Preliminary Design of Bridges for Architects and Engineers

Civil and Environmental Engineering

1. **Preliminary Design of Bridges for Architects and Engineers**
   Michele Melaragno
2. **Concrete Formwork Structures**
   Awad S. Hanna

Additional Volumes in Production

Preliminary Design of Bridges for Architects and Engineers

MICHELE MELARAGNO
University of North Carolina at Charlotte
Charlotte, North Carolina

MARCEL DEKKER

MARCEL DEKKER, INC. NEW YORK • BASEL • HONG KONG

Library of Congress Cataloging-in-Publication Data

Melaragno, Michele G.
Preliminary design of bridges for architects and engineers / Michele Melaragno.
p. cm. — (Civil and environmental engineering series; vol. 1)
Includes bibliographical references (p.) and index.
ISBN 0-8247-0184-4
1. Bridges—Design and construction. I. Title. II. Series.
TG300.M45 1998
624'.25—dc21 98-25608
CIP

This book is printed on acid-free paper.

Headquarters
Marcel Dekker, Inc.
270 Madison Avenue, New York, NY 10016
tel: 212-696-9000; fax: 212-685-4540

Eastern Hemisphere Distribution
Marcel Dekker AG
Hutgasse 4, Postfach 812, CH-4001 Basel, Switzerland
tel: 44-61-261-8482; fax: 44-61-261-8896

World Wide Web
http://www.dekker.com

The publisher offers discounts on this book when ordered in bulk quantities. For more information, write to Special Sales/Professional Marketing at the headquarters address above.

Current printing (last digit):
10 9 8 7 6 5 4 3 2 1

PRINTED IN THE UNITED STATES OF AMERICA

To Fillmore, with whom I have crossed many bridges.

PREFACE

This book is addressed to structural engineers in general practice who are interested in preparing conceptual preliminary plans to be presented at various bridge design competitions. A secondary audience includes architects, landscape architects, environmentalists, and transportation experts interested in acting as consultants for the preparation of such proposals.

For each bridge being designed and built, numerous preliminary designs are made prior to the selection of the final proposal. This is true in the case of open public competitions as well as in the case of in-house design within national agencies. Therefore, the preliminary designs of bridges are much more numerous than the final designs. This clearly indicates the higher order of priorities of the preliminary phase of design over subsequent design phases. In recognition of this, this work concentrates on the conceptual and preliminary stages in the design of bridges, aiming to reach a more widespread audience, larger than the narrow group of specialized bridge designers.

In professional practice, structural engineers who specialize in bridge design constitute a very specific group typically separated from other structural engineers. Yet, due to increased bridge construction, it will be more beneficial to open the field to all structural engineers, especially at the preliminary stage. In this way, new ideas will be available from a larger group of participants, particularly structural engineers with special creative talents that bring a breath of fresh air into the small circle of bridge designer specialists. For instance, an extremely successful contribution in the field of bridge design came from two structural engineers who did not limit their practice to bridges but had a very diversified sphere of interest: Riccardo Morandi and Santiago Calatrava.

In addition to expanding the field of preliminary design of bridges to all structural engineers, present conditions also encourage the participation of other professionals who work as design teams in this field. Potential participants in bridge design teams include architects, landscape architects, transportation experts, and environmental specialists who will benefit from familiarizing themselves with present-day procedures used in the design of bridges. This volume is also aimed at different specialists who could potentially be involved in the team design approach of bridge structures.

This book combines the principles of engineering principles and design with architectural aesthetics and environmental concerns as they relate to bridges with their surroundings. In recent years, tremendous pressure has been applied to the design disciplines of engineering and architecture for considering environmental issues at the top of the hierarchical order of design factors. Furthermore, the public at large has taken an active role in decision making concerning structures such as bridges because of their potential impact on the aesthetics of the environment, in urban sites as well as in the countryside. With regard to the latter, there has been a proliferation of papers in the last few decades aimed at assessing various bridge components with respect to compatibility with surroundings. Working on the definition of structural aesthetics, much of the literature has explored the qualitative aspects of aesthetics as applied to bridges, while many other papers have also been dedicated to the quantification of aesthetic factors applicable to bridges using, for instance, special software referred to as "neuronets." This work, therefore, addresses two major aspects of bridge design: the architectural and the aesthetic.

Considering the dominant role of structural engineering, this book is directed mostly to the average structural engineer who desires to expand his practice to bridge design by participating in design competitions with conceptual and preliminary plans. Realizing the limitations that typical civil engineers have with regard to the architectural components of design and environmental aesthetics, this work expands on these topics, which may not hold equal interest for other professionals. At the same time, all topics related to the preliminary engineering design of bridges will be of major interest to architects and other specialists who intend to join structural engineers in preparing proposals for bridge competitions.

Since this work aims specifically to be a practical working tool, not much emphasis has been given to historical development of bridges, which, although particularly enticing, would not require that much attention in this context. Prolific literature on this topic is currently available; therefore, the reader is referred to other publications. Yet, a very brief introduction to the historical evolution is included as it is considered necessary for understanding the state-of-the art of present-day bridges.

In conclusion, the following fact about the present conditions of bridges in the United States will convey an idea of the extreme relevance of these structures in the context of the current transportation systems on highways and railroads. According to a survey completed in 1989 (Nickerson and Veshosky, 1992), the total number of bridges in the United States is 577,710 of which 42 percent are deficient, 24 percent are structurally deficient, and 18 percent are functionally obsolete. There is great amount of work facing bridge designers.

Michele Melaragno

ACKNOWLEDGMENTS

The author wishes to recognize the contributions of many sources that helped in the completion of the present work. Among the many contributors are: Steve Liao, KC Engineering; Mamoru Kawaguchi, Hosei University, Tokyo; F. W. Klaiber, Iowa State University; Reid W. Castrodale, Portland Cement Association; Ski Brown, Engineering Service Center, California Department of Transportation; Neils J. Gimsing, Technical University of Denmark; R. Douty of MDX Software; Anthony S. Jennings, American Iron and Steel Institute; Joel A. Johnson, North Carolina Department of Transportation; Christian T. Ray, North Carolina Department of Transportation; Jimmy Lee, North Carolina Department of Transportation; Dr. Bijan O. Aalami, ADAPT Structural Concrete Software System; Lee D. Tanase, Leap Software Inc.; C. C. Fu, BEST Center, University of Maryland; Marc Ansuer, Florida Department of Transportation; Libby M. Kop, American Institute of Steel Construction, Inc.; Charles Casey, North Carolina Department of Transportation; Randy Swanson, University of North Carolina—Charlotte; Charles Mitchel, University of North Carolina—Charlotte; Luz M.Aveleyra, University of North Carolina—Charlotte; the staff of the Reference Department of the University of North Carolina—Charlotte Library; Robert Hoganson, Chairman, Wire Rope Technical Board; and Thomas E. Secules, Williamsport Wirerope Works.

Special mention is due to Fuller Moore, Miami University for preparing the manuscript and the illustrations in a camera-ready format, and to B. J. Clark, Executive Acquisitions Editor at Marcel Dekker, Inc.

Finally, I want to recognize the tremendous help by my wife Deborah, who edited the manuscript, plagued by misspellings, omissions, and so many other errors, to reach the stage of acceptance. Furthermore, I want to recognize her skill in taking several of the photographs included in this text. Most of all, however, I want to express my thanks and gratitude for all the support she has given me during the preparation of this volume.

Michele Melaragno

CONTENTS

1

BRIDGE GENERALITIES

FROM THE HISTORICAL INVENTORY OF BRIDGES

In approaching the subject of modern bridges, it is impossible to start without referring to the time frame that describes the historical events through which bridges have developed. The literature of the history of bridges is prolific and it is beneficial to trace the main historical sequence of events in the development of bridge design.

The crossing of barriers of different types requires structures limited only by the availability of materials and by the technological knowledge of the builders. However, another factor that played a major role in the development of bridges was necessity. Thus during warfare conditions the power of organized armies often provided the required energy for accomplishing major engineering tasks that were not possible in peace time. The crossing of rivers by bridges found the most spectacular applications during war. Large applications of bridges built in ancient times refer, for example, to those built during the Persian Wars by Xerxes in 480 BC over the Hellespont; which measures 1.5 kilometers long. Most famous, of course, is the reference to the bridges built by Caesar during the Gallic Wars in 55 BC over the Rhine River (Figure 1-1). This bridge more than any other from an-

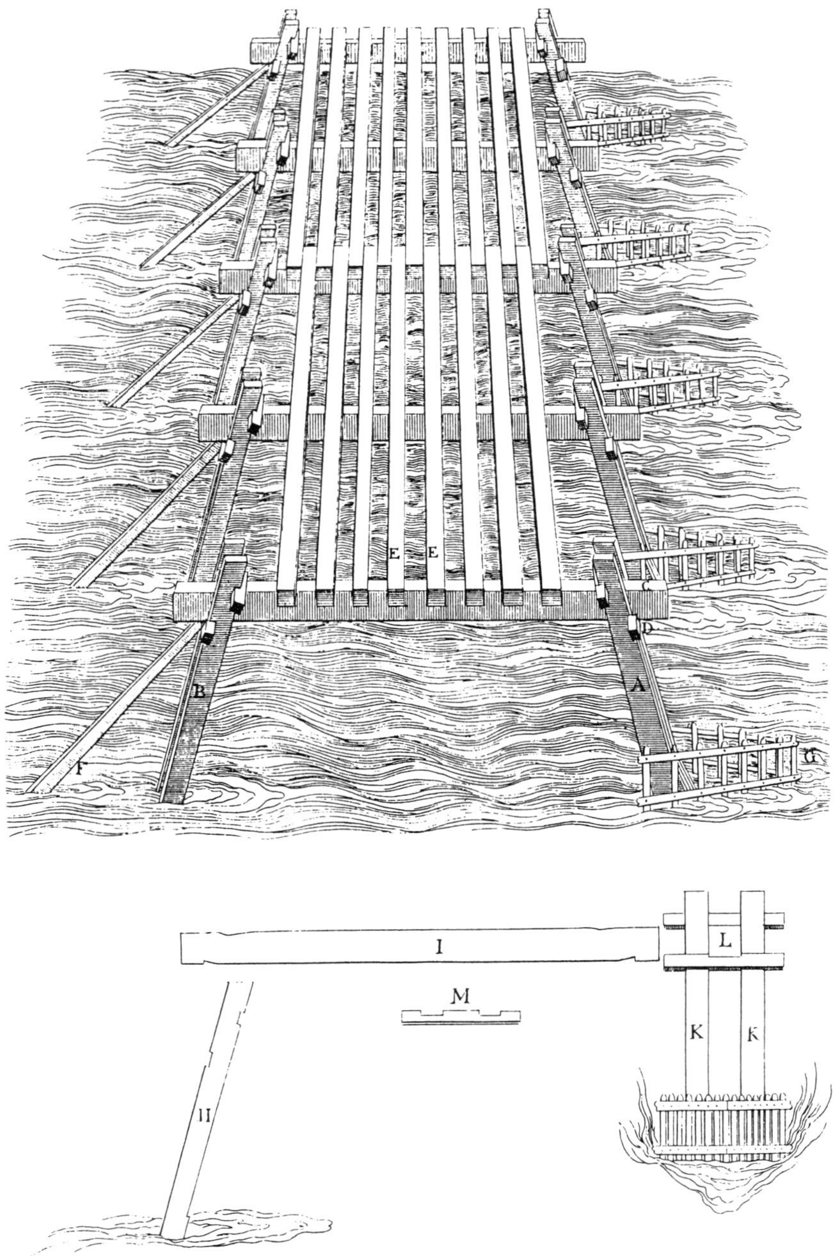

Figure 1-1 Caesar's bridge over the Rhine River (from the reconstruction drawing by Andrea Palladio, from *Four Books of Architecture*, Book III, chapter VI, plate 2).

cient times stands out from the pages of history books for its audacity and its unbelievable execution. Approximately 400 meters long, the structure was erected in record time. In fact, it seems that only ten days elapsed between the time that the order was given and the time when the last man of the Roman legions crossed the river; this included the cutting of the trees, the shaping of the members, the building of the connections, and the erection of the whole structure amid the strong currents of the water course. Obviously the engineering knowledge of Roman legionnaires and the organization of the work had unprecedented standards. Part of their success was also to be found in the exceptional, specialized equipment that the army always had at its disposal, including a variety of forged bronze hardware that allowed for the connection of timber members used for the construction of bridges.

Besides military applications in the old world, masonry bridges for civil structures were also extremely popular in the Roman culture due mostly to the widespread use of arches and vaults. The most famous of these bridges, both in Rome, are the remnants of the Sublicio, as well as Ponte Milvio, built in 110 B.C., which is still in perfect condition, carrying vehicular loads and pedestrian traffic. This is the bridge that gave the name to the battle between Constantine and Massentius, which the former won, resulting in the split of the Roman empire, the recognition of Christianity, and the fall of the Roman civilization. With the fall of the empire and the surge of the Dark Ages, the art of building bridges did not die completely, but survived somehow through the cultural decline.

Toward the Middle Ages, during which the evolution of bridges hardly progressed, a new prototype eventually emerged in Florence: the "Ponte Vecchio" over the Arno River. Rebuilt several times after collapses due to river floods, it was eventually built in its present shape in 1345. Its three arches departed totally from the semicircular Roman prototypes with a much lower rise-to-span ratio.

Another bridge of significant relevance, built originally in 1357 in Prague and eventually restored in 1432, is the Karlsbrucke Bridge named after King Charles IV. Its length measures 515.76 m and includes sixteen arches and particularly large piers; its width measures ten meters and its longitudinal axis changes direction along its length.

Passing from the Middle Ages to the Renaissance, the development of bridges did not produce any major changes except that some masonry bridges, rather than using the semicircular Roman arch, employed arches with a lower rise-to-span ratio. The most significant of the Renaissance arches in Europe is the Ponte Della Trinita, built in 1567 in Florence, that included three elliptical arches with a rise-to-span ratio of 1:5, followed by a few other examples in Florence and Paris. Other examples of bridges in the Renaissance are those few truss-type wooden structures designed by Palladio and described in his *The Four Books of Architecture*. Another landmark in the development of the bridge throughout the Renaissance is the Rialto

Bridge in Venice, consisting of a single arch crossing the Grand Canal and designed by Antonio da Ponte in 1577.

The bridge architecture of Andrea Palladio

The first representative of architecture within the context of bridge design is officially Andrea Palladio, or Andrea Di Pietro della Gondola (1508-1580), or his work includes the design of several bridges. His participation in the development of bridge design is best represented by his writings, *The Four Books of Architecture,* first published in Venice in the year 1570.

More specifically, Book Three, dedicated mostly to bridges, includes 21 chapters, some of which relate specifically to bridge structures: Chapter IV: *Of what ought to be observed in the building of bridges, and of the site that ought to be chosen*; Chapter V: *Of Wooden Bridges, and of the advertencies which ought to be had in the building of them*; Chapter VI: *Of the bridge directed by Julius Caesar over the Rhine*; Chapter VII: *Of the Bridge of Cismone;* Chapter VIII: *Of three other inventions, according to which wooden bridges may be made, without fixing any posts in the water*; Chapter IX: *Of the Bridge of Bassano*; Chapter X: *Of stone bridges, and what ought to be observed in the building of them*; Chapter XI:*Of some celebrated bridges built by the ancients, and of the designs of the bridge of Rimino*; Chapter XII: *Of the Bridge of Vicenza, that is over the Bacchiglione*; Chapter XIII: *Of a stone bridge of my invention*; Chapter XIV: *Of another bridge of my invention*; Chapter XV: *Of the Bridge of Vicenza, that is upon the Rerone*. From this work, the illustrations that accompany his writings, speak more clearly of themselves with a pristine clarity that characterized the drawings made by his own hand. From the large collection of drawings illustrating his bridges, few have been selected to be included in this context (Figures 1-2 through 1-10).

Bridge development

The development of bridges in the New World required the construction of extremely long spans over numerous large bodies of water. Such spans were much larger than those built in European countries and required an audacity not proven before, especially in Europe, where building standards were protected by conservative engineering rules. Among the various structural systems that could promise the construction of long-span bridges, trusses were the most visible choice. Even so, spans that had been bridged were much shorter than those required of the new bridges. Therefore, in the new continent where rules and regulations were practically nonexistent, many builders depended on intuition rather than scientific knowledge. Attempting new design conceptions this way caused several major disasters, but eventually they succeeded in doing what was considered impossible. In conclusion, the new

American truss bridges, initially built of wood and eventually of steel, established a new era in the development of the truss bridges.

Among these pioneer bridge builders was Timothy Palmer, by trade a carpenter. He was the builder of the Essex-Merrimac Bridge over the Essex-Merrimac River and the Pisca Tagua Bridge at Portsmouth, NH using the covered timber truss that protected its structural members from weather deterioration. Another notable American pioneer bridge builder in this first period was Louis Wernwag. With the development of engineering science in America, a second generation of bridge builders included such names as Ithiel Town, Stephen Long, William Howe, Thomas Pratt, Squire Whipple, Wendell Bollman, and Albert Fink.

The glamour of the suspension bridge was slow to appear, after a beginning of small applications both in Europe and in North America, partly because of the limitations of wrought iron chains and bars first used prior to the use of steel strands, and partly due to the limitations of engineering knowledge. Several suspension bridges were eventually built with varying success by different designers, including Charles Ellet. However, the first formidable suspension bridge of long span still standing today was the Brooklyn Bridge connecting Brooklyn with Manhattan over the East River. The Brooklyn Bridge covered a span of 386 m (1,268 ft). Designed by John Roebling who had previously completed in 1866 the Cincinnati Bridge that spanned 322 m (1,057 ft). While John was the designer of the Brook-

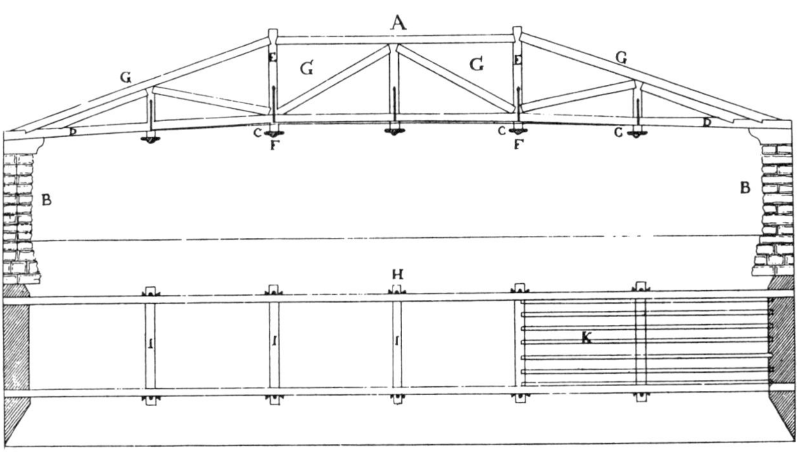

Figure 1-2 Bridge of Cismone (from *Four Books of Architecture*, by Andrea Palladio, Book III, chapter VII, plate 3).

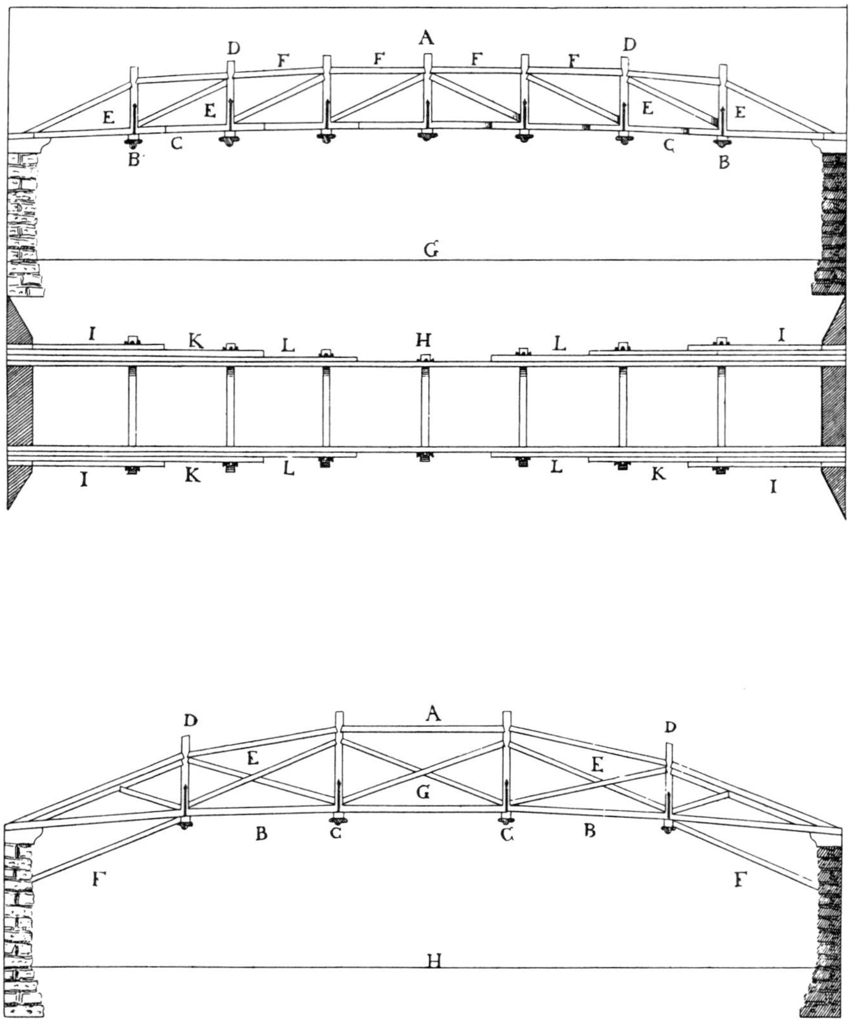

Figure 1-3 Drawings of bridges (from *Four Books of Architecture*, by Andrea Palladio, Book III, chapter VIII, plate 4).

lyn Bridge, his son Washington Roebling built it after the death of his father. After such a success, the suspension bridge continued to present day with, for example, the George Washington Bridge built in New York in 1931 with a span of 1,067 m (3,500 ft); the Golden Gate Bridge built in San Francisco in 1937 with a span of 1,280 m (4,200 ft); the Verrazano-Narrows Bridge, built in 1964 over the New

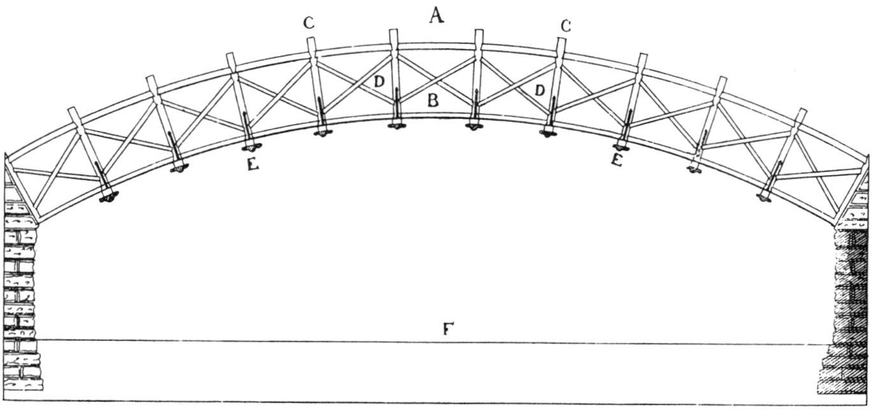

Figure 1-4 Drawings of bridges (from *Four Books of Architecture*, by Andrea Palladio, Book III, chapter VIII, plate 5).

York Harbor with a span of 298 m (4,260 ft); the Fourth Road Bridge built in Scotland in 1964 with a span of 1,006 m (3,300 ft); the Tagus Bridge built in Portugal in 1966 with a span of 1,013 m (3,323 ft); the Severn Bridge built in England/Wales in 1966 with a span of 987.5 m (3,240 ft).

Cable-stayed bridges may be traced back to a small prototype built in Venice and eventually applied to large spans on a few occasions in the 19th century. The modern cable-stayed bridge, however, is really a product of the second part of this century. In 1952, Fritz Leonhardt designed three cable-stayed bridges over the Rhine at Dusseldorf. One of these was the Theodor Heuss, or North Bridge, built of steel and completed in 1958 with two spans of 280 m (919 ft) and 108 m (354 ft). Stayed cables have also been employed in combination with suspension cables as in the case of the Brooklyn Bridge and others. Another cable-stayed bridge that stood out as a new prototype in prestressed concrete was the structure crossing Lake Maracaibo in Venezuela, designed by Riccardo Morandi and completed in 1962. Five and one half miles (9 km) long, the bridge included numerous spans, the largest of which extended 400 m (1,300 ft). Other noteworthy cable-stayed bridges include the Annacis Bridge built in 1986 in British Columbia over the Fraser River with a span of 465 m (1,562 ft) and the Queen Elizabeth II Bridge built in England in 1991 over the Thames River with a span of 812 m (2,664 ft).

Much has been published about the history of bridges in technical, architectural, and laymen's literature. To repeat the history of bridges would be redundant and would require more space than could be given to such a topic. The brief outline about the development of bridges throughout the course of centuries has been given only as a guide to follow the many bibliographic references covering the history of bridges up to the present.

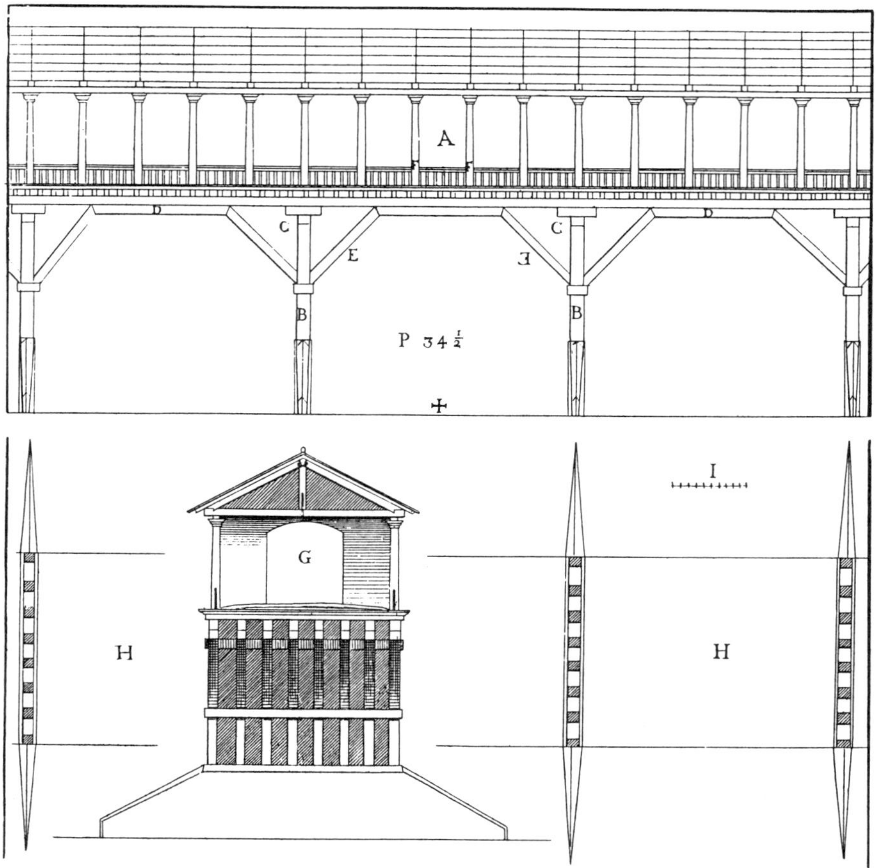

Figure 1-5 Bridge of Bassano (from *Four Books of Architecture*, by Andrea Palladio, Book III, chapter IX, plate 6).

Now consider what significance the reader could find in reviewing the history of bridges in this context. We are addressing structural engineers for the most part, who desire to enter into the field of preliminary design of bridges. Therefore, the material covered herein is mostly topics that have not been covered in the standard educational curricula for civil engineers, such as architectural history as pertains to bridges. To this effect it should be pointed out first that history has to be considered an applied discipline, used not only to give a general knowledge of a subject but also to be used as case studies. Just as architectural history is a fundamental part of architectural education, similarly the history of bridges is necessary for the structural engineer engaging in the design of these types of structures.

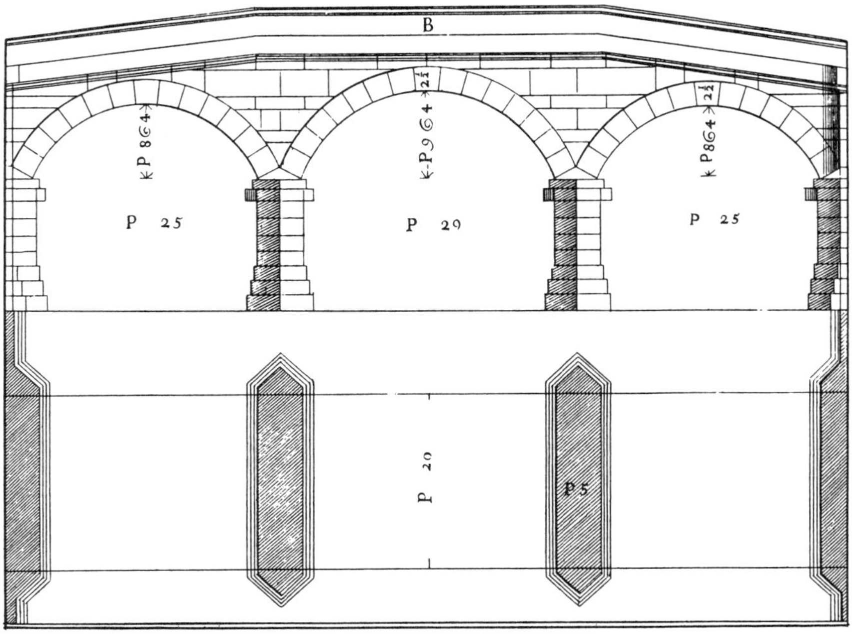

Figure 1-6 Bridge over the Bacchiglione River of Vicenza (from *Four Books of Architecture*, by Andrea Palladio, Book III, chapter XII, plate 8).

What can we learn from the history of bridges? To begin with, the reader can create a memory bank of solutions in his or her mind that, as in a kaleidoscope, will furnish the basic material to articulate creative new solutions. Indeed, it would be impossible to develop any design from an absolutely blank memory bank. At this point there are two phases in the process of learning from case studies: (a) a first phase consisting of examining each individual structure in its overall configuration; and (b) a second phase consisting of comparing each of these individual structures to one another. The reader would definitely discover a sense of understanding after going over and over each individual artifact. However, this also could be achieved in an easier manner, under the guidance of an instructor who could point out the essentials, eliminating the laborious approach of self discovery. Yet, self-education is a natural process that everybody goes through during his or her professional life, and therefore one assumes that this would also apply to the case in question.

Consider a practical case that one could derive as an example. Having established that bridges, like buildings, can be architectural artifacts, two of them have been selected in an attempt to understand some of the reasons why they have achieved a place of relevance in architectural history. The two structures are the

Figure 1-7a Stone bridge of Palladio's invention, elevation (from *Four Books of Architecture*, by Andrea Palladio, Book III, chapter XIII, plate 9).

Rialto Bridge in Venice and the Ponte Vecchio in Florence. These two structures stand out not only in the books of the history of architecture, but also in most of the guidebooks illustrating the architectural highlights of these two cities. Indeed, from the structural point of view, they are definitely not particularly noticeable. Probably the only interest from the engineering point of view could be the foundation system for the Rialto Bridge, which required twelve thousand piles (six thousand on each side) to sustain the foundations for the abutments. However, this was not an exceptional characteristic for the city of Venice, since all buildings of a certain size depended on a similar kind of foundation.

So, what other qualities do these bridges have to justify their place in architectural history? First, both structures blend with the surrounding urban stage in that neither overwhelms the surrounding buildings, nor are they dwarfed by them, but coexist harmoniously at the same scale. These considerations are especially significant today when bridges are judged aesthetically for their ability to harmonize with their background both in the urban context as well as in the countryside. Also, both bridges have spontaneously become a public space for people to gather. The structures no longer function exclusively as typical pedestrian bridges but have become public places where commerce is transacted as in a public market.

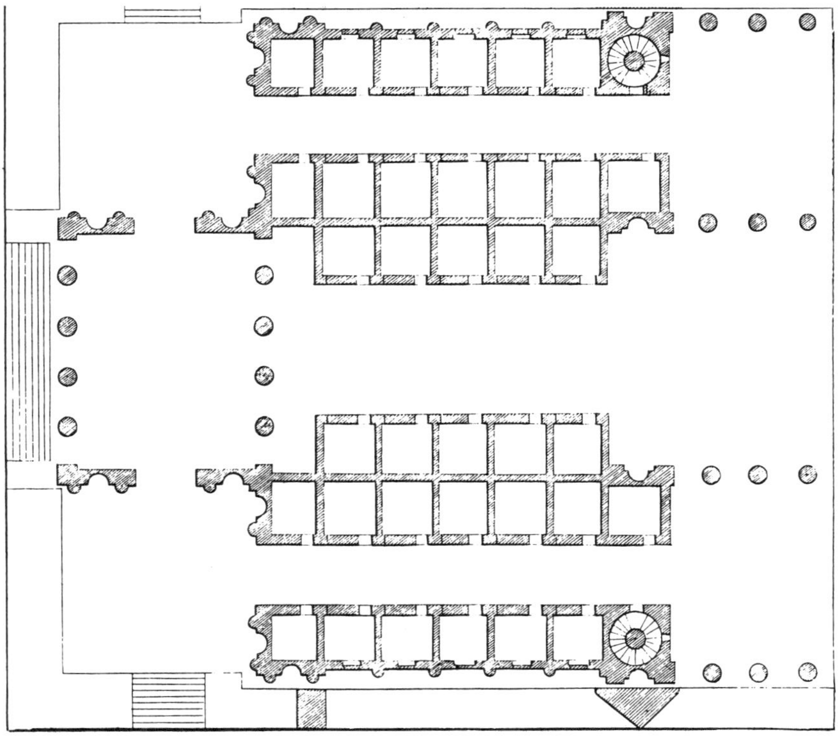

Figure 1-7b Stone bridge of Palladio's invention, plan (from *Four Books of Architecture*, by Andrea Palladio, Book III, chapter XIII, plate 9).

Shops line both sides of the bridges where special merchandise is sold,such as jewelry, silver, and artifacts. Moreover, the process of transforming these bridges into such busy, crowded urban spaces was not only the response of the public to their preferences but also the expression of the regulatory institutions controlling urban spaces that allowed commercial structures to be built on the bridges.

Indeed, these two bridges represent two unique prototypes that did not have similar competitors in Europe for years until the mid-1950s. At this time the Autostrada del Sole (the "Highway of the Sun") in Italy included a few overpass bridges dedicated only to commercial activities such as shops and restaurants for the motorist stopping at rest areas on each side of the highway. Such a communality of uses between the two bridges is not a result of a similitude of forms, dimensions, or other physical characteristics since they do not share them at all, but evidently they must have expressed positive aspects that attracted the public to

Figure 1-8a Stone bridge of Palladio's invention, elevation (from *Four Books of Architecture*, by Andrea Palladio, Book III, chapter XIII, plate 10).

gather on these spaces. One could argue that a bridge is a selective site where pedestrian traffic merges and therefore it constitutes a strategic point for vendors to display their merchandise to a captive public. Yet, if this would be the case, one wonders why all of the bridges in the urban site have not followed a similar course. Evidently there must be other reasons such as: the peculiarity of the site itself; the orientation of the bridge; the near and distant views of the cityscape visible from the bridge; the interest of the water traffic visible from the bridge; and the particularity of the areas on each side of the bridge.

Considering each of the structures separately, let us now observe them from an architectural point of view. Ponte Vecchio over the Arno in Florence was erected in 1345, and designed by Taddeo Gaddi. The period is prior to the Renaissance, more specifically toward the end of the Middle Ages during the time of Dante Alighieri, Francesco Petrarca, etc. The city at this point was the expression of the times, i.e., when simplicity and severity characterized the architecture of the city. The bridge is a direct response of this austere architecture where sobriety and simplicity were the essence of a lifestyle still controlled by the mysticism of the feudal era. The Rialto Bridge over the Grand Canal in Venice is also a product of its times. Its architect, Antonio Da Ponte, won the competition for the design of the bridge, erected in 1591. This was the same competition that Andrea Palladio en-

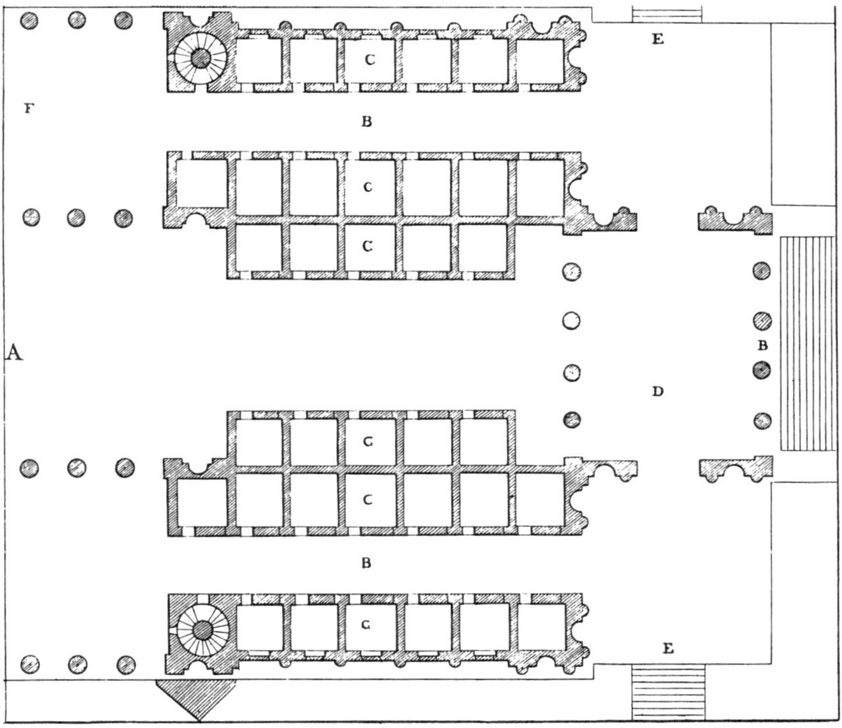

Figure 1-8b Stone bridge of Palladio's invention, plan (from *Four Books of Architecture*, by Andrea Palladio, Book III, chapter XIII, plate 10).

tered but did not win. However, the bridge is very much in a Palladian style expressing feelings of serenity, using purposely the same adjective to which the city was usually referred: "La Serenissima." Like most of Palladian architecture in the city and surrounding territories, this bridge,which reflects the same style, seems to represent an expression of static placidity in counterpoint to any dynamic state. Indeed, this is a full expression of the late Renaissance prior to the insurgence of the Baroque period.

Noticeable, too, is another characteristic that both bridges have in common: each of the two structures was designed by an architect. This is particularly significant considering that the majority of bridges were designed and built anonymously. Particularly so is the case of Ponte Vecchio, before the advent of an era when architects eventually emerged. Thus, the fact that these two bridges carried an architectural signature is a hint justifying the fame that they eventually gained. Furthermore, the Ponte Vecchio, a simple and severe expression of the architecture at the time, had an opportunity to be modified. When it was reconstructed

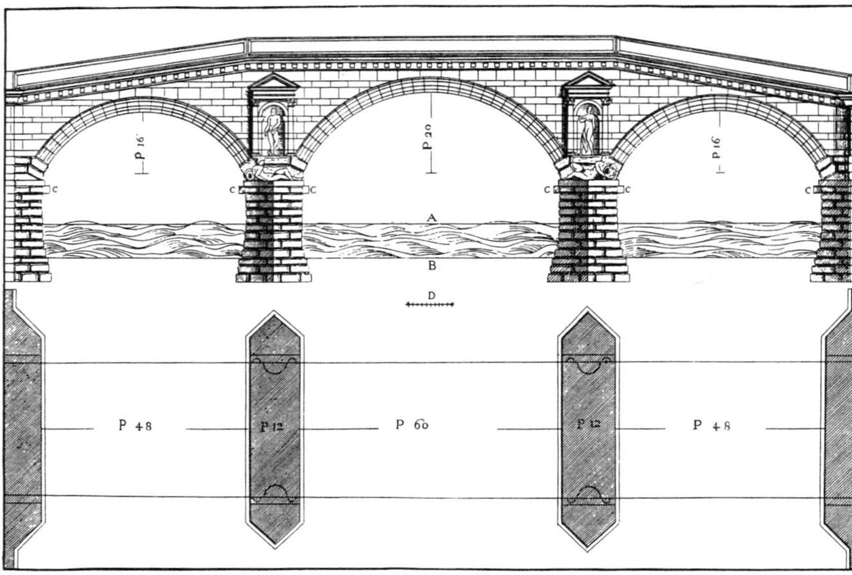

Figure 1-9 Stone bridge of Palladio's invention (from *Four Books of Architecture*, by Andrea Palladio, Book III, chapter XIV, plate 11).

after being destroyed by German troops, however, the bridge was rebuilt exactly as it was in all its details, proving that the original form of the bridge was exactly what the people of Florence wanted.

Another theme from the history of bridges is a unique prototype that has become popular in these last few decades. Specifically, reference is made to a new family of bridges spanning from building to building in several cities of the United States. Such bridges that respond to a specific need in the urban circulation have a similar counterpart in one of the bridges of Venice usually referred to as the Bridge of Sighs. This structure connects the Doge's Palace with the Piombi. The first building included the court house of the old Republic and the second was the famous prison. This explains the name given to the structure used by so many prisoners condemned to reclusion in the upper stories of the building. The idea of connecting one building with another above the street level, which in the case of Venice is the water level in the canal, is an innovation in the use of bridges that has not found similar application before recent times. These present-day structures,which do not represent any major engineering accomplishments, are nevertheless linked to this original prototype in Venice that stood out in the history of bridges.

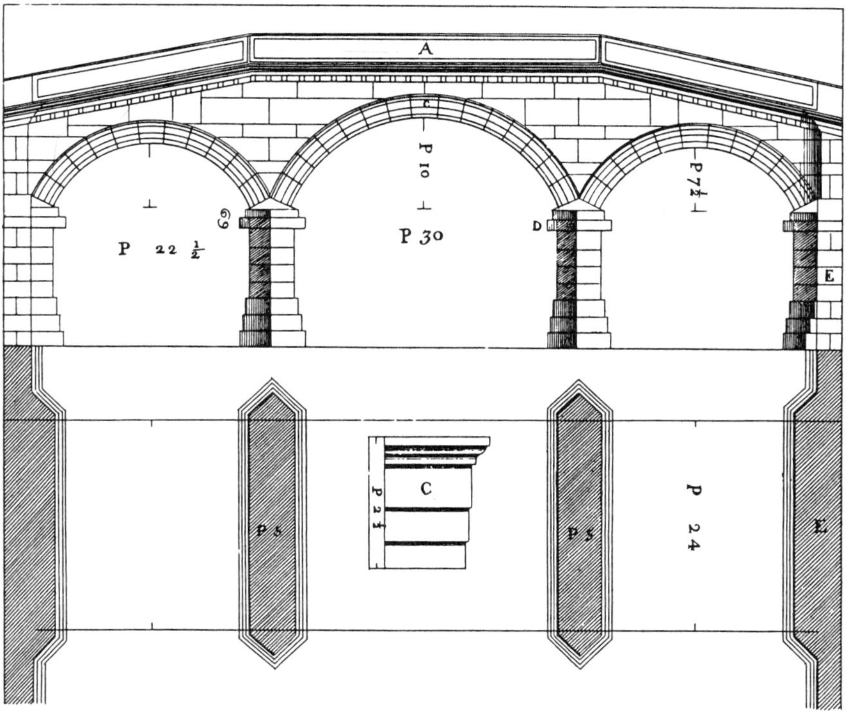

Figure 1-10 Bridge at Vicenza over the Rerone (from *Four Books of Architecture*, by Andrea Palladio, Book III, chapter XV, plate 12).

THE MEGASTRUCTURES

Like the constant state of tension that energizes and propels forward the development of science and technology, the engineering of bridges is continuously reaching toward the next goal of spanning larger and larger gaps between two points. This recalls the typical aphorism in the words of LeRilaise, "Toward structures of zero weight and infinite span." Currently several proposals exist for megastructures spanning over distances never reached before, including the following case.

The Bridge of the Strait of Messina

Connecting the mainland of the Italian peninsula at Reggio Calabria with Messina in Sicily has been a paradoxical idea that started a long time ago and is now ready to be made real. The first proposal for connecting the two points of land with a permanent bridge was launched by David Steinman and consisted of three suspended spans, the largest of which in the middle would have had a length of one mile spanning between two intermediate towers which had to reach a depth of 120 meters. After rejection of the proposal, the Italian state agency of national roads (ANAS) proposed a competition for a bridge in 1969. Among the various submissions, Fritz Leonhardt presented a proposal for a cable-stayed structure that was also rejected.

Yet the idea of constructing such a bridge did not lose momentum and continued gradually to progress. In 1976 a specially formed committee developed a study program for the bridge. In 1981 the Messina company, sponsored by the Italian government, started a study about a feasibility program that was eventually completed in 1986. Among the various proposals presented was that of Riccardo Morandi (Figure 1-11) in 1986 which consisted of a suspended hybrid structure. Eventually a last proposal presented in 1992 consisted of a single-span suspended structure with a length of 3.3 kilometers between the two supporting towers.

This structure will have a particularly wide deck with a width of 60.4 m with an aerofoil cross-sectional shape almost twice the typical width of most other suspension structures. A description of the roadway system for the bridge reported by Brown (1993) as follows: "The 60.4-m deck is intended to accommodate a pair of rail tracks in the center plus two service roadways, a triple carriageway with a service lane on each side of these and, cantilevered beyond the line of the bridge hangers, at the far extremity on each side, an emergency road lane positioned some 2 m below the level of the remainder of the deck."

Still in the preliminary design stage, the project will include two suspension towers sited on land which will support in turn an intermediate span of 3.3 kilometers that will supersede the longest on-record span of 1.9 kilometers of the Akashi-Kaikjio Bridge in Japan to be completed in 1998. In addition to the main span, two other spans, one on each side, will project out of the towers for a length of 180 meters each. The cost of the structure is estimated to be 2.4 billion lire and the date of completion is anticipated to be in 2004.

Aerodynamic studies of the structure have already been conducted at the Poletecnico di Milano and at the Danish Maritime Institute. Here an aeroelastic model of the bridge has been built with a length of 13 m, representing the structure at a scale of 1:250 and tested in a wind tunnel.

Further data about the preliminary design of this bridge include a parabolic shape of the main supporting cables that would have a sag-to-span ratio of one to eleven. The clearance above the sea level would vary between 64.3 and 70.0 m according to tide changes. The cross-section will accommodate three highway traffic lanes, two of which will be in opposite directions, and the third would be used alternatively. The railroad traffic will include two lanes in opposite directions, plus another two lanes in opposite directions for future use by railroad vehicles with rubber tires. In addition there will be two pedestrian lanes in opposite directions. The predicted traffic will include 4,500 vehicles per hour in each direction and also 200 trains per day in each direction. The deck cross-section will be 60 m wide. The overall weight of the structure, including paving and complete railway tracks, will be 23.3 tons/m. Total weight of structural steel will be 70,500 tons. The maximum lateral displacement during service operation is to be 9.9 m. Maximum thermal expansion and contraction is to be 3.4 m±. Tower height will be 376 m, and the tower foundation on the Messina side will include two concrete plinths 55 m in diameter each.

The Akashi-Kaikyo Bridge

While the Messina Bridge is still in the design stage, the Akashi-Kaikyo Bridge is in the ultimate stage of construction and will be ready for use in 1998. When this bridge opens to service it will have the world's longest span. The bridge is part of the Honshu-Shikoku Bridge project and will connect the islands of Honshu and Shikoku. Its total length will be 3,910 m with a central span of 1,990 m, and two adjacent spans of 960 m each. This suspension bridge will include two towers, 297 m high, with an X-configuration bearing on caissons 80 m across and 70 m deep. The two suspension cables have an extraordinary diameter of 1.1 m, carrying the upper structure that includes a deck that carries six highway lanes of traffic.

Particularly difficult for the construction of this structure are the site characteristics, which include deep waters, strong currents, and high winds. Such currents affecting the stability of the caissons are in the order of 4.5 m/s. With regard to the wind forces acting on the upper structure, the design velocity is in the order of 46 m/s. With regard to seismic loads, the technical literature indicates that the design criteria guarantees a survival of the bridge in the event of an earthquake having an intensity of 8.5 on the Richter scale and its epicenter being at the distance of 150 km from the bridge.

Critical description of the bridge in the technical literature points out that the structural configuration of this bridge follows very much the traditions of typical American suspension bridges. In other words, the Akashi Kaikyo Bridge does not follow the typical European configurations of suspension bridges that have been built since the 1960s which include slender aerodynamic decks. Instead, this structure includes decks with deep trusses typical of the American prototypes. This could be explained by the general attitude adopted in Japan which, as discussed later, depends upon the preference for conservative forms.

BRIDGE SIGNATURES

Unlike many works of art that carry the name of the artist who created them, many architectural and engineering artifacts stand by themselves without any recognition for the man who conceived and built them. This seems to be particularly the fate of many bridge structures, including many famous ones used daily by millions of people who have never been told about those heroic men who created them out of their own knowledge and spirit of inventiveness.

To remedy such a situation, it seems particularly useful to fill this vacuum for the benefit of the reader by including in this section some biographical profiles of some of the outstanding bridge designers who have left a strong mark on the history of bridges.

Robert Maillart

In discussing modern bridges it is impossible not to include Robert Maillart, who established in the first part of this century a new typology, i.e., the reinforced concrete bridge. Indeed, his bridges not only established a new family of bridge structures but also established the potential of a new building material capable of creating efficient plastic forms that surpassed masonry artifacts, making then totally obsolete.

Robert Maillart was born in Bern on February 6, 1872, the son of a Swiss-German mother and Belgian father. His educational background includes his studies at the Federal Technical Institute at Zurich under Professor Ritter. There his education included structural engineering and bridge design, which he completed in 1894. After his graduation he acquired a broad experience by working in several engineering offices in Switzerland: first, an engineering company from Bern specializing in railway design, then the public works department of the city of Zurich working on road and bridge design, and lastly the office of a structural engineer working on design and construction of bridges and buildings.

After this first period of apprenticeship he opened his own firm in 1902 by the name of Maillart & Company, specializing in reinforced concrete structures. By this time, reinforced concrete had come of age mostly under the pioneering work of French engineer François Hennebique and the German engineer G. Wayss.

One could definitely say that his work revolutionized with great impetus the architecture of bridges, which up to this point were usually built with traditional forms typical of masonry structures. In a sense Maillart's work was really half a century ahead of the tastes of the times, since masonry bridge forms were too radical in the mind of architects and the public in general. Naturally Maillart had grasped the potential of reinforced concrete in terms of its ability to resist all kinds of stresses such as tension, compression, shear, and torsion. Furthermore, he capitalized on the moldability of this material that could be cast not only in rectilinear formworks but also in curvilinear shapes, in two- and three-dimensional applications. Arches and vaults in reinforced concrete took another morphological identity in comparison to masonry structures strapped within the limits imposed by masonry itself.

His bridges, more than forty, were not only the expression of daring engineering forms, but also they reached the realm of a new form of art that has often been referred to as "structural art." All of his work is to be found only in Switzerland within the surroundings of a rugged landscape. The challenge he confronted in the execution of his work, is not only demonstrated in the unique configuration of each of his bridges but also in overcoming the difficulties imposed by the site. Most of his work spans over extremely difficult terrains, where scaffolding for supporting the formwork was almost impossible to be erected under conventional methods of construction.

Riccardo Morandi

Who is Riccardo Morandi? His recogition in the United States has not reached that of Nervi, but in Europe and South America, his name and his work are much better

Figure 1-11 Riccardo Morandi (photo courtesy of Edizioni Di Comunita).

known. Thus, a few biographical notes are definitely helpful to introduce the man and his work. He was born in Rome in 1902, eleven years after Nervi, which places them in the same historical period. Like Nervi, he graduated with a degree in engineering, not at the University of Bologna, but at the University of Rome in 1927, which at the time, bore the specific name of *Scuola di Applicazione per Ingegneri* at S. Pietro in Vincoli, where the school still exists today. From 1928 to 1931, as one of his first professional assignments, he worked in the earthquake zone in Calabria on concrete structures involving the construction of new churches, after which he returned to Rome. Rome is his home and the site of his office, where he has worked throughout his professional life. From his headquarters, he has reached far distant places with numerous jobs throughout Italy and in other countries, especially in South America. From 1936, his main interest went to pre-stressing, and he succeeded in obtaining several patents. As in the case of Nervi, Morandi, too, was called into academics, and in 1959 he was teaching structural morphology with special emphasis on bridges in the School of Architecture at the University of Florence. Morandi and Nervi came together on other occasions to work in their professional endeavors; in 1962 they shared the award from the Associazione Italiana Tecnico Economica del Cemento. Fame eventually reached Morandi late in life in his sixties, when a series of awards finally recognized his work, which include the following:

1963	Member, Academia di San Luca, Rome
1963	Honorary Fellow, Royal Society of Arts, London
1966	IN/ARCH Prize, Rome
1967	Columbus Prize, International Institute of Communications
1969	IN/ARCH Prize, Liguria
1970	Freyssinet Prize, Federation Internationale de la Precontrainte
1976	XIII Samoter Prize, Verona
1979	Honor Award, Technische Hochschule, Munich
1980	Gold Medal, Institute of Structural Engineers, London
1983	Architecture Prize, Academia Nazionale dei Lincei, Rome
1983	Plaque of Honour, Venezuelan Association of Engineers
1983	Hon.D.Arch., University of Reggio, Calabria

As shown from his personal archives, his work, definitely astonishing for its volume in the course of his life, includes 30,000 documents related to ideas and projects that, for the most part, were actually built at different sites around the world. The listing which follows indicates the vast accomplishments attributable to this man:

1931	House, Viale Vaticano, Rome
1933	Piccini Car Park, Via Vittorio Emanuele, Rome
1934	Augustus Cinema, Rome
1935	Parish Church, Colleferro, near Rome
	Giulio Cesare Cinema, Via Giulio Cesare, Rome
1939	House, S. Felice Circeo, Latina, Italy
1941	House, via Terme Deciane, Rome
1946	Astoria Cinema, Via Stoppani, Rome
1947	Alcyone Cinema, Via Lago di Lesina, Rome (with G. Gandolfi)
	S. Nicolo Bridge, over the River Arno, Florence
1948	Bologna Cinema, Rome
	Bridge over the River Fortore, near Serracapriola, Fiuggi, Italy
	SRE Thermo-Electric Centre, S. Paolo, Rome
	Valentino Bridge enlargement and restoration, River Calore (project)
1949	Calci e Cementi di Segni Cement Factory, near Pescara, Italy
	Grillo Bridge, over the River Tevere, near Rome
1950	House, Via Martelli, Rome
	Espero Cinema, Via Nomentana, Rome
	Bridge over the River Elsa, Tuscany, Italy
	Match Factory, Zaule, Trieste
	Auditorium, Academy of S. Cecilia, Rome
1951	Di Tommaso Barracks, Rome
	SST Thermo-Electric Centre, Civitavecchia, Italy
	Bridge over the River Arrestra, Genoa/Savona Motorway, Italy
	Bridge over the River Tuy, Venezuela
1952	Eden Cinema, Rome
	BPD Study Centre, Colleferro, near Rome
	Bridge over the Lupara Stream, Genoa/Savona Motorway, at Arenzano, Italy
	Aircraft hangars, La Guaira, near Caracas, Venezuela (project)
1953	New Republic Bridge, Paguita, Caracas, Venezuela
	Bridge over the River Storms, Cape Town, South Africa
	Compere Bridge, over the River Liri, Sori, near Frosinone, Italy

1953 Bridge over the River Adda, Lecco, Italy (project)
Verona Arena reinforcement
Bridges over the Agro and Fiumedinisi streams, Sicily
1954 Gornalunga Bridge, Enna, Sicily
Cermami Bridge, Galliano Castelferrato, Enna, Sicily
Lazzi Car Park, Florence
BPD Metallurgic Centre, Colleferro, Italy
Mooring Towers for electric cables across the Straits of Messina
BPD Fibre Plant, Castellaccio, Anagni, Frosinone, Italy
1955 Bridge over the Tambura stream, Vagli di Sotto, near Lucca, Italy
Footbridge over the Lussia stream, Vagli di Sotto, near Lucca, Italy
Hydraulic Reservoir Aqueduct, Livorno (project)
Underground Reservoir, Caracas, Venezuela
SRE Thermo-Electric Centre, Fiumicino, Rome
1956 Maestoso Cinema, Via Appia, Rome
Genoa/Po Valley Motorway alterations, Bolzaneto, Genoa
Oreto Bridge, Palermo, Sicily (project)
Santa Barbara Thermo-Electric Centre, S. Giovanni Valdarno, near Florence
Amerigo Vespucci Bridge, over the River Arno, Florence
1957 Metronio Covered Market and Car Park, Via Magna Grecia, Rome
Calci e Cementi di Segni Cement Factory, Savignano, Italy
Bridge over the Sambro stream, Autostrada del Sole, near Bologna
Bridge over Lake Maracaibo, Venezuela (Figure 1-12)
Fiumicino International Airport Terminal, Rome (competition project), with A. Zavileeri, V. Monaco, and A. Luccichenti
SENN Nuclear Establishment, Garigliano, near Latina, Italy
1958 Bridge over the Tevere River, Magliana, Rome
1959 ABCD Factory, Fagusa, Sicily
Underground Pavilion, Valentino Park, Turin
Palazzo del Lavoro, *Italia* '61, Turin (project)
Bridge over the River Tevere, Tor di Quinto, Rome (project)

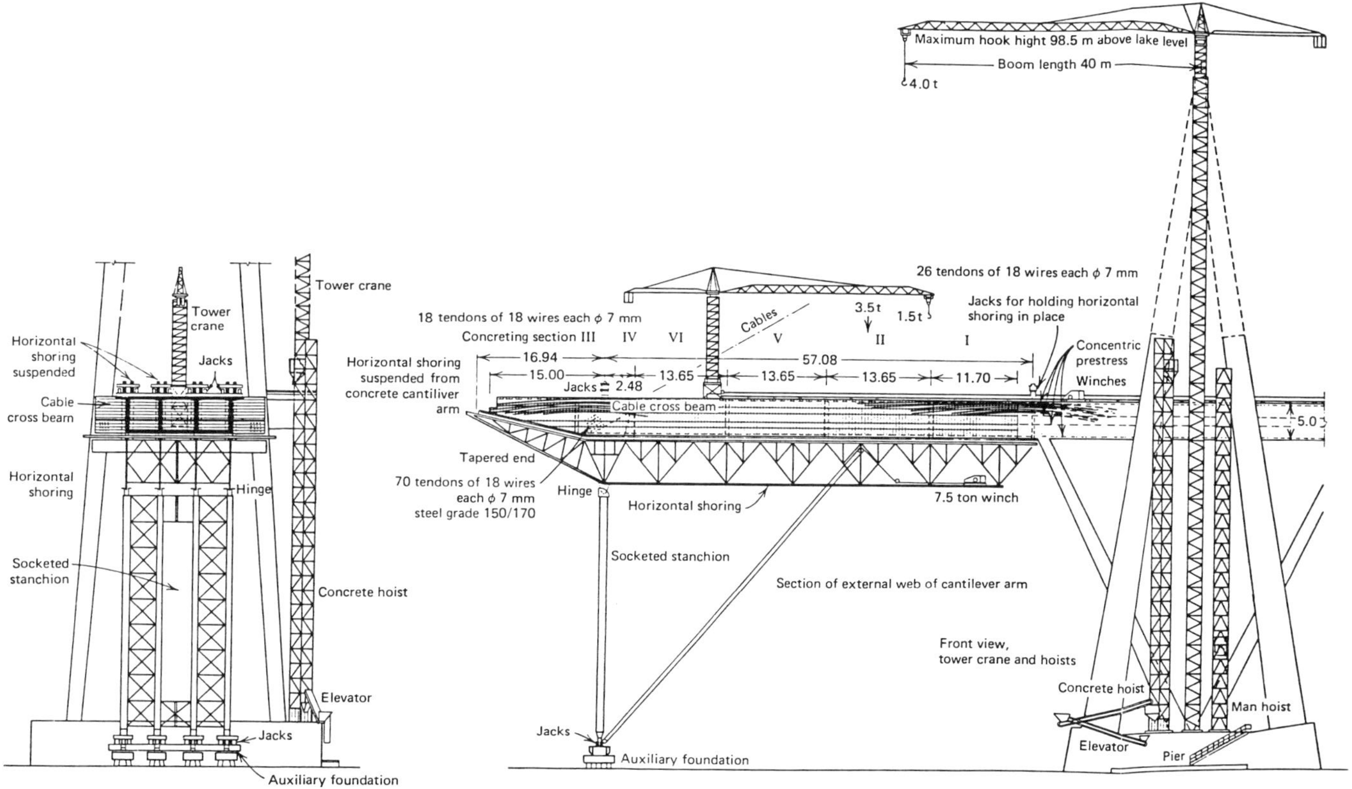

Figure 1-12 Cantilever girder construction, Lake Maracaibo Bridge, by Riccardo Morandi (reproduced from Anonymous, *The Bridge Spanning Lake Maracaibo in Venezuela,* Bauverlag GmbH., Wiesbaden, Berlin, 1963, by permission of Julius Berber, Bauboag Aktiengesellschaft).

1960 Viaduct, Via Olimpica, Rome

Plan for raising of the Abu Simbel Temples, Aswan, Egypt (project; with G. Colonetti and P. Gazzola).

Bridge over the River Tartaro, Canda, Udine, Italy (project)

SELT Thermo-Electric Centre, Valdarno, Livorno

Bridge over the Columbia River, Canada

Overpass, CISA Motorway, Italy

1961 Olympic Stadium, Tehran (project)

Peugeot Building, Buenos Aires

Viaduct, Fiumarella Valley, near Catanzaro, Italy

Bridge over the River Tevere, Autostrada del Sole, Viterbo, Italy

Monorail, *Italia '61*, Turin

1962 Aircraft hangars, Alitalia Building, Fiumicino International Airport, Rome

Bridge over the Port of Goteborg, Sweden (project)

Bridge over the Vella Stream, Sulmona, Italy

Piezometric Reservoir, EUR Zone, Rome (project)

Bridge over the River Schelda, Antwerp (project)

1963 Bridge over Lake Paola, Sabaudia, Latina, Italy

1964 Sports Stadium, Munich (project)

Thermo-Electric Centre, Bastardo, Foligno, Italy

Zeppieri Garage, Frosinone, Italy

1965 Bridge and Viaduct over the Polcevera Valley, Genoa/Savona Motorway, Italy

Solferino Bridge, over the River Arno, Pisa (project)

Bridge over the River Parana, Corrientes, Argentina (project)

S. Nicola Bridge, Benevento, Italy

Bridge over the River Riachuelo, Buenos Aires

Entella Viaduct, Genoa/Savona Motorway, S. Luigi, Italy

Viaduct, Modica Valley, Ragusa, Italy (project)

Calci e Cementi di Segni Cement Factory additions, Colleferro, Italy

Bridge over the River Paranda, Zarate, Argentina

Bridge over the Salerno/Reggio Calabria Motorway, Sicily

Hotel/Restaurant, S. Felice Circeo, Latina, Italy

1966 Viaduct, Latte Valley, near the Franco-Italian border (project)
Wadi-Kuf Bridge, Libya
Overpass, Via Cristoforo Colombo, Casal Palocco, Rome
Viaduct over the Favazzina stream, Salerno/Reggio Calabria Motorway, Sicily
Bridge over the Rome/Fiumicino Airport Motorway, Magliana, Rome
House, Salto di Fondi, Latina, Italy
1967 Bridge over the River Adige, Trento, Italy
Viaduct over the River Ontrano, Salerno/Reggio Calabria Motorway, Sicily
1968 Bridge over the River Salso, Licata, Agrigento, Sicily
1969 Bridge over the River Guyas, Guayaquil, Ecuador
Sports Centre, Milan (Ortensi, Mosco, Musmesi and Stegher)
1970 Viaduct over the Tangential, Naples
Alitalia Maintenance Centre for Boeing 747 Aircraft, Fiumicino International Airport, Rome
1970 Catania Airport, Sicily
Viaduct over the Valley of Temples, Agrigento, Sicily
Bridge over the River Magdalena, Barranquilla, Colombia
1971 ENEL Thermo-Electric Centre, Piombino, Italy
Underground car park, Turin
ENEL Thermo-Electric Centre, Milazzo, Italy
Underground car park, Brescia, Italy
Lecco Motorway, Italy
Bridge, Stmigliano, near Rome (project)
1972 Bridge, Barranquilla, Colombia
Naples Port and Motorway alterations
Bridge, Rande, Spain
1973 Focaro, Ruzzi and Arsella Viaducts, near Chieti, Italy
Arvivo Viaduct, Italy
1974 Servia Bridge, Macedonia, Greece
Mededil car parks, Naples
Thermo-Electric Centre, Termini Imerese, Italy
Bausan Wharf, Naples

1975	Pyramid Building, Abidjan, Ivory Coast
	Carpinteto Viaduct, Potenza, Italy
	Aircraft hangars, at Fertilia, Alghero, Olbia, and Elmaś Cagliari, Sardinia
	Air Force Barracks, Olbia, Sardinia
1976	Ergife Hotel, Via Aurelia, Rome
1977	Irminio Viaduct, Sicily
	Cornice Bridge, Jeddah, Saudi Arabia
	Apartment/office buildings, Jeddah, Saudi Arabia
	FATA Building, Turin (with Oscar Niemeyer)
	Bridge, Jeddah, Saudi Arabia (project)
1978	Bridge, Abidjan, Ivory Coast
	Bridge, over the River Creek, Saudi Arabia (project)
1980	Ohburn Bridge over the Jeddah-Medina Road, Saudi Arabia
	Luxury Hotel, Tripoli, Libya
	Passenger Hall and Canopies for the new Central and East Rail Stations, Baghdad, Iraq (project)
1981	Two Tower Buildings for INAIL, Palermo, Sicily
	Large Water Tanks in Italy and Nigeria (projects)
	Air Terminal, Venice (project)
1982	Banca d'Italia Operational Centre, Frascati, Italy (project)
	Posts and Telecommunications Building, Caltanissetta, Italy (project)
	Al-Ain Airport, Persian Gulf (project)
1983	Cortes Bridge, near Madrid (competition project)
	Footbridge, Algiers (project)
	Sports Palace Building, Bari, Italy (project)
	Railway Bridge, Attigliano, Italy
1984	Motorway Viaduct over the Rio Colorio, near Stresa, Italy (project)
	Motorway Viaduct over the River Fuimetta, near Stresa, Italy (project)
	Motorway Viaduct over the River Scoccia, near Stresa, Italy (project)
	Motorway Viaduct over the River Vallone, near Stresa, Italy (project)
	Motorway Viaduct over the River Roddo, near Stresa, Italy (project)
1986	Proposed Bridge over the Strait of Messina, Italy consisting of a single span suspension structure crossing over the strait 3.3 km wide.

Of these projects, not all have been actually built and have remained at the proposal stage as indicated in the above compiled list. As the records show, Morandi's work, like that of many exceptional men in the arts and sciences, is of unbelievable magnitude that separates this man from most in everyday life. Of course, the volume of his work is nothing more than one component that only compliments the significance of his work, much more valuable in terms of ingenuity, creativity, and unbelievable skills in his art.

Since it is impossible, of course, to analyze or even present with a minimum of proficiency all the work mentioned above, we shall use only some of the various projects to illustrate key points that are fundamental to appreciate the innovative parts of his work. Continuing in the traditions of large applications of reinforced concrete structures built particularly in France, where concrete originally came to be the new material of the new era, he has spread his work on a broader scale than Nervi himself, by expanding not only in bridges, but in buildings, viaducts, industrial structures, tanks, etc., gaining, just like Nervi, the practical recognition as an architect even if his formal education was in engineering. Also, like Nervi, one of the fundamental contributions in the design of reinforced concrete structures was the multi-faceted range of skills that included also the priceless talent of knowing how to build.

Nervi applied this concept all the way down to interpreting the role of builder through his own construction company. Morandi was no less a participant in construction, not by being directly involved with a construction company of his own, but by designing and supervising each step of the most innovative constructive solutions for his unorthodox structures. It is this particular skill in which he is an leading figure worldwide that needs to be explained in detail. Especially where bridge construction was concerned, the design of scaffolding and even more so the design of a feasible methodology for the erection of his unique structures, was a task he included in the design of the primary structure itself, not leaving it to a construction company to solve. It is important to recognize is the strict correlation between the theoretical design of these bridges and the methodology to erect them. The description of some of his bridges illustrates Morandi's excellent facility in solving both problems.

Let us use the word *tectonics*, so much misused in recent years, to refer instead appropriately to that exceptional skill that we find in Morandi's work so well demonstrated in the erection of the structures he designed and supervised. Especially in the viaducts and bridges built in difficult terrains, we find amazing solutions that only a great sense of intuition and skill would have been able to produce.

In the footbridge over the Lussia at Vagli di Sotto, Garfagnana, the structure included a 3-hinged central arch with a span of 230 ft and two additional spans of 72 ft-4 in. The central arch spanning over the Lussia was going to be built without the use of centering that could have interfered with the filling of the basin. The solution adopted in this case involved the construction of the two halves of the bridge with two precast prestressed sections, each one resting on its respective bank and then rotated on the abutments until they reached the desired ultimate position, resting temporarily on a support consisting of a scaffolding tower erected at midspan and supporting one end of the arch at the crown. The operation was carried individually by rotating first one of the spans and then the other connecting the two halves of the temporary tower. The prestressing induced in the structure was determined only by the additional stresses that would take place during the erection process, therefore allowing this unique construction process.

The erection of the following structure is particularly significant for its originality: the bridge over the Storms River on the Port Elizabeth-Cape Town Road. It involves a prestressed reinforced concrete arch, constructed into four segments, rotated in sequence and finally assembled into an integral unity. More specifically, the structure consists of an arch fixed on both ends spanning 328 ft with a rise at the crown of 65.5 ft, and includes a cellular cross-section involving four longitudinal ribs and the top and bottom slabs forming respectively the extrados and the intrados.

The depth of the cross-section at the springline is obviously thicker than at the crown due to the fixed-end type of supports; in fact, the overall depth of the arch at the supports is 23 ft, while at the crown, it is reduced to 20.5 ft. The thickness of the concrete slabs also varies from 17 to 12 in, while the four ribs vary also in thickness from 8 ft at the springing to 4 ft at the crown. The arch itself includes four segments, of which the first two starting at the springline, have a length of 31.5 ft and then the two intermediate ones have a length of 147.5 ft each. The first two sections, relatively short, were cast in place with conventional centering. The two intermediate sections, measuring 147.5 ft each, were cast vertically and then, by means of temporary hinges, were simultaneously rotated to be joined together at the crown. The casting of the intermediate sections in a vertical position involved the construction of a temporary concrete frame cast in situ which was eventually demolished at the end of the operation. The temporary hinges were eventually fixed at the completion of the erection process.

Another innovative technique involves a system applied to a triad of prestressed reinforced concrete bridges based on the same structural concept, which includes the Via Olimpica viaduct in Rome, the prototype at Sulmona, and a proposed one over the Tiber at Tor di Quinto, near Rome, which remained only a proposal that was never built. The concept consists of adopting longitudinal beams carrying the bridge deck that are supported at a certain distance from the end, leaving a cantilevered section on each side of the support. The two cantilevered sections with their negative moments reduce the positive moment at midspan of the central section. Up to this point, the structure would not have displayed any special innovations, following traditional concepts of design; but the main innovative aspects of Morandi's creative design is to increase the negative moments of the cantilevered spans, leaving these unchanged in length. To increase such moments, he introduces some prestressing ties that extend from the end of the cantilevers to the bottom of the piers, which add an additional negative bending moment on the cantilevers. With such a methodology, he succeeds in attaining a very slim structure, which proved to be efficient in each of the applications.

The proposed bridges on the Cisa highway consist of hollow, rectangular piers carrying a two-level bridge with a lower and an upper deck, carried in turn by precast Gerber beams. The hollow piers are constructed in sequence using some sort of a telescopic formwork which allows the various segments of the piers to be cast on top of one another as the formwork is lifted for each casting. The upper deck of the bridge is then erected with scaffolds bearing on the lower deck; therefore, the key point is the construction of the lower deck, therefore the that was meant to be done without centering. This consists of precast Gerber beams lifted and positioned by cranes without any centering and placed on concrete brackets cantilevering from the hollow columns.

The bridge over the Lake Maracaibo in Venezuela is a structure over 5.4 miles long which includes a central span of 1,312 ft with a height of 164 ft plus five additional spans of 492 ft each with a height of 164 ft and the remaining part with unspecified limits. The upper structure consists of prestressed concrete, cantilevered cable-stayed girders with suspended spans having a total length of 771 ft.

The towers carrying the stays extend 303.5 ft upward, while the pier foundations have a width of 113.5 ft required by the eccentric loading during erection of the central span. Morandi was the winner of the competition to design this bridge and his proposal was the only one to consider concrete; all other entries proposed steel structures. Concrete was chosen because of the predictable corrosive action of salt water on steel in a tropical region which would have demanded an expensive maintenance program. Most of the elements constituting the spans and the supports were in prestressed concrete, cast in special sites along the shores of the lake and then eventually placed on floats.

They were tugged on site and then lifted in place. Indeed, the description of the various operations and the sequence that they required for the construction and the erections of the entire structures, is quite difficult to present in this context for the complexity of the details involved, yet, just reading parts from the few sources available on this topic, it is quite amazing to comprehend all that was involved. For instance, quoting from a 1965 text *The Concrete Architecture of Riccardo Morandi,* we read: "The whole operation of loading, transporting and launching of each pair of beams took on average only two hours," which speaks for itself in terms of the efficiency of the construction system that is part of Morandi's work.

From other information concerning this work, a major parameter in the execution of the job was to provide free passage to the busy traffic of oil tankers and other ships, which are constantly passing through the channel within the lake, so that no interference should take place within the erection operation and the marine traffic involved.

The suspension system used by Morandi in his concrete bridges is also an inventive combination of concrete, steel cables, and prestressing action that eventually will result in an efficient system that is easy to maintain. The cables consisting of different strengths are housed in a concrete envelope or sheath, which are not bonded to one another, but are free within the encasement. The first operation includes the casting of the concrete sheath, then after its hardening, the cable is stretched and the concrete is consequently compressed. When the vertical load is applied to the cable, its elastic elongation reduces the prestress in the concrete sheath of a certain amount, never relieving completely the concrete from compression. A more complete discussion of this process can be found in the literature describing Morandi's proposal for the construction of the bridges over the river Tartaro at Canda, near Udine.

The Nueva Republica Bridge in Caracas, Venezuela (1953) is a prestressed concrete structure with a central cellular arch spanning 295 ft and a rise of 24.3 ft extended by two rectilinear spans on each side, going from the central piers to the abutments.

The Viaduct over the Fiumarella Valley, near Cantanzaro, is another bridge, including an arch with a span of 760 ft and a rise of 216 ft, supporting a horizontal deck on inclined columns. The whole bridge consists of two separate arches side by side, rigidly connected and having a cellular cross-sectional configuration. The overall depth of the arches is larger at the springline where it is 82 ft and, diminishes at the crown where the depth is 33.5 ft. Each arch at the springline forks into two branches with an innerspace of 72 ft. The cellular cross-section included a transverse member in correspondence of each support, extending to carry the deck above. The upper structure includes a deck slab reinforced by four longitudinal ribs with a variable thickness, having also a variable depth from 4.8 to 7.75 ft. The deck itself includes a slab 6 in thick, with a stiffening diaphragm 6 in wide and 3 ft 3 in deep.

Fritz Leonhardt

Born in 1909 in Germany, Fritz Leonhardt acquired prominent recognition among bridge engineers in Europe. Among his works, was the bridge over the Rhine at Rodenkirchen (1938) with a span of 1,240 ft and a plate-girder of 11 ft. It was the first example of a plate-girder suspension bridge in Europe, but it no longer exists because it was destroyed during World War II. Other work by him includes a group of three bridges of the cable-stayed type over the Rhine in Dusseldorf. The first one, the Theodor Heuss, or North Bridge, had a main span of 919 ft and side spans of 354 ft. The second, named the Kniebrucke, has a span of 1,050 ft.

Most interesting about Leonhardt's thinking on the dynamic instability of bridges is his interpretation of the disaster that occurred in the United States when the Tacoma Narrows Bridge collapsed in 1940 (Figure 1-13). He concluded that to avoid aerodynamic instability caused by the so-called Karman effect of producing rhythmical vortices, the most effective way would have been not to stiffen the bridge with trusses but rather by elimination of the eddies by using a narrow aerofoil deck suspended by only one cable at the center, supported by A-shaped towers.

Figure 1-13 The Tacoma Narrows Bridge pictured during its collapse in 1940 (from Melaragno, M., 1992, *Wind in Architecture and Environmental Design.* New York: Van Nostand Reinhold, by permission of the author, holder of copyrights).

Tung Yen Lin

A leader in the area of bridge design, Tung Yen Lin represents today the most authoritative exponent worldwide. His name is currently associated with the most ambitious project that will probably be developed in the next century as the avant-garde of all megastructures. Biographical data from the present literature, specifically from Who's Who in the U.S., are reported herein for the benefit of the reader, who should definitely be informed about the major representatives in this area of design. Born in Foochow, China in 1911 Lin married Margaret Kao in 1941 and had two children. He was educated at Chiaotung University, in Tangshan, China, in 1931 (where he received his Bachelor of Science in Civil Engineering); University of California-Berkeley in, 1933 (where he received his Master of Science); LLD, Chinese University in Hong Kong,in 1972; Golden Gate University in San Francisco, in 1982; Tongji University in Shanghai, in1987; and Chiaotung University in Taiwan, in 1987.

Lin's postions have included the following;1933-1946 he was chief bridge engineer, chief design engineer for the Chinese Government Rys.; 1946-1955 assistant and later associate Professor at the University of California; 1955-1976 Chairman of Div. Structural engineering; 1960-1963 Director of a structural lab as well as Chairman of the Board of T.Y. Lin International; 1968, President of Inter-Continental Peace Bridge, Inc., 1968; Consultant to the State of California, Defense Department, also to industry; Chairman of World Conference Prestressed concrete, 1975; Western Conference Prestressed Concrete Buildings, 1960; Chairman of the Board of Lin Tung Yen, China, 1993 to the present.

He is member of the following professional associations: Honorary Professor at the Chiaotung University, 1982; Honorary Professor at Tongji University,1984; Honorary Professor at Shanghai Chiaotung University, 1985; Member of ASCE (which awarded him the Wellington award and the Howard medal); National Academy for Engineering; Academia Sinica; International Federation Prestressing (from whom he received the Freyssinet medal); Honorary member of the American Concrete Institute; and Prestressed Concrete Institution (from whom he received a medal of honor).

Lin was a recipient of the following honors and awards: Berkeley citation award, 1976; NRC Quarter Century award, 1977; AIA Honor award, 1984; President's National Medal of Science, 1986; Merit award American Construction Engineers Counsel, 1987; John A. Roebling Medal Bridge Engineering, 1993; Lifetime Achievement Award, Asian American Archives and Engineering Association, 1993; Outstanding Achievement award AAAE Association of Southern California; named Outstanding Alumni of the Year by the University of California Engineering Alumni Association, 1984; University of California Alumnus of the Year award, 1994.

A major outstanding bridge project is the proposed bridge between Spain and Morocco, (1991). This project is a proposal concerning a project aiming to link the two continents spanning from Punta Oliveros, Spain to Pointe Cires, Morocco, with a shortest distance distance of 14 km. The project includes two spans, each 5 kilometers long plus two sidespans of 2.5 kilometers each. The three piers would have respectively a depth of 80 m, 450 m, and 360 m, which would be the deepest piers ever built. The structural system supporting the spans would include a suspension system combined with a cable-stayed system or a cantilevered system which would constitute a hybrid. This proposal deferred from another proposal by others who envisioned a bridge cited on a different location which would have implied a much longer span, reducing the depth of the piers to approximately half.

Also proposed is the Inter-Continental Peace Bridge (mid 1980's) spanning the Bering Strait from Alaska to Siberia along the Arctic trade and transit corridor between the two continents. The bridge would connect two intermediate islands, the Little and Big Diomede Islands. The whole structure would include two spans of 550 m over the two navigational channels and in addition to 220 m of the cable-stayed type adding a span of 366 m each. The overall length of the series of bridges would include over 80 km. In comparison with the proposed bridge between Spain and Morocco, this would include shorter spans and shallower piers, yet the geographical location would definitely be much more challenging in terms of construction and maintenance.

Santiago Calatrava

An engineer, an architect, and a sculptor, Calatrava combines within himself these different vocations which are all reflected contemporaneously in his creations. He has especially brought a creative new approach to the field of bridge design, standing at the leading edge of a new generation of bridges that are taking shape now and definitely will continue on through the next century. Biographical notes about Calatrava derived from the current literature, Who's Who 1996, include the following data. Born in Valencia in 1951, he married Robertina Maragoni in 1976 and had three children. He was educated at an art school in Valencia and the Spain; Escuela Tecnica Superior de Arquitectura de Valencia; Swiss Federal Institute of Technol in Zurich where he studied civil engineering and received a doctorate in technical science.

From 1979-81, he was an assistant instructor for Building Statics and Constructions and for Aerodynamics and Light Weight Structures. In 1981 he established his professional practice: principal of an architectural and engineering consulting office in Zurich. In1989 he had an architectural practice in Paris, and in 1991 a professional practice in Valencia.

Calatrava is a member of the following professional associations: Union of Swiss Architects, 1987; International Academy of Architecture, 1987; European Academy, 1992; Honorary of FRIBA, 1993; and Honorary member of the Union of German Architects, 1989.

He was a recipient of the following honors and awards: International Association for Bridge and Structural Engineering Award, 1988; Gold Medal, IStructE, 1992.

The numerous list of bridge projects as reported by Frampton, et al., (1993), includes the following works presented in chronological order:

- Acleta Alpine Motor Bridge in Disentis, Switzerland; 1979 diploma project; River Rhine; total length 262 m, maximum span 100 m; height above riverbed 50 m, above support foundation 30 m; Prestressed, post-tensioned concrete bridge; cantilevered construction.
- Letten Motorway Bridge in Zurich, Switzerland; 1982 project; River Limmat; river crossing between two tunnels; total length 161 m, maximum span 70 m; Continuous beam with open box-girder cross-section; concrete and diagonal steel truss member; competition with Marbach & Ruegg.
- Rhine bridge in Diepoldsau, Switzerland; 1982 project; River Rhine; motor and pedestrian bridge; total length 180 m, maximum span 140 m; Cable-stayed with concrete pylon and carriage way; competition with Wenaweser & Wolfensberger.
- Caballeros footbridge in Lerida, Spain; 1985 project; River Segre; total length 210 m, maximum span 140 m; depth of pylon 28m, height above riverbank 7 m; cable-stayed bridge; steel pylon with concrete abutments; light steel and aluminum deck structure; competition.
- Bach de Roda - Felipe II bridge in Barcelona, Spain; 1985-1987; Railway crossing; motor and pedestrian bridge; total length 129 m, maximum span 46 m; depth of arch 10 m, height over railway 8 m; concrete approach spans, steel arch with compound steel beam structure for carriage way.
- Raitenau overpass in Salzburg, Austria; 1986 project; pedestrian bridge; Residenz Square; temporary exhibition entrance; total length 24 m, maximum span 15 m; cantilevered box-girder construction in timber; competition.
- 9 d'Octubre bridge in Valencia, Spain; 1986-1989; Turia Riverbed; motor and pedestrian bridge; extension of the Campanar Boulevard over proposed lake and park; total length 2,144 m with regular spans of 7.2 m; height above foundation 4.9 m; viaduct type structure, concrete carriage way supported on concrete and steel columns.
- Oudry - Mesly bridge in Créteil/Paris, France; 1987 - 1988; motorway crossing; foot bridge connecting two housing estates; total length of crossing 120 m, maximum span 55 m; depth of arch 8 m; steel arch.
- Passerelle de Thiers in Thiers, France; 1987 project; "Creux de l'Enfer" waterfall; pedestrian bridge; total length and span 8 m; steel construction.

- Pontevedra Bridge in Pontevedra, Spain; 1987 project; River Mino; motor bridge; total length 510 m, maximum span 130 m; cable-stayed with box-girder and pylon of concrete; pylon head and cables in steel; competition.
- Alamillo Bridge and Cartuja viaduct in Seville, Spain; 1987 - 1992; River Guadalquivir; pedestrian and motor crossing; total length 250 m, maximum span 200 m, height of the mast over carriage way 142 m; cable-stayed bridge with inclined pylon, steel box structure filled with concrete; ribbed steel deck structure; viaduct of concrete.
- Casine footbridge in Florence, Italy; 1987 project; River Arno; footbridge to Le Cascine Park; depth of arch 10 m, height above river 21 m; total length 125 m, maximum span 74 m at deck and 104 m between piers; project; concrete abutments supporting a steel arch.
- Leimbach footbridge in Zurich, Switzerland; 1988 project; River Sihl; footbridge leading to Leimbach; railway station; total length 49 m, maximum span 41 m; concrete pylons with steel stays; competition.
- Lusitania Bridge in Merida, Spain; 1988 -1991; River Guadiana; motor and pedestrian bridge; total length 465 m, maximum span 189 m; depth of arch 32 m; Mid-span steel arch with pre-stressed continuous box girder; cross post-tensioning.
- Wettstein Bridge in Basel, Switzerland; 1988 project; River Rhine; tram, motor and pedestrian bridge; total length 199 m, maximum span 66 m; depth of arch 6 m; steel arched space frame structure; counter-proposal to official version.
- Gentil Bridge in Paris, France; 1988 project; River Seine; pedestrian, motor and metro bridge; total length 186 m, maximum span 137 m; depth of arch 30 m; steel box-girder carriage way with inclined steel arch; competition.
- Ruess Steg in Flüelen-Seedorf, Switzerland; 1989 project; River Teuss; rural footbridge; total length 40 m, maximum span 40 m; depth of arch 5 m; arch and deck of timber construction.
- Mira Flores Bridge in Cordoba, Spain; 1989 project; River Guadalquivir; motor and pedestrian bridge; motorway crossing; total length 132 m, maximum span 130 m; stiffened box-girder deck with steel arch.
- Gran Via bridge in Barcelona, Spain; 1989 project; motorway bridge; total length 258 m, maximum span 126 m; double steel arch bridge with twin box-girder deck.

- Ria bridge in Ondarroa, Spain; 1989; River Artibay; motor and pedestrian bridge; total length and maximum span 74.5 m; asymmetric steel arch dividing the box-girder carriage deck from the cantilevered pedestrian deck.
- La Deves footbridge in Ripoli, Spain; 1989-1991; River Ter; pedestrian bridge; total length 65 m, maximum span 44 m; Depth of arch 6.5 m; inclined steel arch and ribbed deck structure, timber deck.
- Port de la Lune swing bridge in Bordeaux, France; 1989 project; River Garonne; Motor and pedestrian bridge; total length 300 m, maximum span 100 m; cable-stayed swivel bridge; proposal for an exhibition of ideas.
- Urevitarte footbridge in Bilbao, Spain; 1990; River Nervion; pedestrian crossing; total length 73 m, maximum span 71m; depth of arch14.5 m; concrete abutments and ramps; Inclined steel arch with steel box-girder deck.
- East London river crossing in London, England; River Thames; motor and pedestrian bridge; Maximum span between abutments 510 m, span of arch 300 m; depth of arch 25 m, height above abutments 103 m; conventional box-girders for approach spans, steel central arch and box-girder with concrete pylons.
- Nouveau Pont Sur le Vecchio bridge in Corsica, France; 1990 project; River Vecchio; motor bridge; total length 91 m, maximum span 65 m; height above riverbed 23 m, above support foundation 5 m; concrete construction; competition.
- Grand Pont bridge in Lille, France, 1991 project; motorway bridge over railway lines; total length 132 m, maximum span 108 m; steel arch and deck construction.
- Médoc swing bridge in Bordeaux, France; 1991 project; River Garonne; motor and pedestrian bridge; total length 416 m with two spans of 120 m each; height of pylon 100 m, over water approximately. 120 m; cable-stayed swivel bridge with steel pylon; competition.
- Ronprinzen bridge in Berlin, Germany; 1991; River Spree; motor and pedestrian bridge; total length 74 m, maximum span 44 m; approximate height above river 7.5 m; steel construction; competition.
- Oberbaum bridge in Berlin, Germany; 1991; River Spree; motor, pedestrian and underground bridge on two levels; total length and maximum span of 22 m; steel construction; renovation of existing defunct structure with introduction of a new central span.

- Klosterstrasse railway viaduct in Berlin, Germany; 1991 project; urban street crossing; railway bridge; total length 23 m, maximum span 15 m; steel bridge with railway deck in composite steel.
- Alemeda bridge in Valencia, Spain; 1991; Turia Riverbed; motor and pedestrian bridge with access to Alameda underground station; total length 164 m, maximum span 130.5 m; depth of arch 14 m; steel arch and steel box-girder carriage way.
- Solerino footbridge in Paris, France; 1992 project; River Seine; pedestrian bridge between Louvre and Musée d'Orsay; total length 130 m, maximum span 76 m; steel construction; competition.
- Serpis bridge in Alcoy, Spain; 1992; River Serpis; motor and pedestrian bridge; total length 520 m, maximum span 210 m at deck and 400 m from base to base; depth of arch 30 m; steel arch, compound steel beam structure for carriage way.

Among the various projects listed above, the Bach de Roda - Felipe II bridge in Barcelona stands out in particular for its bold delineation emerging from the urban landscape of the city with a unique identity that differentiates it from most any other bridge structure ever built. The structure crosses over a railroad right of way in the city connecting two areas, the Sant Andrea in the north to the San Marti in the south on the outskirts of Cerda. In so doing, this project was intended to revitalize these two impoverished areas, a task that the structure eventually accomplished.

The incentive for direction of this bridge came from the major plan for the face-lifting of the city in connection with the Olympics that the city hosted in 1992. Plans for the structure started in 1985 and the bridge was eventually completed in 1987, way ahead of the Olympic games. Since its completion, the bridge has become a definite focal point in the urban texture of Barcelona. For instance, the two pair of arches on each side of the roadway are painted white, standing out considerably visually. Both arches are tied by the main girders supporting the deck. These girders support the cross beams, which in turn carry the concrete roadbed and the pedestrian walkways as well.

Other structures besides the previously mentioned bridges include: Stadelhofen Railway Station, Zurich, 1983-1989; Lucerne Station Hall, 1982; BCE Place, Toronto; Montjuic Communication Tower, Barcelona, 1991; Lyon Airport Railway Station, 1989-1994; Kuwait Pavilion Expo '92, Seville, 1987-1992.

Exhibitions include the Museum of Architecture, Basel, 1987; Design Museum, Zurich, 1991; Museum of Modern Art, New York, 1992; Retrospective, RIBA, London, 1992; Deutsches Museum, Munich, 1993; Valencia, Lubeck, and Copenhagen, 1993.

Jörg Schliach

Particularly versatile throughout the wide scope of structures, the professional practice of Jörg Schliach puts him on the leading edge of structural engineering and at the avante-garde of architectural morphology.

Schliach was born in 1934 in Stetten/Remstal, 17 km east of Stuttgart, which is within that region of Germany named Schwaben. He attended primary school in Stuttgart and then because of the events of World War II he moved to the countryside with his family.

After the war he eventually finished his studies at Stetten. He then continued in his professional development by studying at Technische Hochschule in Stuttgart where he acquired a two year degree. Then, because of his particular interest in engineering and architecture he continued to study in these directions. In 1955, after recieving his two year degree he moved to Berlin to continue his studies, attaining a Diplom at Technische Universitat. Along with his techinical cirriculum biograpahers point to the humanistic component of his culture, which in combination with engineering and architecture made him quite close to the profile of a renaissance man.

After graduating in Berlin in 1959, he then moved back to Stuttgart to work on a doctoral thesis under Leonhardt. Then he attended Case Institute of Technology in Cleveland for a masters degree. The year after, he attained a scholarship to return back to Stuttgart to complete his original doctoral thesis under Leonhardt from 1960 to 1963.

Upon his graduation, he joined the consulting group of Leonhardt und Andra,which marked the beginning of his professional career. He is particularly interested in a variety of structures, and his career includes major fields in structural engineering, including: towers; concrete thin-shells; cable net structures; glass grid roofs; textile membrane roofs; highway, railway, and pedestrian bridges; suspended buildings; and solar power plants. Although it would be of major interest to cover all of Schliach's work, we can only focus on the railroad and highway, pedestrian bridges that carry his signature.

His bridges include:

- Calcutta Hooghly River Bridge, 1971-1993 (Figure 1-14)
- Akkar Bridge in Sikkim, 1988
- Evripos Bridge, 1992
- Obere Argen Crossing, 1990
- Kirchheim Overpass, 1993 (Figure 1-15)
- Kelheim Westtangente Bridge, completed1988
- New York Williamsburg Bridge proposal, 1988
- Prince Edward Island Link proposal, 1988

- Strait of Gibraltar Crossing proposal, 1990
- High-Speed Train Viaducts, 1990

Figure 1-14 The second Hooghly bridge in Calcutta by Jorg Schlaich (reproduced with the permission of the copyright holder).

Figure 1-15 The overpass at Kirchheim, Germany by Jorg Schlaich (reproduced with the permission of the copyright holder).

- Berlin Havel and Humboldthafen Railbridge projects, 1994
- Ingolstadt Danube Bridge, 1994
- Stuttgart Nesenbachtal crossing, 1994
- Macau-Taipa Crossing, 1994
- St. Paul Mississippi Bridge proposal, 1992
- Hong Kong Ting Kau Bridge, 1994

Pedestrian bridges by Dr. Schlaich include:

- Stuttgart BGS Rosenstein I, Rosenstein II, Cannstatter Straße, 1977
- Neckarsulm I and II, 1985
- Stuttgart Max-Eyth-See, 1989 (Figure 1-16)

Figure 1-16 Pedestrian bridge at Max-Eyth-See Park in Stutgart, Germany by Jorg Schlaich (reproduced with the permission of the copyright holder).

- Kelheim, 1987
- Suttgart Neckarstraße, 1990
- Stuttgart IGA, Nordbahnhof, Heilbronner Straße, Lowentor, 1992
- Pforzheim LGS, LGS I, LGS II, and LGS III, 1991
- Stuttgart BGS, Heinrich-Baumann Straße, 1977
- Sindelfingen, 1986
- Unterturkheim, 1991
- Stuttgart IGA, Pragsattel I and II, 1992
- Nantenbach, 1990
- Bad Windsheim, 1988
- Herrenberg, 1992
- Kiel, 1995

For more information about the prolific work of Dr. Schlaich, see *The Art of Structural Engineering* by Alan Holgate and published by Axel Menges in 1997.

MATERIALS IDENTITY

From a historical perspective the evolution of bridges depended not only on the ingenuity of designers, but also on the availability of materials that have gradually been developed through the course of centuries. The dominant role that materials have played in the anthropological evolution of humanity is a historical reality that has been present throughout the eras such as the old and new Stone Ages, as well as the ages of Iron and Bronze. From this perspective, it is perfectly justified to underline the role of materials even in the history of bridges, without the danger of overstating it. Besides the practical aspects of different technologies that each material requires, another major point that any designer must understand, including those involved in bridge structures, is that each individual material speaks a language of its own that the designer needs to understand. Such observations that may sound dictated by poetic overtones are by no means abstract concepts, but are intended to be practical suggestions for designers aiming at submitting meaningful proposals in bridge competitions.

From the very inception of the design phases, the selection of material to be used for the design of a bridge is among the factors to be selected first in the hierarchial order of topics. The choices include: reinforced concrete; prestressed or posttensioned concrete, steel, aluminum, timber (for special occasions), and masonry, which is mostly used only for architectural reasons today. Each one of these materials implies a number of factors to be considered during the selection process. Intuitively, one could include the following: availability of materials within the region or a reasonable distance; availability of skillful labor for the construction and erection of the structure; resistance to weather and corrosive actions of the elements; ease of construction; time for construction and erection; durability; aesthetic factors; harmonious interaction with surroundings; public acceptance.

Limiting the discussion to a strict minimum, a few remarks are included in the following sections for each of the basic materials mentioned above.

Reinforced concrete

The dominance of reinforced concrete as a structural material for buildings as well as for bridges is an obvious fact, clearly indicated by the large number of bridge structures built throughout the world using reinforced concrete, prestressed or posttensioned concrete, cast-in-place or precast. The major advantage of reinforced concrete, prestressed or not, is definitely an economic factor, which is particularly significant for bridges built in geographical areas, where structural steel is particularly expensive.

Also, presently, even in countries such as the Unites States, Canada, and Germany, reinforced concrete is extremely popular due to its reduced maintenance costs in comparison to structural steel.

Makes the use of reinforced concrete over structural steeleven more desireable the present increased efficiency of the materials available. These include, high strength concrete mixtures that have reached extremely high values in the range of 10,000 psi and reinforcing steel grade 60 that has a yield strength of 60,000 psi.

At first glance concrete may seem to have an indestructible capacity that could extend through the centuries like some masonry structures that have survived through centuries. Yet such an impression is definitely false. It is hard to believe in fact, that a structure like the Ponte Milvio in Rome is still in service after 2000 years requiring minimum or no maintenance, while many concrete bridges have a life-time less than 50 years (Figures 1-17 and 1-18).

Concrete, which may seem to be relatively inert, is very vulnerable to atmospheric agents which produce an active chemical process that progresses inward, reaching the reinforcement and exposing it also to corrosion and failure. Some of the causes of concrete degeneration, either exposed to air or water, are: chemical attack; cavitation; movement of abrasive material by waterflow; traffic; wind blasting; and the impact of floating ice.

Formation of soluble products is a process that attacks concrete surfaces degenerating the quality of concrete, in a slow progressing attack. Organic and inorganic acids create soluble products in concrete surfaces which are

Figure 1-17 Ponte Milvio crossing the Tiber River in Rome. Built in 109 B.C. The bridge stands today as it was originally built, having withstood centuries of exposure to weather (photograph by Deborah Melaragno).

removed by leaching. Specifically, the substances in question include: inorganic acids, acetic acid, various sulfates, hot distilled water, and others.

Some substances combined with cement create certain compounds that disrupt the concrete because they have a larger volume than that of the cement paste from which they derived. White alkali made of the sulfates of sodium, magnesium, and calcium are the most aggressive. The sulfates reacts with the hydrated lime and calcium in cement paste forming calcium sulfate and calcium sulfoaluminte. During such a reaction the cement paste expands and eventually is destroyed.

Crystallization is another destructive process that occurs when concrete is alternately wet and dry. In this case sodium carbonates are produced which crystallize in the pores of the concrete disintegrating the concrete surface. Spalling of the concrete is another effect of chemical attack that occurs when destructive agents reach the concrete reinforcement corroding it. At present there are no additives that could prevent such erosion processes generated either in air, water, or soil. On this basis we must recognize that all of these processes attacking concrete doom any concrete structure to a slow process of degeneration resulting in until final failure.

Figure 1-18 Ponte Milvio, showing the gateway to the bridge on the south side of the river (photograph by Deborah Melaragno).

Structural steel

The various types of structural steel identified in the *Standard Specifications For Highway Bridges* include the types which are respectively designated according to AASHTO and the ASTM as follows:

- M 270 grade 36 (AASHTO); A 709 grade 36 (ASTM)
- M 270 grade 50 (AASHTO); A 709 grade 50 (ASTM)
- M 270 grade 70W (ASHTO); A 709 grade 70W (ASTM)
- M 270 grades 100/100W (ASHTO); A 709 grades 100/100W

The designer will base the selection of the steel type upon the mechanical characteristics of the material, including elements of strength and durability and the resistance to weathering agents. The modulus of elasticity for all kinds of steel is E = 29,000,000 psi.

Standard specifications for the structural analysis and fabrication of steel members and their connections cannot be included in the present context for reasons of space. Consequently, the reader is referred to major references such as the *Standard Specifications for Highway Bridges* followed in the United States. However, for preliminary design, the following few recommendations could be useful.

The ratio of depth-to-span for beams and girders is suggested to be not less than 1:25. For trusses, the ratio of depth to span should be not less than 1:10. Maximum deflection for flexural members due to live loads and impact should be not larger than the span length/800. Maximum deflection allowed for cantilevered flexural members, due to live loads and impacts, should be not larger than span length/300, but in case of pedestrian load the limit should be not larger than span length/375. The slenderness ratio for primary members should not exceed 120, while for secondary members and bracing members it should not exceed 140.

Considering steel bridges on the basis of their overall behavior, with regard to durability, one should consider the major causes that generate the deterioration of steel bridges. The aging of steel structures exposed to the weather is mostly due to corrosive effects from water, chemicals, and also the effect of fatigue which generates cracks. To this effect, the reader is referred to the literature concerning fatigue stresses especially to those articles that consider fatigue in the structural members of steel bridges. For instance in the AASHTO *Standard Specifications for Highway Bridges* the effects of fatigue are mentioned in Division I, Sec. 10.58 Fatigue. In this section, the analysis of the probability of fatigue of steel members or connections under service loads and the allowable range of stress for fatigue shall conform to article 10.3.

Such article, in turn, refers to repetitive loading and toughness considerations. Under this section members and fasteners subject to repeated variations or reversals of stress should be designed so that the maximum stress should not exceed the basic allowable stresses. Furthermore, the range of stresses should not exceed the allowable fatigue stress range. These in turn are given in Table 10.3.1A. This table includes eight categories and five categories of cycles that include: 100,000 cycles; 500,000 cycles; 2,000,000 cycles; and over 2,000,000 cycles. The categories are, in turn, a reference to Table 10.3.1B.

Furthermore, slab reinforcement should be designed in accordance to Article 10.58.2.1 and shear connectors should be designed in accordance to Article 10.38.5.1, while hybrid beams and girders should be designed according to Article 10.3.

Masonry

Masonry of present day, applied mostly to buildings, still has its place in architectural artifacts, because of its character in the majority of cases and because of its practical aspects in areas wheremasons still exist. More than stone masonry, current applications include bricks as well as the relatively modern concrete blocks.

The application of masonry to the bridges of today in the industrialized countries is definitely an abandoned-building methodology; however it could eventually still be alive in those countries with low levels of technology, especially concerning extremely short spans or culverts. The reason for expanding on this subject is not really for considering any potential applications to any new structure. Instead, the subject is considered for comparing old masonry structures with present-day reinforced concrete equivalents.

When examining the longevity of reinforced concrete bridges, it has been seen that the average lifespan is extremely short. This is true not only in terms of what one may intuitively have expected, but also when reinforced concrete is compared to masonry. We have seen that the process of decay of concrete bridges is usually a chemical attack the material by water or other chemicals present in the atmosphere. The surface in a concrete mass is an area of highly intensive chemical attacks which at first examination would not be expected, considering it as an artificial stone. But, such an assumption is definitely not true because of the cement paste that interconnects the lapideous materials (i.e., the fine and course aggregates). The cement paste, in fact, is not an inactive product and it undergoes chemical deteriorations that does not affect stones.

Inspection to reinforced concrete bridges shows that the deterioration of concrete is a progressive process that eventually reaches the steel reinforcement, and exposes it to corrosion. In the case of masonry, one cannot explain how the mortar in the joints survives for centuries. In other words, how do we explain the resistance of mortar joints in between bricks or stones that have allowed masonry structures to survive for thousands of years, while concrete bridge structures have failed after only a few decades? Undoubtedly, the strength of present-day concrete is much superior to the quality of mortars used in Roman times.

The pozzolana used in the mixing of mortars with calcium hydrate in substitution of the inert sand had produced what is usually referred to as Roman cement. But definitely such a compound was not assumed to have better qualities than modern cement pastes. Furthermore, the mortar joints between bricks seems so vulnerable in comparison to a mass of concrete, both exposed to the same atmospheric agents; but the tangible reality is that mortar joints have a greater longevity than concrete masses.

Tangible examples supporting this discussion are available among the numerous remnants of Roman structures. Most amazing among them is the Pantheon in Rome, considering the exterior as well as the interior state of preservation. However, since the topic in question is that of covering bridges, the most astonishing example is the Ponte Milvio, previously mentioned in historical context of bridge evolution. The bridge, Ponte Milvio, as it is presently referred to, had also been called Pons Mulvius or Ponte Molle, and was built in 109 BC, historically significant for major events. The bridge includes several arches spanning from pier to pier, and additional smaller arches in between, that allow the flow of water in case of floods.

The structure clearly verifies the state of preservation of the bridge which has been in service since its construction and still carries pedestrian traffic. Most astonishing is that up to a few decades ago the bridge carried a trolley service including two cars travelling on rails transferring live loads as well as vibrations throughout the entire structure. This was line number one, connecting Piazzale Flaminio with Piazza di Ponte Milvio that operated for several years.

CONCEPTUAL DESIGN PHASE

The conceptual design phase for the submission of bridge design proposals is discussed in part three, yet at this point after attaining a historical perspective of the subject, conceptual principles of the structure start to germinate. Before any delineation starts to take shape, ideas need to be organized along logical patterns that will progress, subsequently creating a logical foundation for the artifact that will eventually emerge from this process.

Any artifact produced by humans that aims to attain a personality or an identity of its own, needs as a prerequisite to have a meaning developed through rational thoughts, which are eventually shaped by emotional components of the human mind. This obviously applies to all kinds of human products, from the small to the large, including in this context large structures like buildings and bridges. In practical terms this implies that any bridge proposal should start from a phase that could be referred to as "conceptual." This may sound like an exaggerated expression of formality to many structural engineers, yet it will be readily understood by any practicing architect due to differences between the engineering and architectural point of view.

The design of bridges, like that of any architectural structure, is a holistic package produced by a professional with a broad personality capable of handling the product with a multifaceted approach. The artifact being produced has to satisfy quite a number of design goals, including the approval of the client and that of the users of the artifact. But, while architects have been trained to articulate such an approach to design, structural engineers acting as principals of a design firm for bridge proposals definitely need to learn. Therefore, engineering firms should be adequately prepared to elaborate conceptual designs that have meaning and offer a sense of appeal that will capture a design jury.

As the word implies, the conceptual design phase is based on the rationality of the project expressed in descriptive terms. A vocabulary should be established using terms clearly understood or capable to be defined in simple terms. This is a key point, especially when such a vocabulary must use abstract terms which are not clear in themselves and could be easily misunderstood. In this phase, therefore, the focus is not on the physical description of a project in a drawing or model format, but it is mostly in terms of a verbal presentation of ideas to be stated in a clear form that any juror can grasp whether the jury is composed of professionals, experts or by the public at large.

The format need not be a stereotyped, mechanical presentation of facts, but it could be elaborated in an informal way by anyone with clear ideas in mind. Therefore, the various goals and objectives definitely need to be included and many other points could be individually presented in form of a long checklist. Such a checklist could be prepared following some of the data presented later in when the phase of preliminary design is discussed. However, regardless of the topics presented in the conceptual phase, it is essential to make sure that the bridge being proposed has a meaning that will capture the imagination of the jury with clear simple logic, avoiding confusion, repetitions, and over emphasizing.

A conceptualization of structure is not part of the problem solving techniques used later in the preliminary and in the final design, but it is rather a process of introspection. In other words after having procured any data concerning the project and after a period of assimilation of such data, a designer should be able to react and eventually be able to generate an output of ideas. This is the typical process of creating objects that eventually will gain meaning and personality of their own. It could probably be the most demanding and frustrating phase of design, which nevertheless cannot be avoided because it is definitely the very key to the solution of the problem, i.e., submitting a proposal that will be the winner.

AN INVENTORY OF SIGNIFICANT BRIDGES

The number of bridges in the world is amazingly large. In fact, in California alone, one could count several thousand bridges. Therefore, it will be logical to have some sort of an inventory to identify and classify them in a way that they could be identified with ease. Such as inventory should be organized in terms of categories such as: (1) urban or suburban; (2) along primary or secondary roads; (3) for highways or for railroads; (4) in accordance with materials such as wood, masonry, reinforced concrete or steel; (5) according to structural systems such as suspension, cantilever, simple truss, steel truss, continuous truss, etc.; and (6) according to country.

The use for such inventories will be multifaceted. Designers, of course, can have their own part in using such classifications for identifying bridges according to their own interest. For instance, designers may want to make their own inventory of bridges that are more significant to them for different reasons, such as to verify special solutions in terms of structural systems or in terms of aesthetic factors as well as in terms of special details.

Other purposes for bridge inventories include reasons of national pride when bridges constitute a national investment in the country's highway system, railroad system, and urban viability. Besides the capital investment that bridges represent, they are an object of civic pride. In this respect bridges are greatly admired when they push forward the limits of engineering achievements. Consequently, bridges covering the longest spans are admired most, and they are often listed according to the length of the span they cover.

For all of these reasons, an inventory of bridges is included in the appendix. This appendix comprises two lists, i. e., one for the United States and another for the rest of the world. As prepared by the Survey of State Highway Engineers in 1995, both lists include the year of construction, the name of the bridge, the location, and the longest span. In turn, we notice that the way in which the list is organized considers various structural groups such as suspension, cantilever, truss, etc., placing individual bridges in each category in the order of span length. The appendix identifies the most notable bridges, for bridges of historical meaning, special literature must be consulted.

CURRENT DESIGN FIRMS

Knowledge of the most significant bridges built in the last decade has particular value for understanding design trends, which is essential for any designer entering into an area of presenting new proposals. Nothing would be more efficient in this respect than examining bridges that have received national attention through awards such as those given by the American Institute of Steel Construction and the Portland Cement Association. The AISC in fact has given awards to bridges through a sequence of competitions being offered every two years since 1928.

Divided into several groups, bridge submissions have been judged by juries of experts. Since the first competition the bridges receiving awards by the AISC have been listed in various issues of *Modern Steel Construction*. The information given for each project being recognized with an award include the category, name of bridge, location, designer, consultant, general contractor, fabricator, erector, owner, steel weight/sq. ft. of deck, cost, and steel tonnage.

For reason of space the category, the bridge name, and the location have been given in the appendix for each bridge that received awards by the AISC in the 1996, 1993, and 1991 competitions only. More detailed information about these bridges and others awarded in previous competitions, is provided the works cited in the bibliography. In this context, however, we limit the list of names of the designing firms and the names of the consultants for the bridges being recognized by the AISC in the 1996, 1993, and 1991 competition.

1996 AISC Competition

Designers: Hanson Engineers, Inc.; Greiner Engineering, Inc.; WSDOT - Bridge and Structures Office; HNTB Corporation; Sverdrup Civil, Inc.; Greiner, Inc.; Steinman Boynton Gronquist & Birdsall; Beiswenger, Hoch & Associates, Inc.; Howard Needles Tammen & Bergendoff Corporation; Clark Engineering Corporation; Hardesty & Hanover, Llp.; Bucher, Willis, & Ratliff Corporation; LONCO, Inc.; Teng & Associates, Inc.; Kunde, Sprecher & Associates, Inc.; De Leuw, Cather & Company; HNTB Corporation; Hayes, Seay, Mattern & Mattern, Inc.; Petratrovich, Nottingham & Drage, Inc.; Mackin Engineering Company; and Collins Engineers Inc.

Consultants: Figg Engineers, Inc.; Leonhardt, Andra & Partners; Ford, Powell & Carson, Inc.; MK Centennial; Portuondo Perotti Architects, Inc.; and Group III Design.

1993 AISC Competition

Designers: Alfred Benesch & Company; HNTB Corporation; Idaho Transportation; ENTRANCO; Grieger, Inc.; Meheen Engineering Corporation; Sverdrup Corporation; California Department of Transportation; Ayres Associates; Modjeski and Masters, Inc.; Pickering, Corts & Summerson, Inc.; Zimmer Gunsul Frasca Partnership/Skidmore Owings & Merrill; Bergmann Associates; Kohn Pedersen Fox Associates; Robinson Engineering; Tensor, Inc.

Consultants: The Hastings Group; Rowan Williams Davis & Irwin; Sasaki Associates; and Weiskopf & Pickworth.

1991 AISC Competition

Designers: Howard Needles Tammen & Bergendoff; STV/ Seelye Stevenson Value & Knecht; New York State Department of Transportation; T. Y. Lin International; McTish, Kunkel & Associates; Minnesota Department of Transportation; Bakke Kopp Ballou & McFarlin; Tennessee Department of Transportation; Parsons, Brinckerhoff Quade & Douglas, Inc.; HDR Engineering, Inc.; Hazelet & Erdal, Inc.; Hardesty & Hanover; Zimmer Gunsul Frasca Partnership; Envirodyne Engineers, Inc.; Modjeski and Master; and Sverdrup Corporation.

Consultants: SEA Engineering; Jupiter Engineering; Alfred Benesch & Co.; Howard Neddles Tammen & Bergendoff; Teng & Associates; KPFF Consulting Engineers.

Similar to the AISC, the Portland Cement Association (PCA) also established a Biennial Bridge Awards Competition in 1988 for the purpose of recognizing excellence in design and construction of concrete bridges. While more detailed data for each winning project are listed in the appendix, the list of the designers are presented herein for the 1996 and 1994 competitions as follows:

1996 PCA Competition

Designers: Sandwell, Inc.; Nathan L. Jacobson & Associates, Inc.; HNTB Corporation; Ontario Ministry of Transportation; Oregon Department of Transportation, Bridge Engineering Section; LoBuono, Armstrong and Associates; Edwares and Kelcey, Inc.; Wiss, Janney Elstner and Associates, Inc.; Wells Engineers Minnesota, Inc.; Dillion Consulting, Ltd.; Nicholls Engineering.

1994 PCA Competition

Designers: Fred Wilson & Associates; Wilson Okamoto and Associates; J. Muller International; Libby Engineers; Law Crandall; Eastern Federal Lands Highway Division, Federal Highway Administration; Figg Engineering Group; California Department of Transportation; SEQAD Inc.; Burkett & Wong, Inc.; A. A. Williams & Associates, Ltd.; McCormick Rankin & Associates, Ltd.

Consultants: DRC Consultants, Inc.

EFFICIENCY OF MODERN STEEL BRIDGES

The efficiency of steel bridges among the last generation of structures is definitely a major element to characterize the technological progress in this field. Obviously with the increasing strength of materials and the availability of computers for more accurate analysis, a growth in efficiency that can easily be measured considering the structural weight in pounds per square foot of deck area is to be expected. From available data about bridge winners of AISC competitions, it is possible to attain a general overlook of the efficiency of the new structures. Organizing a list in order of efficiency expressed in pounds per square foot of deck area we attain the following:

For the period 1994-1996

- Rainbow arch bridge (short span) 20 lb/sf.
- Woodlands water crossing at Mall Ring Road (short span) 22.5 lb/sf.
- Fred Hartman Bridge (long span) 29 lb/sf.

- Quivira Road viaduct (grade separation) 29.12 lb/sf.
- Pulaski Road over I-55 Spudi (reconstructed) 35 lb/sf.
- 8th Street bridge over the Sheboygan River (moveable span) 40 lb/sf.
- South Station connector ramp (short span) 40 lb/sf.
- United States Naval Academy Bridge (medium span) 45 lb/sf.
- Ramp Y at I-95 Davie Boulevard Interchange (medium span) 47 lb/sf.
- Route 19 - Interstate 80 interchange (grade separation) 48 lb/sf.
- Great River bridge (long span) 53 lb/sf.
- Hoffstadt Creek bridge (long span) 68 lb/sf.
- 20th Street HOV Structure / Denver Union Terminal ramp (grade separation) 70 lb/sf.
- Bob Michel bridge (medium span) 74 lb/sf.
- New Brickell Bridge (movable span) 75 lb/sf.
- Tudoe Road trail crossing (special purpose) 83 lb/sf.
- I-526 Bridge over the Cooper River (long span) 84 lb/sf.
- Union Pacific Railroad Joppa Bridge (railroad) 95 lb/sf.
- Market Square walkway (special purpose) 102 lb/sf.
- Endicott Causeway Breach (special purpose) 120 lb/sf.
- Aviation / Rosecrans Railroad underpass (railroad) 275 lb/sf.

For the period 1991-1993

- Flaming Geyser Bridge no. 3024 (medium span, low clearance) 30 lb/sf.
- American Falls, SH39 over UPRR (grade separation) 34.5 lb/sf.
- Florida's Turnpike/I-595 interchange (grade separation) 34.7 lb/sf.
- Bridge No. 13010 over Paulins Kill Creek (short span) 35 lb/sf.
- I-15 Tropicana flyover (elevated highway) 37.7 lb/sf.
- I-70 Eastbound Mainline approach structure (elevated highway) 38 lb/sf.
- Olmstead Island footbridges (special purpose) 45 lb/sf.
- Lawyer's Canyon bridge (medium span, high clearance) 46.5 lb/sf.
- Skyway at the St. Paul Companies Headquaters (special purpose)

54 lb/sf.

- Potato Slough Bridge (moveable span) 66 lb/sf.
- Pine Street Bridge over Palmetto Canal (short span) 70 lb/sf.
- South River corridor pedestrian bridge (special purpose) 70 lb/sf.
- Alsea Bay Bridge replacement (medium span, high clearance) 86 lb/sf.
- I-255 over the Mississippi River (long span) 95 lb/sf.
- Veterans Administration Skybridge (special purpose) 300 lb/sf.

For the period 1989-1991

- Paducah tied arches (reconstructed) 19 lb/sf.
- McCarran Boulevard Bridge (medium span, high clearance) 25.5 lb/sf.
- St. Peter Street bridge (short span) 30 lb/sf.
- Bridge "B" over MCO entrance road (grade separation) 30 lb/sf.
- SR-386/I-65, CSX railroad, Two-Mile Pike (elevated highway or viaduct) 39.93 lb/sf.
- Geneva Spur pedestrian bridge (special purpose) 44.7 lb/sf.
- Susquehanna Street bridge (short span) 49.1 lb/sf.
- Ramp B - Turkey Lake Road over I-4 (medium span, low clearance) 49.4 lb/sf.
- U.S. 10 over St. Croix River (movable span) 58.8 lb/sf.
- The Link (special purpose) 65 lb/sf.
- Soo Line Railroad Bridge over Marshall and Snelling Avenues (railroad) 138 lb/sf.
- Roosevelt Lake Bridge (long span) 145 lb/sf.
- Greenpoint Avenue Bridge (movable span) 150 lb/sf.

It is particularly important to notice that the above listing of the various bridges in the order of specific weight in pounds per square foot of deck surface is not an absolute index of efficiency. This is based on the fact that the comparison does not take into account the length of the span and the difference in loading that varies, for example, between highway bridges and railroad bridges. Thus, when comparisons are made between two structures, the reader should take into account the different set of circumstances characterizing each individual bridge.

BRIDGES IN THE REGIONALISM OF THE WORLD

Regionalism in the architecture of large structures such as bridges is indeed a reality, even if designers all over the world belong to one large family where everyone is aware of the work being done internationally. As an example, let us consider the work being conducted in Japan as discussed by M. Ito, 1996. Bridge construction in Japan in the last thirty years has been on the avant-garde, with a large number of structures being built at the same time technological progress in engineering put Japan on the leading edge in this field.

In general, the typical scenario of bridge construction in Japan seems to impinge on four major points which include the following: (a) bridges in that country seem to be somehow anonymous in the sense that they are not associated with the designers that have created them but are instead totally controlled by the government and other public institutions. This is the main authority that controls the work and hires engineers and other professionals as consultants only; (b) local physical conditions dominate the freedom of design because of their overwhelming influence. Such conditions include high winds, earthquakes, and specific geological factors; (c) typical characteristics of structures in Japan seem to be extremely conservative, restricting architectural forms and construction techniques to old standards.

However, some of these conditions appear to be changing in recent years. For instance, bridge consultants to the governmental agencies have gained more authority than they had in the past. An example of such an attitude, in which consultants had more to say about the design of a major bridge, is the Katsushika Harp Bridge in Tokyo. This structure, the world's first curved, cable-stayed bridge, had already been designed from the structural point-of-view, yet the intervention of private consultants had a major role in determining the aesthetic values of the bridge.

The basic approach to design of major engineering structures, including bridges, depends very strongly on the use of committees. These usually include professionals and faculty members from local universities. Specifically where aesthetics is concerned, architects, landscape architects, industrial designers, representatives of public and private sectors, and artists are definitely invited to join such committees right from the preliminary stage.

A singular situation seems to characterize the operation of such committees. Such a situation involves the use of more than one committee for the planning and design of a single bridge or group of bridges. One committee may be responsible for the structural system, while the second committee may be charged with the aesthetics of the structure. Often, in fact, the second committee may make drastic changes to the structure originally designed as indicated in several case studies. More detailed information on the process controlling the various stages of design of bridge structures in Japan is available in the the work of M. Ito, 1996, from the University of Tokyo, which provides specific data from actual case studies.

Taking Poland as another example, one notices that the majority of bridges built in that country from 1956 to1990 consist of prefabricated concrete girders, yet in the latest years these type have been gradually loose in popularity. More specifically statistical data on the percentage of different categories of bridges in poland indicated that bridges consisting of precast concrete elements amounted to 18% of bridges in the country. On the other hand, the same bridge types amounted to 67% of the number of bridge structures built between 1956 and 1990. Other data indicate that the total number of prefabricated concrete bridges built is 5,300 (Biliszczuk, et al.,1997).

INNOVATIONS IN BRIDGE DESIGN CONCEPTS

New ideas of particular interest for their uniqueness in the design of bridge structures have recently been brought up by Potzl (1996).

A reflection of great insight presented by this author is the consideration that, prior to the design of steel and reinforced concrete bridges, bridge structures consisting of masonry did not include any joints or bearings, but acted as monolithic structures. Joints and bearings, are the cause of expensive maintenance costs in typical present-day structures. Obviously, monolithic structures which do not allow movements such as deformations and rotations, create additional stresses that have to be resisted. Yet even with the additional costs that these stresses imply, the concept can be proved valid considering the reduced cost of maintenance.

The author starts from the assumption that the cost of maintaining a bridge during its lifetime is approximately the same as the initial cost of constructing the bridge. It is interesting with record a monolithic bridge fixed on both sides, to consider the different behavior according to two different conditions: (1) the bridge is fixed on both ends but it includes a rectilinear axis; and (2) the bridge is fixed also on both ends but its axis is curvilinear in the horizontal plane. Their behavior under thermal expansion and contraction, for instance, are much less pronounced in the second case with respect to the first. Supporting this conclusion, the author calculated the effect of temperature drop for a curved bridge, finding that the tension stresses resulting from this thermal load were so small that they could have been absorbed by the concrete itself without any steel reinforcements. Specifically, the bridge in question was 15 m wide with a slab of 30 cm average thickness, an abutment distance of 350 m, and a curvature radius of 500 m, which was exposed to a temperature reduction of 50°K.

Also relevant for monolithic concrete bridges, fixed on both ends and curved in the horizontal plane, is that, in comparing them with a similar bridge having no horizontal curvature, curved bridges experience a smaller horizontal deflection when the curvature of the bridge is within an angle of aperture contained within 80° and 110°. On the other hand, bending moments at the supports may be reduced when the aperture is larger than 110°.

An innovative concept also brought up by Potzl in his paper is the concept of "robustness," which he advocates should be applied to structures such as bridges. The term is clearly described in his paper and is briefly discussed herein. The reader should consult the original source mentioned in the bibliography to attain more information about this subject.

Robustness in the design of the structure aims at the following five major points: (1) avoidance of unfavorable impacts; (2) ruling out of unforeseen failure and early recognition of safety-relevant damages; (3) amicable reaction of structures to catastrophies; (4) slight deviations from calculations may not lead to disproportionately large damages; and (5) restriction of impact in extent and intensity during the lifetime of the structure.

Basic criteria for the implementation of the concept of robustness for bridge structures, include the following: (a) structural redundancy in general; (b) prevention of failure through redundancy of members; (c) stabilization through tensile members which are more resilient to impact loads; (d) overall ductility of the whole structure; (e) monolithical construction of the whole bridge; (f) allowance of deformability, even for statically indeterminant structures by adjusting the shape; (g) stress flow using a minimum amount of deformation energy; (h) compactness of form that reduces the surface area exposed to wind forces and weathering agents; (i) replaceability of components that will extend the lifespan of the structure; (j) adaptability of bridges to changing requirements in vehicular traffic demands according to volume and loads; (k) over-design of component parts to avoid the possibility of damage or deterioration due to faulty fabrication of components.

2

PRELIMINARY DESIGN

PRELIMINARY DESIGN CHECKLIST

In the realization of a bridge structure, a conceptual design procedure (see Chapter 1) is recommended. A second stage that would logically follow is the *preliminary design*.

As previously mentioned, the conceptual stage that precedes the preliminary design stage established a program that would allow design contributors to participate in a competition by submitting design proposals in agreement with certain conceptual data. The preliminary design stage would allow different experts in bridge design to produce documents that could be judged using the same set of criteria, making the competition a valid means from which a winner could eventually be established. The process might be repeated if the examining board of judges could not select a winner for that competition. Some of the topics that are considered in this phase include aesthetics, construction, structure and maintenance.

It should be noted at the beginning of this preliminary design phase that a major goal is the approximation of the proportions of the members constituting the bridge. To achieve this, the candidates who seek to receive a commission for the bridge should depend on their own knowledge of theory, practice and judgment ability. Therefore, a structure is eventually delineated and proportioned so that it would constitute the basis for a structural analysis to be performed and costs to be estimated. These are steps that could be reiterated as often as desired especially now when computer programs allow a quick revision of data for different solutions. At the end of this phase, different solutions have been tested to obtain unique prototypes concerning aesthetics and costs that would characterize each individual entry submitted by the various competitors.

At this point, a checklist could be presented to try to organize the preliminary phase of design into a sequence of parameters considered more influential than others. To list the most important relevant parameters is by no means an assurance that such a list is comprehensive, and notice that the organization of such lists of parameters could follow any format with equal or more validity.

Loads

- See Appendix 2 for list of loads to be considered.

Geometry

- Alignment
- Section
- Total length
- Individual span lengths
- Depth-to-span ratio
- Horizontal clearances to be spanned
- Vertical clearances
- Obstacle type to be crossed over
- Characteristics of supports (type and geometry)
- Traffic detours during construction
- Accommodation for utilities
- Approximate proportioning of members
- Overall proportioning

Soil conditions for foundations

- Soil-bedding capacity
- Foundation options: spread footing, piling, caissons
- Potential settlement of pier foundations
- Potential settlements of abutments and embankments

- Longitudinal movements at embankments
- Active soil pressure on abutments

Initial cost of construction

- Construction time
- Construction details
- Cost of approaches (earth moving).
- Anticipated cost for future expansion (additional lanes)

Maintenance costs

- Cost of anticipated repairs
- Inspection costs

Safety aspects

- Personal safety during construction
- Protection of adjacent properties during construction
- Protection of existing utilities during construction
- Safety of traffic adjacent to construction
- Safety of traffic on bridge in operational conditions
- Structural safety: use of redundant members
- Night lighting

Aesthetics

- Compatibility with the surrounding areas: urban, rural
- In harmony with adjacent urban structures
- In harmony with landscape
- Holistic effects on the observer
- Evaluation of details

Requirements for spanning over water bodies

- Width of obstacle
- Vertical clearance
- Minimum pier spacing for potential maximum flow
- Effects of backwater lines for flow restrictions

- Scouring effects on piers and abutments
- Cofferdam requirements
- Protection of piers
- Navigational requirements
- Construction clearance for existing water traffic
- Ice thrust
- Debris load

Special requirements of railroad bridges

- Vertical clearances (overhead)
- Horizontal clearances
- Typical railroad loads
- Maintenance of railway traffic
- Deck details for ballast
- Deck maintenance for ballast

Special requirements for prestressed concrete

- Special additional costs over regular concrete
- Maximum member length in plant
- Maximum member length on-site
- Maximum member weight
- Transportation costs
- Erection costs
- Limits of handling equipment (cranes, trucks)
- Availability of specialized labor

Weathering agents

- Acid rain, salt, others

Materials

- Selection of basic materials: steel, concrete, masonry, timber
- Availability of local materials
- Availability of specialized labor with respect to materials

The above list may have duplications with others since the line differentiating one phase of design from another is somewhat flexible.

BRIDGE TYPOLOGIES

Within the large spectrum of bridges built in recent times, it is possible to differentiate them into several groups according to different factors, including uses, materials, or structural systems as follows:

Types of bridges according to uses:

- Railway
- Highway
- Canal
- Aqueduct
- Utility
- Pipeline
- Pedestrian

Types of bridges according to materials:

- Cast iron
- Wrought iron
- Steel
- Reinforced concrete
- Prestressed concrete
- Timber
- Masonry
- Aluminum

Cast iron and wrought iron, are materials that were used for a short period of time only prior to the introduction of steel. Their inclusion in this list is for historical reasons. Steel, of course, is the material that revolutionized the concept of bridge construction by allowing the longest span presently possible, especially through the structural systems of suspension bridges. Masonry bridges have become obsolete with the advent of reinforced concrete, since the new material creates more efficient structures by reducing the weight and increasing the span.

However, reinforced concrete has its limitations with respect to steel structures, and was surpassed with the introduction of prestressed concrete, which allows more flexibility than reinforced concrete alone, reaching greater spans and lower weights. Timber had its use in the past, but it could be considered obsolete at present unless purposely used for architectural reasons. Present treatments of wood make this material more durable than in the past, not only with preservatives, but also with the lamination of several layers which allow longer spans than possible before. Most typically, a practical application of timber bridges could be for pedestrian uses with limited spans and loads.

Types of bridges according to structural systems:

- Truss
- Arch
- Suspension
- Stringer or girder
- Stayed-girder
- Composite construction
- Hybrid girder
- Continuous
- Cantilever
- Orthotropic (steel deck plate)
- Prestressed
- Segmental (concrete)

Types of bridges according to alignment:

- Square
- Skewed
- Horizontal curved alignments

Types of bridges according to mobility:

- Fixed
- Movable or drawbridges
  - vertical alignment movable (vertical-lift bridge)
  - horizontal alignment movable (swing bridges)
  - bascule bridge (with one or two leaves)

Vertical alignment movable or vertical-lift bridges consist of a movable span which can be lifted by raising the span on a vertical plane while the span remains in a horizontal position. The bridge includes two towers, one at each end of the movable span. The span is lifted by means of ropes and it is balanced by two counterweights. The lifting is accomplished by machinery which can be mounted on the towers and is referred to as "tower drive;" or if located on the movable span, it is referred to as "span drive." Typical examples include the bridge over the Arthur Kill, in the New York harbor with a span of 585 ft and the Stratford Avenue Bridge in Stratford, CT.

Horizontal alignment movable or swing bridges are structures that rotate horizontally around a vertical axis on a pier that is the pivot. They can consist of a truss structure or a plate-girder. Bascule bridges consist of a movable span referred to as a leaf which rotates on a vertical plane. To balance the weight of the cantilevered arm a counterweight is applied. Bascule bridges could consist of one leaf crossing the water body or of two leaves, one on each side of the water body, which meet at the center of the span when the bridge is closed. These bridges can be built using trusses or plate girders. For highway applications two leaves are commonly used with a total span of 175 ft which includes two cantilevered arms of 87.5 ft each. For railroad applications only one leaf is used with a maximum span of 250 ft; in the latter case the bridge consists of one span singularly supported.

Types of bridges according to supports:

- Floating or pontoon bridges
- Trestle bridges (multi-span girders)
- Viaduct (multi-span arches)

Types of bridges according to construction:

- Short spans: solid slabs, and single, parallel beams
- Medium spans: girders and trusses
- Medium-long spans: arches
- Long spans; stayed-girders
- Longest spans: suspension

Types of bridges according to load sharing (substructure, superstructure and deck acting together):

- Composite construction (stringer)

- Hybrid construction (girder)
- Orthotropic bridge (girders or stringers interacting with deck)
- Rigid-frame construction (substructure and superstructure interacting together)

Types of bridges according to obstacles being crossed over

- Navigable waters
- Non-navigable waters
- Rivers
- Canals
- Lakes
- Marshes
- Railroads
- Highways
- Narrow roads

The previous classification groups bridge types into broad categories; the following goes into more detail according to the requirements of structural and construction systems.

Beam bridges

In this category of bridges, main parallel steel beams or girders carry the roadway on top of them. These beams are usually of the rolled type directly available from standard production. The beams span parallel to traffic in single spans from one abutment to the other or in multiple spans with intermediate supports. Economical spans for railroad bridges are in the order of 50 ft, while for highway bridges the equivalent economical span is 100 ft.

Composite beam bridges

In this category of bridges, the main beams or girders interact with the concrete deck superimposed on them forming a T-section. The bond between the top of the beams and the concrete slab is obtained by shear connectors welded on top of the beams extending at least halfway through the thickness of the concrete slab above. These connectors usually consist of rolled shapes such as channel, angle, and I, or spiral bars or studs.

Plate-girder bridges

In this category of bridges, rather than using available hot rolled shape beams, girders of larger cross-sections are fabricated so that they can bridge longer spans. The girders are fabricated by using a deep plate for the web and welding two angles on the top and two on the bottom, forming the flanges in the shape of an I cross-section. For railroad bridges each track is supported by two plate-girder bridges carrying transverse beams on top. Thus a two-track bridge consists of two, single track bridges side by side with a total of four plate girders.

Through-plate girder bridges

This category is similar to the previous one except that the floor system carrying the tracks bears on transverse beams which, in turn, bear on the bottom flange of the plate-girders in order to minimize the overall thickness of the bridge.

Box-girder bridges

In this category the main supporting members consist of fabricated steel beams in a shape made of four plates welded together to form a hollow box (Figure 2-1). The box-girder can be fabricated with different components using plates welded to hot rolled standard steel shapes to create the box. The top flange of the box-girder acts as the deck. The steel deck is then reinforced and stiffened by welding closely spaced stiffeners, corrugated steel plates or honeycombed-type plates.

The such stiffened top plate of the box-girder is usually referred to as "orthotropic deck." On top of the orthotropic surface is a thin layer of asphalt which constitutes the wearing surface. This kind of bridge has been built with spans reaching 850 ft, as in the case of the bridge over the Rhine at Cologne, Germany and 750 ft, as in the case of the San Mateo-Hayward Bridge in California. Typically one box-girder could suffice for narrow bridges, but as the width of the bridge increases more than one box-girder can be used parallel to one another.

Generally speaking, steel box-girders are usually more expensive than I-beam, comparing two structures of equal weight. However, there are more considerations that must be taken into account before deciding which of the two structures is the most efficient. Even in the design process steel box-girders require more effort. Other negative aspects of steel box girders include some difficulties in fabrication and some initial failures that occurred in Europe during the 1960s and local failures along the welding of the plates that occurred in some structures in the United States. Also, problems have been found for the transportation of steel box girders on highways and on barges due to length limits. Still another limitation for steel box girders occurs when the size of the box is too small for an inside inspection.

TYPICAL BOX GIRDER ARRANGEMENTS		
Arrangement		Deck cross-sections
a	Single rectangular box girder	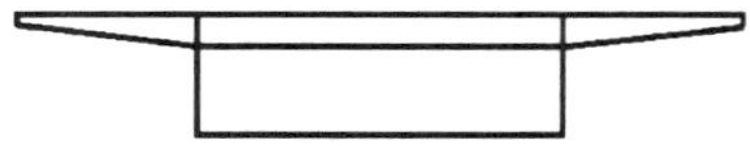
b	Central box girder and side single web girders	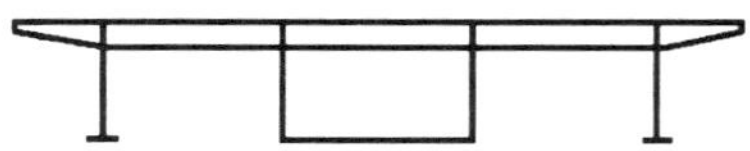
c	Single twin cellular box girder and sloping structs	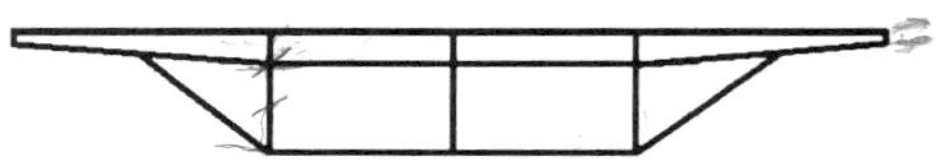
d	Single trapezoidal box girder	
e	Twin rectangular box girder	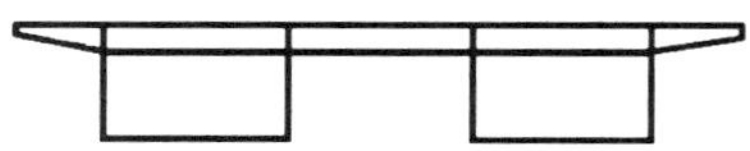
f	Twin trapezoidal box girder	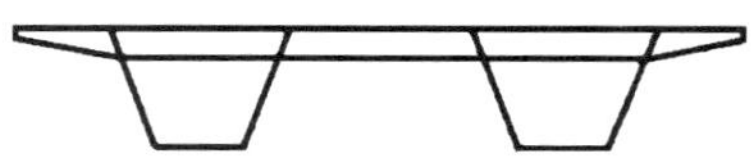

Figure 2-1 Types of main box girders for stayed bridges.

Having stated most of the negative aspects of this kind of structure let us now examine the positive ones that have made them so popular. First of all we must mention the structural efficiency, especially when torsional stresses are involved, including the case of horizontally curved girders. Moreover the steel box-girder performs satisfactorily in terms of being aesthetically pleasing, easy to build, easy to maintain in terms of painting, and easy to be inspected if the interior space is sufficiently large.

The beginning of the steel box-girder is associated in the United States with its acceptance by the AASHTO *Standard Specifications for Highway Bridges* in the early 1970s. Steel box-girders particularly resistant to torsional stresses allowed curved girders to be built with horizontal radius of curvature down to 150 ft, as in the case of two 1960s bridges built in Massachusetts.

There are two types of steel box girders: traditional cross section (consists of four plates welded together to form the box), and "tub" girders which are box beams made of a bottom flange and two vertical webs made of steel plates, plus a top flange made of more than one steel plate interacting with a composite concrete deck that is part of the box.

Another step forward in the design of steel box girders occurred in 1980 when AASHTO *Guide Specifications* included guidelines for horizontally curved bridges. In so doing steel box girders could be designed in terms of their structural resistance to torsional stress similar to the specification for the segmental concrete box girders.

Typical steel box girders in the United States are of the "tub" type with inclined webs forming a trapezoidal cross section. The trapezoidal shape has a narrower bottom flange and a wider top flange. Usually the top flange with the interacting concrete deck cantilevers out considerably from the side of the box. Many bridges consist only of one steel box girder, either straight or curved, but many others include two box girders or more. In terms of supports, each box girder does not require more than one support shaft.

The concrete upper deck of the single, steel box girder or multi-steel box-girder bridge is often very wide and therefore post tensioned across its width. Examples of this are typically the single box girder, curved ramps in many bridges, and the MacMillan Yard Bridge in Toronto. In this latter case the bridge had two, steel box girders carrying an arched concrete deck approximately 100 ft wide that required posttensioning.

What are the typical dimensions of box-girders? The depth which is approximately 5 ft controlled by limits of accessibility due to inspection. This value also represents a minimum for structural reasons. We have discussed previously the relatively higher price of fabrications steel box-beams with respect to I-beams thus to compete and to be even more economical than I-beams, the box-beams must be obviously lighter than an equivalent I-beam carrying the same load. Another point to keep in mind is that the cost of one single steel box-beam is less than the cost of two or more equivalent box-beams.

Curved-girder bridges

This category includes bridges that have curvilinear alignments. For large radius of curvature, the supporting girders can be straight, while for small radius of curvature the supporting girders could be made to be curved along their longitudinal axis. In the latter case girders have to be designed also to carry torsional stresses which at present are less difficult to calculate than in the past due to computer analysis. These kinds of bridges can employ girders, box-girders, both in steel and concrete, carrying the deck directly on top of them.

Through-truss bridges

This category includes bridges whose decks are supported by the lower chord of trusses, differentiating themselves from the deck truss bridges in which the deck is carried by the upper chords of the trusses. The deck truss bridges are more economical than the true truss bridges, because the true trusses can be spaced closer to one another and therefore the transverse supporting beams and the bracings can be shorter. Different kinds of trusses can be used. The most economical depth-to-span ratios vary between 1/5 to 1/8. Besides the truss members and the transverse beams, other members referred to as stiffeners must be added to add lateral stiffness and rigidity for the whole structure. Bracers connect the top chords of the through trusses as well as the bottom chords.

One set of ties perpendicular to the trusses, connects the panel point of one truss with the corresponding panel point of the other, both on the top chord and on the bottom chord. Then to add more lateral rigidity, cross-bracings are also used on the top and bottom chord. The bottom beams carry the steel or concrete deck. The deck also acts as a lateral stiffener. Vertical members of the truss and horizontal bracing members on the upper chord form portals which increase the lateral rigidity of the bridge. Special consideration must be given to the end portals which are directly connected to the abutments and increase their stiffness using additional stiffeners or larger cross-sections of the members forming the portals. In railroad bridges of this kind, the tracks are carried directly by longitudinal stringers which, in turn, are supported by the floor beams. Because of the lateral sway of the passing trains, the stringers must be properly braced.

Deck-truss bridges

This category includes bridges with the deck over the supporting trusses. Most configurations of trusses with parallel chords can be used (Warren or Pratt, for example). As with true truss bridges, the most economical depth-to-span ratio varies between 1/5 to 1/8, including diagonal members that form angles with a horizontal line varying between 45° and 60°. In order to reduce the length of each member, it is convenient to introduce additional members that would subdivide each panel in two.

Arch bridges

Throughout history arch bridges have been throughout history the classical type of bridge built in masonry that constitutes the only permanent material for civil engineering work. Masonry, includes stones and bricks embedded in mortar and is no longer in use in present-day construction. The new materials now are steel and reinforced concrete. Arch bridges at present are considered not economical for spans shorter than 150 ft since other systems are less expensive. Maximum range of steel arch bridges are currently demonstrated by the steel bridge in West Virginia, with a 1,700 foot span, and by the reinforced concrete arch bridge over the Parramatta River in Sydney, Australia, with a span of 1,000 ft.

The old masonry bridges consisted of a masonry vault supporting superimposed walls or earth fill, "spandrel fill," over which the pavement was installed. This system, however, is no longer in use. Modern concrete bridges include two or more parallel reinforced concrete ribs spanning from one abutment to the other. The ribs, in turn, support the upper deck through a system of columns referred to as "spandrel columns" which in turn carry the deck. The inclined surface of the abutment from which the ribs spring and is referred to as *skewback*. Notice that with arch bridges the deck does not have to be on top of the arch ribs all the time, but it can be below the crown of the arch. The part of the deck below the arch is attached by suspenders and the part above the arch is by supported by spandrel columns bearing on the arch. The abutments have to be on solid foundation to absorb not only the vertical but also the horizontal component of the arch thrust. If the sole conditions are not adequate to resist these horizontal components, then steel ties can be used, and in that case the bridge is referred to as a "tied arch bridge."

Basically there are three types of structural arches according to the type of supports: "the fixed arch," where the arch is fixed at both ends at the abutments; the "two-hinged arch," where the arch is hinged at the two abutments; and the "three-hinged arch," where the arch is hinged at the two abutments and also at the crown. The first two are statically indeterminate structures which used to be more difficult to calculate and were critically affected by potential settlements of the supports. The last type being a statically determinate structure, is easy to calculate and is not critically affected by the differential settlements of the supports or by thermal effects.

Continuous-beam bridges

Continuous beam bridges are statically indeterminate structures which are supported by more that two supports. Being statically indeterminate their weight is lighter than the equivalent individual single spans, but on the other hand they are critically affected by differential settlements of the supports and by thermal forces.

They can be built in steel or in reinforced concrete and they can be built as trusses, plate-girders, or box-girders. Usually continuous beams are no longer than three spans, because additional spans would not increase the stiffness but would increase considerably the effect of settlement of the supports and the effects of thermal forces.

"Gerber beams" or cantilever bridges

These kinds of structures, first introduced by Professor Gerber in Germany in the middle of the last century, are statically determinant structures that include continuous beams on more than two supports and also include interior hinges. The number of these interior hinges is equal to the number of spans minus one. Being statically determinant these structures are easy to calculate and are not critically affected either by differential settlements of the supports or by thermal forces. Their advantage over simple beams is that they are lighter and therefore more economical.

Cable-stayed bridges

Included in this category is a new type of bridge consisting of two or more units. Each unit consists of a vertical tower supporting a horizontal beam cantilevering on each side of the tower. The beam is also supported by inclined cables coming down from the top of the tower and by other cables suspended below the top at different levels. Two or more of these units can be placed in series forming a continuous bridge. The beams can also be trusses, plate-girders, or box-girders. When completed, these bridges very closely resemble typical suspension bridges.

Suspension bridges

This category of bridges includes the longest spans ever built. To indicate the order of magnitude of the spans that can be obtained, consider the following examples: the Mackinac Bridge in Michigan with a span of 3,800 ft; the Golden Gate Bridge in San Francisco with a span of 4,200 ft; the Verrazano-Narrows Bridge in New York City with a span of 4,260 ft; the Humber Bridge of England with a span of 4,626 ft; and the Akashi Strait Bridge in Japan to be completed in 1998, with a center span of 6,633 ft.

The structure consists of two towers carrying a main cable that spans from tower to tower, and from the tower to the ground (backstays), where they are firmly anchored. From the main cables, vertical suspenders carry longitudinal trusses which in turn carry the deck. The function of such trusses is to create sufficient rigidity to counteract the aerodynamic forces of the wind and the vibrating forces of the traffic that would make the cable oscillate and consequently set the deck in a wavy motion. Of particular interest are the cables which can reach substantial sizes, as in the case of the main cables of the George Washington Bridge, with an overall diameter of 36 in. In this regard the cables can consist of a series of small strands placed parallel to one another with diameters of approximately 1/4 in, or they may be made of small strands of wires wound form ropes which can eventually be combined together to form larger cables. In this case the ropes are pretensioned in the factory prior to being installed on site.

Pontoon bridges

This category includes floating bridges consisting of a series of interconnected barges supporting a deck. Typically these kinds of bridges have been used by the military for temporary use. In Venice, Italy such a bridge is assembled by the Army Engineers on the occasion of the festivity at the Church of Redentore, allowing pedestrian traffic to reach the church across the lagoon. Stable bridges on barges have been built in the United States, as in the case of the three floating bridges on Lake Washington in Seattle. These include cast-in-place concrete or prestressed concrete barges 360 ft long, 60 ft wide, and14.5 ft deep bolted to one another and anchored to the bottom on each side. A floating draw span can be opened to allow the use of a 200 foot navigable channel.

BRIDGE SPANS

Studies of maximum bridge spans that can feasibly be built within the different bridge typologies have predicted the following values as reported by Brown (1993):

- Truss bridges: 1,640 ft
- Arch bridges: 4,920 ft
- Cables-stayed bridges: 8,200 ft
- Suspension bridges: 6,400 ft

These values have been based on the mechanical characteristics of existing building materials, excluding any potential new material that could probably be available in the near future. Such materials may include high-strength reinforcing fibers embedded within a solid mass of plastics or resins. Such fibers could include, for instance, glass, almost pure carbon, and armada (*Coverall*, an organic material). In this case the maximum span predicted for suspension bridges would be 7.5 miles, as reported from the same source.

BRIDGE COMPONENTS

Substructure

The substructure of a bridge supports the upper structure. The main components of the substructure include piers and abutments. Other components include skewbacks which constitute the springing surface for arches and pile bents for trestles.

Upper structure

The upper structure of a bridge includes trusses, girders, stringers, floorbeams, and decks.

Abutments

Abutments of a bridge constitute the bedding structure upon which one of the ends of the bridge is bearing. Presently, they are usually built in concrete, while older structures were built with masonry. In simple terms, abutments include a footing, a wall carrying the bridge supports, a back wall acting as a retaining wall to support the earth pressure, and wingwalls which also act as retaining walls.

Abutments can be subdivided into two basic categories: open-end and closed-end. Each of them can, in turn, be subdivided into the following categories:

- Diaphragm or integral open-type abutment (Figure 2-2)
- Seat open abutment (Figure 2-3)
- Spill-through open abutment (Figure 2-4)
- Cantilever closed abutment (Figure 2-5)
- Restrained closed abutment (Figure 2-6)
- Rigid frame closed abutment (Figure 2-7)
- Cellular closed abutment (Figure 2-8)
- Semi-gravity (railroad) closed abutment (Figure 2-9)

In the selection of the various types of abutments previously mentioned, the following should be introduced:

- Open-end abutments are usually more economical than closed end ones.
- In the open abutment type, approach embankments can be erected before construction of the abutment and therefore be compacted with minimum settlement.

- In the closed abutment type, the back-filling can produce settlement that is higher than in the previous case.
- When open type abutments are not constructed with sidewalls, they are more flexible for expansion for future widening of the bridge.
- The diaphragm open-end type is preferred over the seat-open abutment type.
- The diaphragm type eliminates expansion bearings and deck joints.
- The diaphragm type is limited to use on bridges no longer than 300 ft due to limitations of expansion movements due to thermal changes.

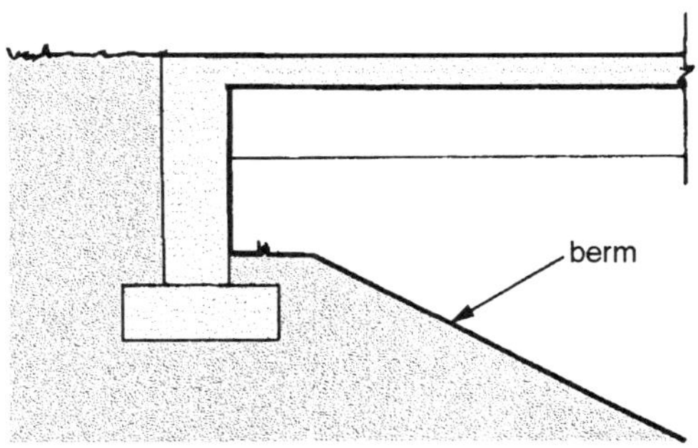

Figure 2-2 Diaphragm or integral open abutment (reproduced from *Analysis and Design of Reinforced Concrete Bridge Structures,* 1988 by permission of American Concrete Institute).

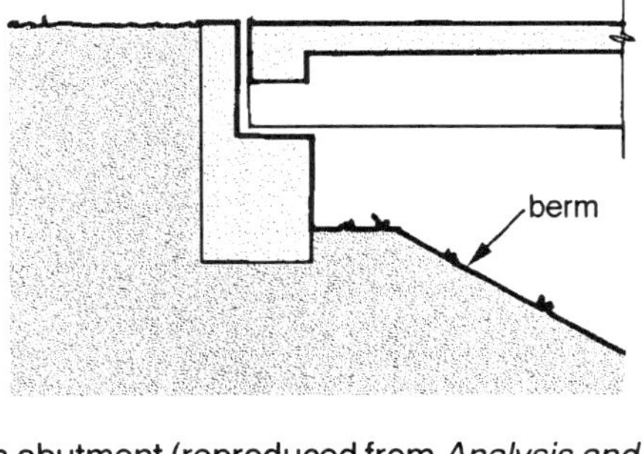

Figure 2-3 Seat open abutment (reproduced from *Analysis and Design of Reinforced Concrete Bridge Structures,* 1988 by permission of American Concrete Institute).

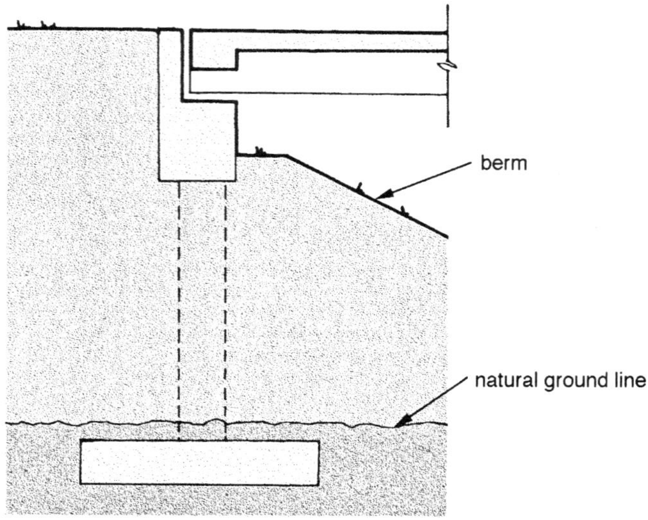

Figure 2-4 Spill-through open abutment (reproduced from *Analysis and Design of Reinforced Concrete Bridge Structures,* 1988 by permission of American Concrete Institute).

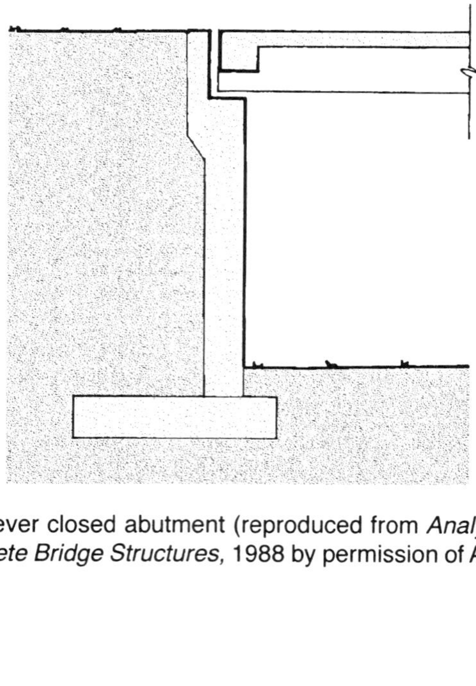

Figure 2-5 Cantilever closed abutment (reproduced from *Analysis and Design of Reinforced Concrete Bridge Structures,* 1988 by permission of American Concrete Institute).

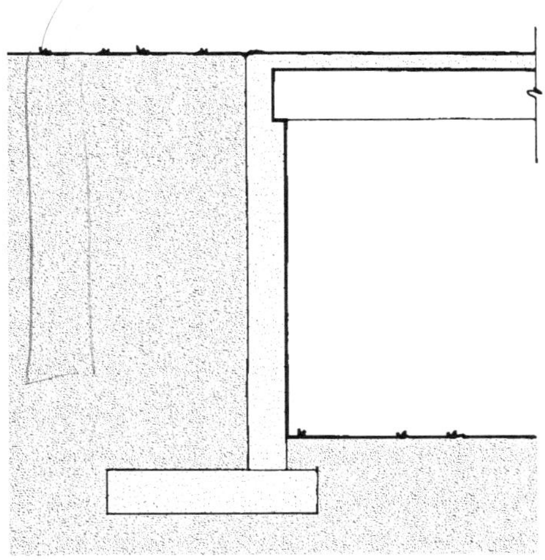

Figure 2-6 Restrained closed abutment (reproduced from *Analysis and Design of Reinforced Concrete Bridge Structures,* 1988 by permission of American Concrete Institute).

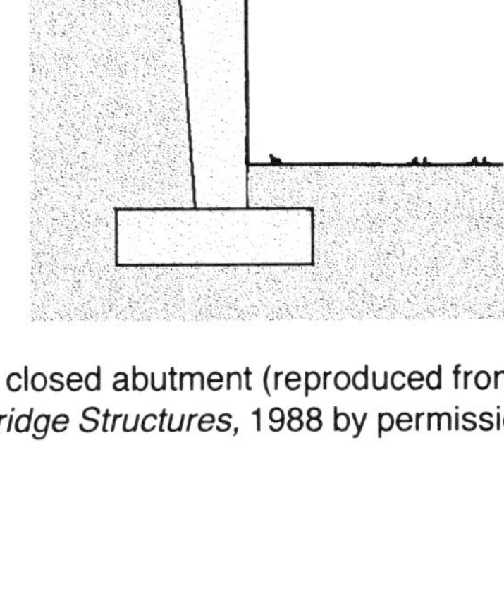

Figure 2-7 Rigid-frame closed abutment (reproduced from *Analysis and Design of Reinforced Concrete Bridge Structures,* 1988 by permission of American Concrete Institute).

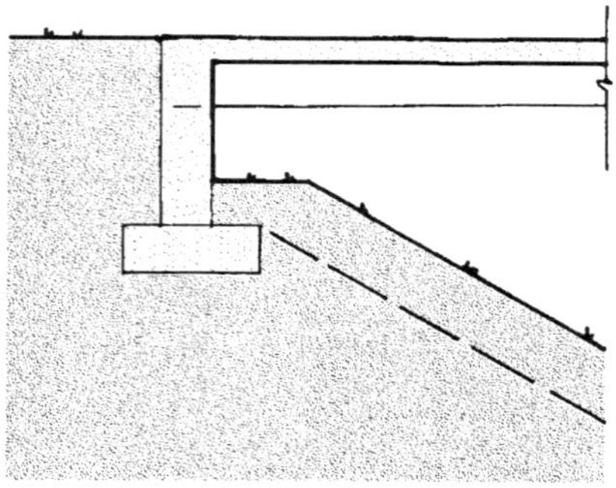

Figure 2-8 Cellular closed abutment (reproduced from *Analysis and Design of Reinforced Concrete Bridge Structures,* 1988 by permission of American Concrete Institute).

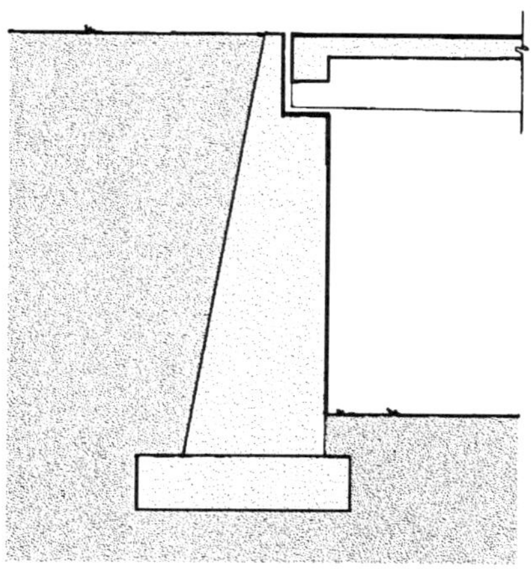

Figure 2-9 Semi-gravity (railroad) closed abutment (reproduced from *Analysis and Design of Reinforced Concrete Bridge Structures,* 1988 by permission of American Concrete Institute).

The footing can be one solid concrete base bearing directly on soil, referred to as a "spread footing." The spread footing can also be supported by piles if the soil bearing capacity is not adequate. In such a case different kinds of piles could be used: point bearing piles, frictional piles, and a combination of the two. With regard to the materials used for piles, these may include timber, steel, precast concrete, cast-in-place concrete, or prestressed concrete. The installation of piles may include the following processes: driven by a pile driver or augered, jetted, prebored, predrilled, and cast-in-place.

Piers

Piers are the intermediate supports placed within the two abutments for reducing the span of the whole structure. They can be built over dry land or in water. Their shape is determined mostly by the soil conditions and by the forces that they have to resist. Most typical types of piers include solid shafts, T or hammerhead, multiple columns and portal, separate columns, and cantilever types. The basic forces that they are intended to resist include dead loads, live loads, braking, wind, ice, earthquake, and stream flow forces. Piers in water are usually protected with stone, steel, and other protecting elements below the water level to guarantee the integrity of the pier against scour, ice and from the impact of floating materials. Further protection of the piers against the impact of transient vessels include fenders made of clusters of timber piles.

Cofferdams or caissons

Cofferdams or caissons are used for the construction of piers and abutments below the earth's surface or below the water surface. Their function is to excavate the soil within the interior areas of the cofferdam pumping the water out so that the work can be performed in the dry. Notice that cofferdams in water have a tendency to be lifted due to their buoyancy and therefore they must be anchored downward using a concrete seal ring at the bottom, referred to as a tremie seal. This is constructed by using vertical steel pipes (tremies) through which concrete is poured from the top up to a level whose weight is adequate.

CONCRETE BRIDGES

As is well known and for obvious reasons, reinforced concrete bridges have replaced almost completely the older masonry structures. Reinforced concrete is much more versatile than masonry, because the new material can resist stresses that masonry could not, allowing new structural concepts as well as a new structural morphology. It also allows the reach of spans of a much larger order of magnitude than those that could be achieved with masonry. But, most of all, the new material allows for more economical structures. If a masonry bridge were to be built today it would probably, and with few exceptions, not be for any reason other than architectural aesthetics.

Reinforced concrete bridges definitely constitute the structures of the present and future and, generally speaking, they are in most cases always more competitive in price to steel counterparts. These must be considered in relative terms, because reinforced concrete bridges cannot compete with the longer spans that steel bridges can achieve.

However, when a span can be covered with a concrete or steel equivalent structure, concrete is usually more economical. Such a condition may at times be reversed for areas, such as the United States and Germany, where structural steel is cheaper than reinforced concrete. For the most part, concrete from other nations could be made out of local material such as clay and limestone, which are much more abundant than iron ore. The following sections give a general overview of reinforced concrete bridge structures that could be sufficient to cover the needs for basic design proposals.

Concrete bridge types

Concrete bridges constitute a large family of structures that could be classified in the following ten structural types. Some of these types are further divided into subtypes as indicated.

Non-prestressed concrete slab bridges

- Solid slabs (Figure 2-10)
- Cored or voided slabs (Figure 2-11)
- Waffle soffit slabs

Non-prestressed concrete girder bridges

- T-beams (Figure 2-12)
- Box girder (Figure 2-13)
- Precast I-girder (Figure 2-14)
- Precast spread box beam (Figure 2-15)
- Precast butted box beam (Figure 2-16)
- Voided slab

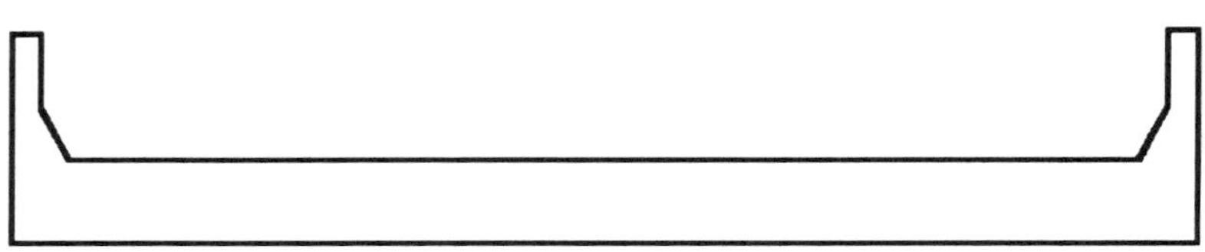

Figure 2-10 Cross section of solid-slab bridge (reproduced from *Analysis and Design of Reinforced Concrete Bridge Structures,* 1988 by permission of American Concrete Institute).

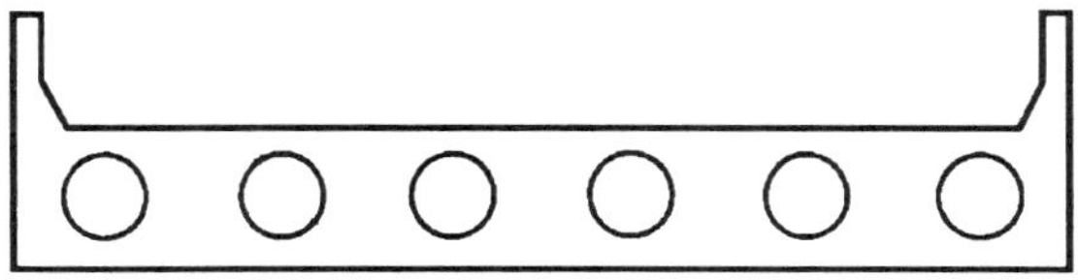

Figure 2-11 Cross section of voided-slab bridge (reproduced from *Analysis and Design of Reinforced Concrete Bridge Structures,* 1988 by permission of American Concrete Institute).

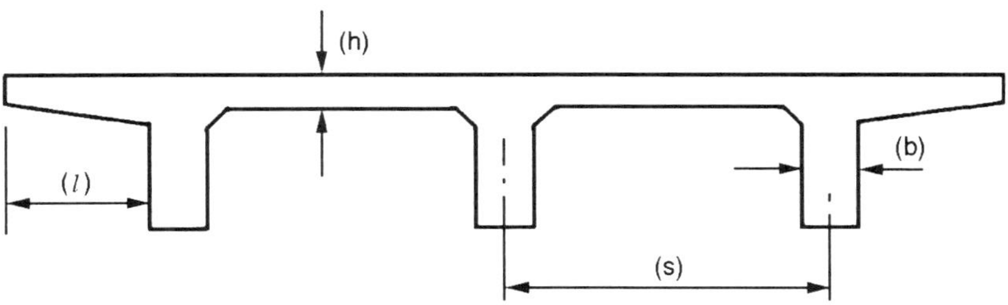

Figure 2-12 Typical concrete T-beam bridge configuration (reproduced from *Analysis and Design of Reinforced Concrete Bridge Structures,* 1988 by permission of American Concrete Institute).

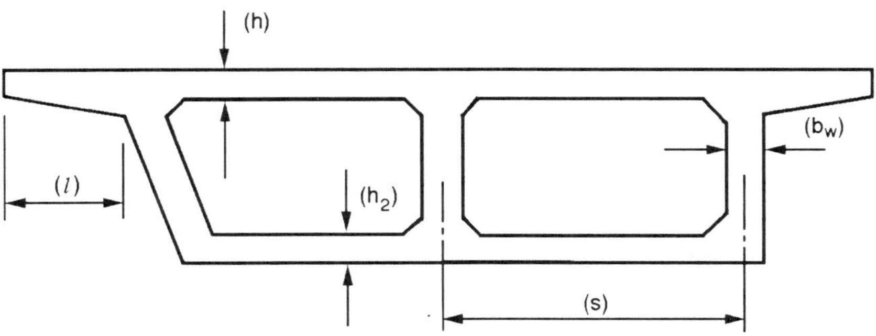

Figure 2-13 Typical concrete box-girder bridge configuration (reproduced from *Analysis and Design of Reinforced Concrete Bridge Structures,* 1988 by permission of American Concrete Institute).

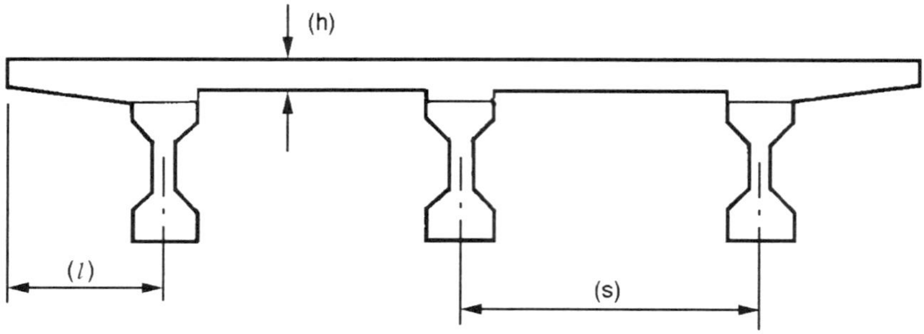

Figure 2-14 Typical precast-concrete I-girder bridge configuration (reproduced from *Analysis and Design of Reinforced Concrete Bridge Structures, 1988* by permission of American Concrete Institute).

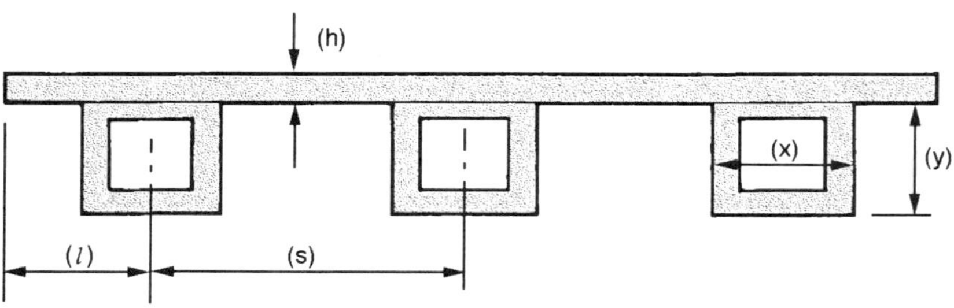

Figure 2-15 Typical precast-concrete spread box-beam bridge configuration (reproduced from *Analysis and Design of Reinforced Concrete Bridge Structures,* 1988 by permission of American Concrete Institute).

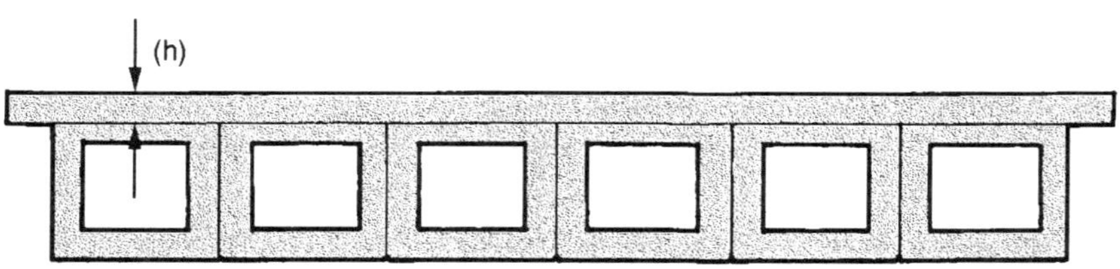

Figure 2-16 Typical precast-concrete butted box-beam bridge configuration (reproduced from *Analysis and Design of Reinforced Concrete Bridge Structures,* 1988 by permission of American Concrete Institute).

Prestressed concrete slab bridges

- Cast-in-place posttensioned
- Precast pretensioned

Prestressed concrete girder bridges

- Cast-in-place posttensioned
- Precast T-beam
- Precast prestressed I-girder
- Precast prestressed box girder
  - spread box girder
  - butted box girder

Posttensioned segmental bridges

Rigid frame bridges

Arch bridges

Spandrel-filled arch

- Barrel arch
- Two-hinged rib open-spandrel arch
- Fixed rib open spandrel
- Tied arch

Truss bridges

Cable-stayed bridges

Suspension bridges

Span ranges and span-to-depth ratios

Typical span ranges and span-to-depth ratios for the individual types of concrete bridge types are:

Non-prestressed concrete slab bridges

- Solid slabs
  - typical span ranges for solid slabs used for highway bridges vary from 14 to 44 ft.
  - typical span ranges for solid slabs used for railroad bridges vary from 12 to 24 ft.
- Cored or voided slabs
  - spans for cored or voided slabs for highway bridges are 40 to 65 ft.
  - spans for cored or voided slabs for railroad bridges are 20 to 40 ft.
- Waffle soffit slabs
  - spans for waffle soffit slabs for highway bridges are 40 to 80 ft.
  - span-to-depth ratios vary with support condition:
    - for simple-span slabs the ratio is 15± for highway bridges.
    - for continuous spans the ratio is 20 to 24.

Non-prestressed concrete girder bridges

- T-Beams
  - typical span for T-beams used for highway bridges are 30 to 40 ft.
  - typical span for T-beams used for railroad bridges are 20 to 40 ft.
  - span-to-depth ratio depends on support conditions:
    - for simple T-beam highway bridges the ratio is 14±.
    - for continuous T-beam railroad bridges the ratio is 15±.

- Box-girders
  - typical spans for highway bridges are 80 to 200 ft.
  - typical spans for railroad bridges are 50 to 100 ft.
  - span-to-depth ratios depend on support conditions:
    - for simple spans the ratio is 16 ± for highway bridges.
    - for continuous spans, the ratio is 18± for highway bridges.

Precast-concrete I-girder bridges

Precast-concrete spread box-beam bridges

Precast-concrete butted box-beam bridges

Precast-concrete voided-slab bridges

Prestressed Concrete Slab Bridges Bridges

- Cast-in-place, posttensioned
  - spans for solid slabs used for highway bridges vary up to 80 ft.
  - spans for solid slabs used for railroad bridges vary up to 20 ft.
  - spans for cored or voided slabs used for highway bridges vary up to 150 ft.
  - spans for cored or voided slabs used for railroad bridges vary up to 40 ft.
  - spans for partial voided slabs for highway bridges are over 150 ft.
  - span-to-depth ratio depends on support conditions:
    - for simple slabs the ratio is 20± for highway bridges.
    - for continuous spans the ratio is 24 to 40 for highway bridges.
- Precast pretensioned
  - spans for solid slabs for highway bridges vary from 12 to 20 ft.
  - spans for solid slabs for railroad bridges vary from 12 to 20 ft.
  - spans for cored slabs for highway bridges vary from 30 to 80 ft.
  - spans for cored slabs for railroad bridges vary from 24 and 40 ft.
  - span-to-depth ratio depends on support conditions:
    - for simple spans, the ratio is 22± for highway bridges.
    - for continuous spans, the ratio varies between 25 to 33.
    - for haunched spans, the ratio varies between 33 to 50.

Prestressed concrete girder bridges

- Cast-in-place, posttensioned
  - spans for constant depth for highway bridges are 80 to 500 ft.
  - spans for constant depth for railroad bridges vary up to 110 ft.
  - spans for haunched for highway bridges vary up to700 ft.
  - span-to-depth ratio depends on support conditions:

- for simple spans the ratio is 22± for highway bridges.
- for continuous spans the ratio is 25 to 33 for highway bridges.

- Precast T-beam
  - spans for simple T-beams for highway bridges is 30 to 160 ft.
  - spans for simple T-beams for railroad bridges varies up to 110 ft.
  - span-to-depth ratio depends on support conditions:
    - for simple spans, the ratio is 18± for highway bridges.
    - for continuous spans, the ratio is 20± for highway bridges.
- Precast Prestressed I-Girder (see precast T-beamabove).
- Precast Prestressed Box Girder
  - Spread-type: span-to-depth ratio depends on support conditions:
    - for simple spans, the ratio is 18± for highway bridges.
    - for continuous spans, the ratio is 22± for highway bridges.
  - Butted-type: span-to-depth ratio depends on support conditions:
    - for simple spans the ratio is 25± for highway bridges.
    - for continuous spans the ratio is 28± for highway bridges.

Posttensioned segmental bridges

Rigid-frame bridges

Arch bridges

- Spandrel-filled arch (spans vary between 50 and 100 ft)
- Barrel arch
- Two-hinged rib open-spandrel arch
- Fixed-rib open-spandrel arch
- Tied arch (spans variable from small all the way to 1,000 ft)

Truss bridges

- Spans vary between 120 and 250 ft.

Cable-stayed bridges

- Spans vary between 600 and 1,200 ft.

Suspension bridges

- Spans up to 680 ft.

Besides the range of spans typically used and the range of the span-to-depth previously given, basic data for each individual type are as follows:

Non-prestressed concrete slab bridges

The three different types of slab bridges, previously described, share all the same basic characteristics which can be briefly summarized as follows. The main appearance is that of extreme simplicity and the system is commonly applied to short spans. In terms of construction, the process is particularly fast in comparison to any other system.

Non-prestressed concrete girder or stringer bridges

For all the different six cases in this category (Figure 2), one notices the simplicity of the elevation, while at the same time the bottom view of the structure is somewhat awkward especially near the supports where the stem widens. Considering the central box-girder (Figure 2-1b), its appearance is very simple and attractive from all points of view. These types are also useful for concealing utilities that run through the interior space.

The formwork can be left relatively rough on the interior faces, but on the exterior surfaces, which are exposed, the formwork should be carefully finished, and therefore the construction time may be a little longer. With regard to maintenance, this type does not require particular attention. In the case of the T-beams in Figure 2-1a, the formwork is somewhat more complicated and consequentially even the construction time will also be longer. With regard to maintenance, the requirements are not that demanding; however, attention must be paid to the possibility of damage by floating debris when the clearance is low.

Prestressed-concrete slab bridges

This category applies particularly to short spans and is characterized by the simplicity of its appearance. With regard to maintenance the requirements are very small, but transverse joints may require more attention. For the case in the design shown in Figure 2-12, where prestressing is applied, the process requires special attention and special workmen for the prestressing. In the case shown in Figure 2-13, the prestressing is done in the factory; therefore limiting the operation in the field, which can be very rapid by means of a crane that also eliminates the need for any falsework.

Prestressed-concrete girder or stringer bridges

From the aesthetic point of view the box-beam girder is always the most attractive for its simplicity and for the smoothness of the surfaces.

The T-beam is also aesthetically pleasing except when viewed from below. In general, however, all the structures in this category are pleasant when looked at from the front elevation. In the design shown in Figure 2.12, the posttensioning will require more efforts than the typically reinforced girders, and the cast-in-place operation will require falsework that the precast would not need. The construction time would be longer in this case than in the other cases when precast is used. The effect of longitudinal and transverse stresses and deformations will reduce the possibility of cracks. For all other cases, (Figures 2-13 and 2-14), the prestressing obviously adds more time to construction in comparison to regular reinforced girders, especially for long spans. These add particular difficulties in handling, not only in the factor but also at the erection site. Precast girders can be either simply supported or continuous. In combination with precast girders it is possible to use cast-in-place decks without the need for falsework. Maintenance is usually minimal.

Posttensioned segmental bridges

Bridges of this type are built in segments, either cast-in-place or precast, and then eventually they can be posttensioned all at once. Precast segments are usually more economical since they can be built in the plant even in long sections and then be erected by crane without any need of falsework.

Rigid-frame bridges

This kind of structure may include single, double, or triple portals according to the structure laying underneath. The single portal may be used for narrow water courses, while the multiple portals can usually be employed when the bridge spans over highways having different lines of traffic with median strips that will accommodate the intermediate supports. Simple in appearance, the portals usually include variable sections of different depth as required by the modern diagrams.

Arch bridges

The concrete arch of short span is usually built with a constant cross-section, while for medium and long spans the cross-section of the arch is typically variable. Naturally the rise of the arch is a factor determined by the clearance necessary for the traffic below. As the rise of the arch increases, the horizontal truss at the base is significantly reduced. The thickness of the arch varies according to the end supports whether fixed or hinged. When fixed, usually the thickness of the arch at the spring line is twice that at the crown. From the construction point of view, the mass between the spandrel arches and the deck are usually for short span bridges, while for longer arches the area between the deck and the arch is open. In this case, if the deck is always above the arch, it is carried by vertical struts in compression.

On the other hand, when through bridges are used, the deck is suspended by tendons from the arch rising above. The appearance of the arch is usually very dynamic, especially when the span-to-rise area is very large. Slender arches are very pleasant, whether spanning wide water bodies or spanning over steep ravines. The silhouette of the arch reflects in all cases accepted standard morphological configurations extending throughout history. However, the present-day concrete arches are extremely more slender than the old typical masonry counterparts. Usually the construction time required is longer, but it could be drastically reduced by using precast segments and eventually adopting posttensioning with the assembly.

In designs where the roadway is carried directly on fill, this type is used for small spans instead of for long spans where the fill could add substantially to the dead load. It is important that the field be well drained.

In designs where the spandrel area above the barrel is open and the deck carried by the walls, the arch in itself is as wide as the deck above or slightly less and it is usually fixed at the supports but at times it could also be hinged. The deck rests just over the thickness of the arch at the crown. Minimum maintenance is required except for the hinges themselves.

Designs which include two hinges of the supports require special attention and maintenance. In this case, usually the thickness of the arch is constant; however, the arches constituting the whole bridge need to be laterally supported with one another. The deck above the arch is supported by means of vertical struts as in the previous case, while when the structure is a through arch bridge, the deck is suspended by hangers.

When the arch is fixed at the base, the cross-section is usually variable, decreasing gradually toward the ground. The variation of the cross-section produces a benefit by reducing the thermal forces development in the arch.

The tied arch is used when a tension member can be applied between the supports for the purpose of absorbing the horizontal thrusts when the soil conditions do not allow the abutments to do so. The tied bridge can be built only as a through arch bridge, since this is the only type that would allow for the installation of ties. In a tie bridge, the deck is carried by hangers suspended from the bridge above. Notice that tied arches can be built as isolated structures or as a series of individual structures. In terms of maintenance, tied arches require more attention than other structures due to the inspection necessary for the ties and the hangers.

Truss bridges

Concrete truss bridges in category 8 can have different kinds of support conditions. These may be simply supported spans, simply supported spans with cantilevers, or continuous supports, as well as Gerber structures of continuous spans with interposed internal hinges. Bridges supported by trusses can be of the deck type, when the supporting trusses are below the deck, or they can be of the true type when the deck is carried by the lowered chord of the trusses. In the deck bridge type it is possible to use chords or trusses with the upper chord horizontal or the lower chord sloped; for through truss bridges the deck rests on the horizontal lower chord while the upper chord can be either parallel or sloped.

Construction wise, trusses may be precast and prestressed and joined in the field or precast and posttensioned. In comparing the truss-type with the girder-type bridge, the truss type is definitely more efficient and more rigid when the amount of material is the same. It is interesting to note the fact that when parallel chord trusses are used, it is more efficient to design the truss having the panels at both ends terminating with an inclined member.

Concrete truss bridges with diagonal members have been built in so many structures around the world that it has been pointed out that pouring the diagonal members on an incline may at times be difficult. If this is the case, inclined diagonal members for complete trusses can be substituted by complete steel members as a potential option. Another alternative is that of Vierendeel trusses that are much easier built cast-in-place, or built with precast members assembled on the site.

Due to the formwork and to the required falsework, this category of bridges is considered more expensive to fabricate than other types and is considered to require more maintenance. While some sources describe the concrete truss bridge as less attractive than other forms, this is particularly questionable, especially when sensitive designers may interpret the theme with special configurations that give plasticity and articulation to the various components. As a matter of fact, variable cross-sections with nonprismatic geometries and sharp angularities have at times produced vibrant sculptural compositions of high morphological values.

Cable-stayed bridges

These types of bridges in concrete follow their similar counterparts in steel. They are used for long spans and usually are employed for spans that do not necessarily require the use of suspension bridge *per se*. In appearance these structures are quite light and delicate due to their slenderness. They are usually built with systems specifically mentioned elsewhere in this text that do not require the use of falsework. Special attention must be given to the protection of the stays, especially at the connection. Similarly, the structures require a larger maintenance time in comparison to others, due especially to the complexity of the suspension members.

Suspension bridges

Bridges in this category are intended to span the largest distances that can be covered. They include steel towers and cables and concrete decks and girders. Their time of construction is longer than that of stayed bridges, and also like these, suspension bridges do not require falsework. The concrete part of suspension bridges is constituted by the decks and girders which span the distances equal to the spacing between hangers. The hangers are in turn carried by the two main cables spanning from tower to tower.

Loads for concrete bridges

The following discussion on different loads applicable for the design of concrete bridges is derived from the AASHTO, the Ontario Highway Bridge Design Code (OHBDC), and data from different studies such as "Recommended Design Loads for Bridges" in the *Journal of the Structural Division ASCE*.

Typical loads to be considered include the following: structural dead loads and superimposed dead loads; construction, handling, and erection loads; deformation effects which include: settlements of supports, shrinkage and creep, axial load deformations, thermal effects, prestress effects, and frictional forces; environmental loads which include wind loads, snow loads, earthquake loads, earth pressures, buoyancy, stream flow pressure, ice loads, debris loads, wave action and ship impact; pedestrian bridge live loads which include deck live loads, railing live loads and provision for overloads; highway bridge live loads which include standard vehicular live loads, special truck loads, and fatigue; railroad bridge live loads which include design live loads and provision for overloads; rail transit bridge live loads; airport runway bridge loads which include landing gear loads, impact, application of wheel loads, braking forces and provisions for crash landing loads; pipeline and conveyor bridge loads which include fluid loads, solid loads, and equipment loads; and load combinations.

Not all the loading conditions mentioned above are going to be examined herein due to their general nature, and for these the reader is referred to the ACI-ASCE Committee 343, *Analysis and Design of Reinforced Concrete Structures*, Chapter 5. Only those that apply specifically to bridge structures are briefly discussed hereafter.

Structural dead loads

Structural dead loads include the weight of structural members themselves plus the weight of other structural members that they support, including rails, sidewalks, slabs, and beams.

Superimposed dead loads

Superimposed dead loads include the weight of the following components: wearing surface, earth fill, permanent forms, tracks and ballast, barrier walls and railings, waterproofing, signs, architectural ornamentation, pipes, conduits, and any other appurtenances installed on the structure. For preliminary estimates of superimposed dead loads the ACI specifications recommend the following:

- for highway bridges:
  - barrier rail (New Jersey curb): 450-600 psf (6.6-8.8 kN/m)
  - allowance for future overlay: 25 psf (1.2 kPa)
- for railroad bridges
  - ballast and tracks: 3500 lb/ft (51.0 kN/m)
  - tracks only: 600 lb/ft (8.8 kN/m)
  - walkway: 500 lb/ft (7.3 kN/m)

Construction, handling, and erection loads

See: ACI-ASCE Committee 343, *Analysis and Design of Reinforced Concrete Structures,* Chapter 5, p. 343R-59.

Deformation effects

See: ACI-ASCE Committee 343, *Analysis and Design of Reinforced Concrete Structures,* Chapter 5, p. 343R-59.

Wind loads

Bridges with lengths up to 400 ft are considered in sensitive to dynamic wind effects and can be designed using the standard procedure indicated in ACI-ASCE Committee 343, *Analysis and Design of Reinforced Concrete Structures,* Chapter 5, p. 343R-260, that consider climatological site conditions, gradient winds, gustiness, shape coefficients, and orientation with respect to the mean wind direction and natural frequency of the structure.

Bridges with lengths larger than 400 ft, including some lightweight pedestrian bridges, are considered to be sensitive to dynamic wind effects, more specific analysis is required, including wind tunnel testing.

Snow loads

Snow loads are based on the typical weight of 5 to 12 lb/ft^3 or 0.9 to 1.9 kN/m^3. Compacted snow is based on the typical weight of 15 to 20 lb/ft^3 or 2.4 to

3.1 kN/m^3. For covered bridges effects of snow drift accumulation must be considered. For unsheltered bridges snow loads can be considered in combination with reduced live loads. If the snow load exceeds the live load, the latter could be disregarded.

Stream flow pressure

Stream flow pressure on piers is calculated by:

$$SF = Kv^2 \, (0.515\ Kv^2)$$

where:

SF = stream flow pressure in psf (kPa)
v = velocity of water in ft/s (m/s)
K = 1.4 for square pier ends
= 0.5 for angled ends where the interior angle < 30°
= 0.67 for circular pier ends

Seismic loads

See ACI-ASCE Committee 343, *Analysis and Design of Reinforced Concrete Structures,* Chapter 5, p. 343R-62.

Earth pressures

See ACI-ASCE Committee 343, *Analysis and Design of Reinforced Concrete Structures,* Chapter 5, p. 343R-64.

Buoyancy

See ACI-ASCE Committee 343, *Analysis and Design of Reinforced Concrete Structures,* Chapter 5, p. 343R-64.

Ice loads

See ACI-ASCE Committee 343, *Analysis and Design of Reinforced Concrete Structures,* Chapter 5, p. 343R-65.

Debris loads

See ACI-ASCE Committee 343, *Analysis and Design of Reinforced Concrete Structures,* Chapter 5, p. 343R-66.

Wave action

See ACI-ASCE Committee 343, *Analysis and Design of Reinforced Concrete Structures,* Chapter 5, p. 343R-66.

Ship impact

See ACI-ASCE Committee 343, *Analysis and Design of Reinforced Concrete Structures,* Chapter 5, p. 343R-66.

Pedestrian bridge live loads

The live loads for pedestrian bridges are divided into the following three categories: deck live loads, railing live loads, and overloads.

Deck live loads

The live load on the deck should be one hundred pounds per square foot on the walk area between the two railings or between the two curbs. This load could be applied continuously over the entire area or discontinuously in accordance to what will produce the maximum stresses in the structural members. Reductions of the live loads could be applied for bridges particularly long and wide. Maximum deflection due to live loads is one thousandth/span. Suspension bridges and "ribbon" bridges should have a natural frequency of vibration higher than that of the pedestrian traffic, more specifically a desirable natural frequency in hz for the structure should be: $f = (600/\text{span in ft})^{-2}$ or $f = (183/\text{span in m})^{-2}$.

Railing live loads

The live loads on the railing are applied vertically and horizontally at the same time. The recommended value is 50 lb/ft (0.75 kN/m). If the railing is higher than 5 ft above the walk surface such load can be disregarded. Posts should be designed to support lateral forces applied at the centroid of the post for the upper rail or for high rails, at 5 ft (1.5 m) minimum above the rails.

Overloads

If pedestrian bridges allow occasional vehicular loads, this must be included in the design loads. If the vehicles are not specifically identified, the recommended load that could be used is: two 4,000 lb (17.7 kN) axles spaced at 12 ft (3.6 m).

Highway bridge live loads

- *Standard vehicular live loads:* standard vehicular live loads plus impacts, centrifugal forces, braking forces, curb loads, railing loads, special alternate live loads, and provisions for overloads can be found for the United States in *Standard Specification for Highway Bridges*, published by the American Association of State Highway and Transportation Officials. For Canada the reference can be found in the Ontario Highway Bridge Design Code.

- *Logging Loads:* special kinds of roads and bridges are allowed in the United States for the logging industry which are designed for much larger loads than those allowed on state highway systems.
- *Military Loads:* loads for military traffic should comply with "Military Fixed Bridges," TM5-312, Superintendent of Documents, Washington, DC.
- *Overload Provisions:* in addition to standard vehicular live loads, bridges should be designed for special, infrequent heavy vehicle loads. If the bridge is designed for a live load less than the standard H20 loading, the design vehicle load increased to twice its value should be applied on any single lane, without concurrent loading of any other lanes (see additional data in the ACI-ASCE Committee 343, *Analysis and Design of Reinforced Concrete Structures*, Chapter 5, p. 343R-67).
- *Design Traffic Lanes:* the traffic lanes on a bridge all have equal widths.
- *Traffic Lane Units:* the truck loads or lane loads are considered to be applied over a width of ten ft or three meters (ACI-ASCE Committee 343, *Analysis and Design of Reinforced Concrete Analysis and Design of Reinforced Concrete Bridge Structures*, Chapter 5, pp. 343R-67).
- *Positioning of Live Loads:* the positioning of vehicular loads should be such as to produce the maximum stresses in the structural members.
- *Reduction in Load Intensity:* due to the improbability that the maximum load would include the simultaneous applications of all possible loads, it is possible to reduce the intensity of the load when calculating the maximum stresses in structural members. To this effect the following percentage of live load can be used: first and second lanes, 100%; third lane,70%, fourth and subsequent lanes, 40%. For design of transverse members, a reduction of intensity loads can also be applied (ACI-ASCE Committee 343, *Analysis and Design of Reinforced Concrete Structures*, Chapter 5, pp. 343R-67).

- *Distribution of Loads to Beams:* see ACI-ASCE Committee 343, *Analysis and Design of Reinforced Concrete* Chapter 5, p. 343R-67.
- *Fatigue:* the design of reinforcing steel and prestressing steel at cracked sections of concrete bridges should include the effects of fatigue (ACI-ASCE Committee 343, *Analysis and Design of Reinforced Concrete Structures,* Chapter 5, p. 343R-67).

Railroad bridge live loads

Specifications for the design live loads for railroad bridges can be found in the following reference: *The Manual of the American Railway Engineering Association,* Chapter 8. Overloads must be anticipated in the design of railroad bridges. When such overloads and impact are applied, the maximum stresses should not exceed 1.5 times the values of the allowable stresses. When overloads are applied on the structure at the reduced speed, the impact loads should be adjusted as specified in the ACI-ASCE Committee 343, *Analysis and Design of Reinforced Concrete Structures,* Chapter 5, p. 343R-68.

Rail transit bridge live loads

See reference ACI Committee 358, "State-of-the-Art Report on Concrete Guideways," p. 358R-80.

Airport runway bridge loads

Landing gear loads — Design landing gear loads for runway and taxiway bridges must be designed for the largest aircraft serviced at a given airport. For a more complete list see Federal Aviation Agency's Advisory Circular "Aircraft Data," AC #150/5325-5, 1975, and "Airplane Airworthiness," Advisory Circular. Other references include "Airplane Characteristics" and "Airport Planning" published by individual aircraft manufacturers.

Impact — Vibrations and impact effects of landing gears must be accounted for by increasing the live load of a certain percentage as follows: for parking aprons and low speed taxiways, 30 percent; high speed taxiways and runways, 40 percent; touch areas or runways, 100 percent. For additional data see ACI-ASCE Committee 343, *Analysis and Design of Reinforced Concrete Structures,* Chapter 5, p.□343R-68.

***Wheel loads** (slabs)* — The width of transverse slab resisting the load effects of an individual (or dual) tire should not be assumed greater than the slab span, the spacing of tandem axles, or seven ft or 2.1 meters, whichever is less. The width of the longitudinal slab resisting the load effects of an individual tire should not be assumed greater than the slab span, the tire spacing, or seven ft or 2.1 meters, whichever is less. For additional information see ACI-ASCE Committee 343, *Analysis and Design of Reinforced Concrete Structures*, Chapter 5, p. 343R-68.

Wheel loads (beams and girders — See ACI-ASCE Committee 343, *Analysis and Design of Reinforced Concrete Structures*, Chapter 5, p. 343R-68.

Structural units of the substructure and connections between the substructures and upper structure should be designed to resist longitudinal brakeing force in accordance to reference ACI-ASCE Committee 343, *Analysis and Design of Reinforced Concrete Structures*, Chapter 5, p. 343R-70.

Structures located on the portion of the runway where crash landings may be anticipated should be designed for the failure load of the landing gear. The Federal Aviation Agency requires that landing gears have a reserved strength capable of withstanding the impact of the aircraft dropping at a rate of 12 ft/s or 3.6 m/s.

Pipe Line and Conveyor Bridge Loads

Fluid Loads — Pipe line bridges should be designed to carry the heaviest fluid anticipated in the pipe line. Where a concrete pipe (or channel) acts as a structural member of the bridge, or is not specifically isolated from becoming stressed longitudinally by the live load deflection of the structure, stresses should be checked and reinforcement or prestressing added as necessary.

Solid Loads — Bridges that will carry solids, semifluids and fluid loads should be designed to carry not only continuous loads but also discontinued ones, whichever is critical. In addition they should be able to resist forces derived from starting and stopping the conveyor. The bridges in this category should be designed to carry the live load caused by the material being conveyed plus the weight of any equipment that could be used for maintenance and repairs. Catwalks should be designed for a live load of 60 lb/ft^2 (3 kN/m^2) applied continuously or discontinuously according to whichever load is critical.

Load Combinations — For detailed data see ACI-ASCE Committee 343, *Analysis and Design of Reinforced Concrete Bridge Structures*, Chapter 5, p. 343R-68.

Box girders for concrete bridges

Box girders apply not only to steel structures but also to concrete bridges. In fact, box-beam bridges are the most common types for spans varying between 35 and 250 m. Concrete box girders can be either cast-in-place or precast. Usually the shorter spans are cast-in-place while the longer spans are precast. The efficiency of the concrete box-girder increases as the span increases. In fact, for short girders up to 20 units, the percentage of voids is 40. For spans varying between 20 and 35 the percentage of voids varies between 40 and 55. Finally, for the longest spans varying between 35 and 150, the percentage of voids reaches over 80. Obviously, the larger the percentage of voids, the less the concrete in the cross section, and therefore the higher the efficiency.

When the cross section of bridges is larger than 55 ft in width, it is customary to use at least two box girders. The two girders are interconnected with a closure strip. Each box beam can be individually transversely posttensioned or the transverse posttensioning can also tie together the two box girders extending through the intermediate closure strip. The exterior webs of the box girders can be sloped, making the box girder more efficient. Inclined webs facilitate the stripping of the forms by raising or lowering the form travelers. Box girders can also be built with haunches. For long spans, prestressing of the girder box webs is strongly recommended.

Box girders require stiffening diaphragms to increase their torsional rigidity. They are usually installed near the supports or at the hinges and are usually omitted along the span unless the span is curved and torsion is anticipated to occur.

Cable-stayed bridges

Among the different typologies of bridge structures, the stayed bridge occupies a relevant role, especially for long span structures. More economical than the suspension bridge type, the stayed bridge solution is usually adopted for long span structures which do not necessarily require the suspension type. Although the stayed cable bridge depends on tensile steel members supported by relatively tall towers just like the suspension system, the structural behavior is quite different and the construction methodology used is also significantly unlike that of the classical suspension bridges.

Although the present type of stayed cable bridge has gained popularity since the middle of this century, its origin can be traced back to the 17th century in Venice. There a bridge designer by the name of Verantius originally built a bridge prototype based on a central tower supporting two spans by means of several iron chains. The engineering principle was definitely established for its simplicity of construction, and many structures of this type have been built on the basis of this example.

Yet it was only in 1955 that the cable-stayed bridge in its present form was eventually built in Sweden: the Stromsund Bridge was designed by a German engineer, F. Dischinger, and built by a German contractor, Demag. From the earlier prototype in Venice to the present-day cable-stayed bridge, the structure had developed considerably. One of the main differences was the substitution of chains and rods in tension with present-day steel cables having much higher tensile strength. One colorful example of the stayed bridge which should be mentioned is the Loscher bridge, built in 1784 by a German carpenter from Fribourg was used only wood, even for the stays.

Other early bridges include the King's Meadow footbridge built in 1817 by British engineers using a cast iron tower and cable steel stays supporting a span of approximately 110 ft. However, two major collapses of stayed bridges cast a shadow of doubt over these structural systems. One such structural failure involved a bridge which crossed the Tweed River near Dryburgh-Abbey, England in 1818, and another involved the bridge crossing the Saale River near Nienburg in 1824. With the temporary abandonment of the stayed bridge, the typical suspended bridge gained tremendous popularity for spanning the longest distance. From this point on, suspension bridges dominated the scene and continued to do so, while the cable-stayed bridge reappeared only a few decades ago.

Basic characteristics

The cable-stayed bridge in essence can be defined as the assembly of three basic components: (a) the longitudinal upper structure carrying the deck; (b) a vertical tower; (c) the inclined cable stays suspended from the tower and supporting the upper structure. This is, of course, the simplest form which eventually can be further developed and used as a single structure or as the assembly of several similar structures combined. The simplicity of the concept in structural terms is also found in the ease of construction, which is much simpler than the construction, requirements of typical suspension bridges. In fact, as the tower is first built the longitudinal upper structure can be gradually built without any use of falsework since the upper structure can progress gradually upward in both directions being supported from above by the diagonal stays. This explains the rapid success of the cable-stayed bridge system once the concept had been resuscitated in the 1950s.

The technical literature explains the large popularity these bridges gained by the sudden need in the postwar period to rebuild thousands of bridges destroyed during World War II. The cable-stayed bridge gained its popularity only on the basis of its efficiency in terms of design and construction techniques. Some data proving the fast growth in acceptance of this structural bridge type can be found in the following bibliographic reference referring to the mid 1970's. The distribution of cable-stayed bridges built worldwide by 1974 was as follows: Germany,17; United States, 8; Japan, 7; Canada, 4; Great Britain, 3; Italy, 3; Argentina, 2; Australia, 2, France, 2, Holland, 2, USSR, 2, Austria, 1, Denmark, 1, India, 1, Libya, 1, Sweden, 1; Venezuela, 1; and Zambia, 1 (Podolny, 1976).

Geometrical types

The first classification of stayed bridges includes the following three categories:

- The single tower carries a span to one side and a span to the other side. These two spans may or may not be equal.
- Two towers carry one span in between the two of them, including one span projecting out of the first tower and another projecting out of the second tower.
- Three or more towers carry spans between two consecutive towers. The first tower carries a span projecting out, and the last tower carries a similar span projecting out.

The following three categories constitute a geometrical layout. In structural terms all the spans are continuous with one another. They are supported by several stays suspended directly from the towers themselves at the point of intersection.

Stays

The stays are the diagonal cables that suspend from the towers and support the spans at different points. There are four different types of layouts possible with the stays in accordance with the plane in which they lie.

- All the stays are contained within a vertical plane passing through the longitudinal center line of the bridge and through the towers. This is possible when the bridge is symmetrical with respect to the center lines. In this case the traveler over the bridge can have an open view of the panorama on the exterior side. Stays in this category (i.e., contained within one single plane, are usually found in Europe rather than in the United States). The reason is due to the loading condition required by codes. In fact, AASHTO requires that when the bridge carries multiple lanes, each lane must be loaded with a certain percentage of the load, W, applied in the case of a single load. An example given by Buckland (1990, p. 234) indicates that in the United States in a case of a six-lane bridge, each lane would be loaded with 0.75 if the single lane loading, W, equals 4.5W. The British Code, BS 5400, would require in the same case that two lanes carry the maximum lane load and the remainder carry one third of the maximum lane load. Therefore, the total load is 3.3W. In comparing the total it is obvious that the single plane layer of cables is a more convenient solution in the territories subjected to the British codes than in those of the United States.

- The stays are contained within two vertical planes, each one on the outside of the deck as in typical suspension bridges.
- All the stays are contained in inclined surfaces that pass from the single point of connection at the top of the tower to the points of intersection at the outside edge of the deck. These surfaces are warped in case the stays are not all supported by a single point on the tower, but rather are supported from different points at different heights.
- All the stays are contained within three vertical planes, one passing through the longitudinal center line and the other two passing through each side of the deck. In other words, these three planes constitute a combination of category one and category two.

Stays arrangement

The stays supported by the towers may be radiating all from one point or they may be suspended from different points located at different heights on the towers. Four different categories of stays arrangements are as follows:

Radiating – in this case all the stays converge at one single point at the top of the tower.

Harp – in this case the stays are suspended from different points on the tower, equally spaced vertically so that the stays remain parallel to one another just like the strings in the homonymous musical instrument. The spacing along the horizontal are also equal to one another.

Fan – in this case the stays are arranged in a manner that includes the radiating and harp systems together. More specifically, the stays are suspended from the tower along a vertical line with different spacing. Then they descend diagonally at different points of suspension along the roadway structure at intervals having equal length.

Star – in this arrangement the cables intersect the tower at different heights and then converge on each side of the tower to intersect the roadway structure at a common point.

Cables

In determining the mechanical characteristics of steel cables, the determination of the modulus of elasticity, *E*, is based on tests made on the specimen itself, using the gross metallic area, including the zinc coating which reduces the original value of 29,000,000 psi.

In the determination of the modulus of elasticity for cables, the practical operation is performed on a specimen not less than 100 in. long with a gross sectional area of the strand taken in accordance to ASTM specifications. When the elongation of the strand under stress is measured, the stress should not be less than 10% of the breaking strength or more than 90% of the prestretching load.

It is important to note that the required minimum values of the modulus of elasticity for steel cables are as follows:

- Strands, having a diameter from 0.5 to 2.5625 in have a minimum modulus of elasticity E = 24,000,000 lb/in^2.
- Strands, having a diameter 2.5625 in or larger, have a minimum modulus of elasticity E = 23,000,000 lb/in^2.
- Wire rope having a diameter from 0.375 to 4 in have a minimum modulus of elasticity E = 20,000,000 lb/in^2.

These cable types have a Class A coating of zinc. More precisely, for all cables, the modules of elasticity is less than that of solid steel. There are four basic factors that intervene in the determination of the modulus of elasticity of the cable: (a) the elasticity of the wires; (b) the internal deformation of the cable; (c) the sag of the cable; and (d) a combination of the above. From all of this, the modulus of elasticity of cables remains constant and it is a function of the applied stress.

Equivalent modulus of elasticity of stays

In the design of stays for stayed bridges it is assumed that the cables are straight lines subjected to tensile forces. In reality this is not true because any cable sags under the effect of its own weight and assumes a curvature and an elongation. Therefore, if one assumes the cable to be straight, a compensation must be made. This consists in assuming an equivalent modulus of elasticity, E, which is less than that of the material. To find out such an equivalent, E, there are basically two formulas. One expression is given by H. J. Ernst for locked-coil strand; a similar expression for parallel wire-strands was proposed by Loenhardt (both in Podolny and Scalzi, 1976). As previously mentioned the modulus of elasticity of cables varies according to size of diameter.

Cables and wire rope generalities

Up to this point, it has been assumed that the reader has an intuitive understanding of what a cable is. Yet in practical terms it is possible that some confusion can be made concerning not only cables but other similar flexible tension members. To this effect, it is proper to introduce some definitions of terms to distinguish one from the other to avoid misunderstanding and the misnaming of individual members in question.

The following definitions are derived from *Tentative Criteria for Structural Applications of Steel Cables for Buildings*, published by American Iron and Steel Institute (1966):

- *Cable:* a flexible tension member consisting of one or more groups of wires, strands, or ropes. Of course, diameter varies according to the required tensile forces, yet to give a sense of the order of magnitude, we point to one example involving the largest suspension bridge ever built, the Akashi Kaikyo Bridge in Japan, having a central span of 1,990 m. For such a span the diameter of the two cables are 1.1 m.
- *Fitting:* any accessory used as an attachment to the cable or its components.
- *Zinc-coated:* coated with zinc by either the hot dip or the electrolytic process.
- *Inner Wires:* all wires of a strand except surface or cover wires.
- *Rope:* a plurality of strands laid helically around a core composed of a strand or another wire rope.
- *Strand:* an arrangement of wires helically laid about a center wire to produce a symmetrical section.
- *Wire:* a single continuous length of metal cold drawn from a rod.

In addition to an abstract knowledge of cables, wire ropes, and strands, it is convenient to focus on the numerical values to obtain a more factual idea of these tensile members (Tables 2-1 and 2-2).

A word of caution is due when considering numerical values included in the tables. Such values refer to data being published in 1966 and should therefore be checked to verify whether they still apply or whether they have been modified in the last updated publications. In practical terms the reader should consult the Wire Rope Technical Board or any of the present-day manufacturers included in the appendix. Names, addresses, and telephone numbers, when available, for quick reference.

See appendices for more data about Wire Rope Technical Board and about the individual manufacturers of these products in the United States.

Saddles

The cables, continuous over the towers, are supported by saddles that are either cast or fabricated out of steel plates. As the cables pass over the saddle, they fit into grooves. Such grooves are also fitted with a soft material like zinc or aluminum that deforms under the pressure of the cables and that guarantees a uniform bearing stress. The saddles are either fixed or are supported by rollers or rocker bearings. If the saddle is fixed, the cable supporting the end span and the cable supporting the middle span may have a different tensile force. Vice versa, when the support is a roller or rocker bearing, the tension on both sides is equal. Saddles are not standard and must be designed individually for each bridge (Figures 2-17 and 2-18).

Table 2-1: Mechanical Properties of Zinc-Coated Bridge Rope

Standards Established by the Wire Rope Technical Board

Nominal Diameter In Inches	Min. Breaking Strength in Tons of 2,000 Lbs. Class A Coating	Approx. Weight Per Ft. In Lbs.	Approx. Metallic Area in Sq. In.
3/8	6.5	0.24	.065
1/2	11.5	0.42	.119
5/8	18.0	0.65	.182
3/4	26.0	0.95	.268
7/8	35.0	1.28	.361
1	45.7	1.67	.471
1 1/8	57.8	2.11	.596
1 1/4	72.2	2.64	.745
1 3/8	87.8	3.21	.906
1 1/2	104.0	3.82	1.076
1 5/8	123.0	4.51	1.270
1 3/4	143.0	5.24	1.470
1 7/8	164.0	6.03	1.690
2	186.0	6.85	1.920
2 1/8	210.0	7.73	2.170
2 1/4	235.0	8.66	2.420
2 3/8	261.0	9.61	2.690
2 1/2	288.0	10.60	2.970
2 5/8	317.0	11.62	3.270
2 3/4	347.0	12.74	3.580
2 7/8	379.0	13.90	3.910
3	412.0	15.11	4.250
3 1/4	475.0	18.00	5.040
3 1/2	555.0	21.00	5.830
3 3/4	640.0	24.00	6.670
4	730.0	27.00	7.590

Table 2-2: Mechanical Properties of Zinc-Coated Bridge Strand

Standards Established by the Wire Rope Technical Board

Nominal Diameter In Inches	Minimum Breaking Strength in Tons of 2000 lbs. Class "A" Coating Throughout	Class "A" Coating Inner Wires Class "B" Coating Outer Wires	Class "A" Coating Inner Wires Class "C" Coating Outer Wires	Approx. Metallic Area in Sq. In.	Approx. Weight Per Ft. In Lbs.
1/2	15.0	14.5	14.2	0.150	0.52
9/16	19.0	18.4	18.0	.190	.66
5/8	24.0	23.3	22.8	.234	.82
11/16	29.0	28.1	27.5	.284	.99
3/4	34.0	33.0	32.3	.338	1.18
13/16	40.0	38.8	38.0	.396	1.39
7/8	46.0	44.6	43.7	.459	1.61
15/16	54.0	52.4	51.3	.527	1.85
1	61.0	59.2	57.9	.600	2.10
1 1/16	69.0	66.9	65.5	.677	2.37
1 1/8	78.0	75.7	74.1	.759	2.66
1 3/16	86.0	83.4	81.7	.846	2.96
1 1/4	96.0	94.1	92.2	.938	3.28
1 5/16	106.0	104.0	102.0	1.03	3.62
1 3/8	116.0	114.0	111.0	1.13	3.97
1 7/16	126.0	123.0	121.0	1.24	4.34
1 1/2	138.0	135.0	132.0	1.35	4.73
1 9/16	150.0	147.0	144.0	1.47	5.13
1 5/8	162.0	159.0	155.0	1.59	5.55
1 11/16	176.0	172.0	169.0	1.71	5.98
1 3/4	188.0	184.0	180.0	1.84	6.43
1 13/16	202.0	198.0	194.0	1.97	6.90
1 7/8	216.0	212.0	207.0	2.11	7.39
1 15/16	230.0	226.0	221.0	2.25	7.89
2	245.0	241.0	238.0	2.40	8.40
2 1/16	261.0	257.0	253.0	2.55	8.94
2 1/8	277.0	273.0	269.0	2.71	9.49
2 3/16	293.0	289.0	284.0	2.87	10.05
2 1/4	310.0	305.0	301.0	3.04	10.64
2 5/16	327.0	322.0	317.0	3.21	11.24
2 3/8	344.0	339.0	334.0	3.38	11.85
2 7/16	360.0	355.0	349.0	3.57	12.48
2 1/2	376.0	370.0	365.0	3.75	13.13
2 9/16	392.0	386.0	380.0	3.94	13.80
2 5/8	417.0	411.0	404.0	4.13	14.47
2 11/16	432.0	425.0	419.0	4.33	15.16
2 3/4	452.0	445.0	438.0	4.54	15.88
2 7/8	494.0	486.0	479.0	4.96	17.36
3	538.0	530.0	522.0	5.40	18.90
3 1/8	584.0	575.0	566.0	5.86	20.51
3 1/4	625.0	616.0	606.0	6.34	22.18
3 3/8	673.0	663.0	653.0	6.83	23.92
3 1/2	724.0	714.0	702.0	7.35	25.73
3 5/8	768.0	757.0	745.0	7.88	27.60
3 3/4	822.0	810.0	797.0	8.44	29.53
3 7/8	878.0	865.0	852.0	9.01	31.53
4	925.0	911.0	897.0	9.60	33.60

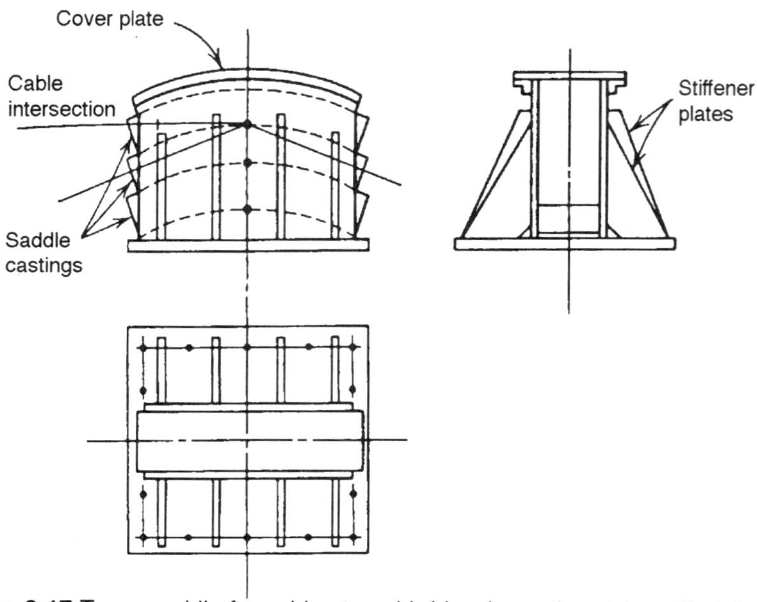

Figure 2-17 Tower saddle for cable-stayed bridge (reproduced from Podolny, W., 1976, *Construction of Cable-Stayed Bridges.* New York: John Wiley and Sons by permission of publisher).

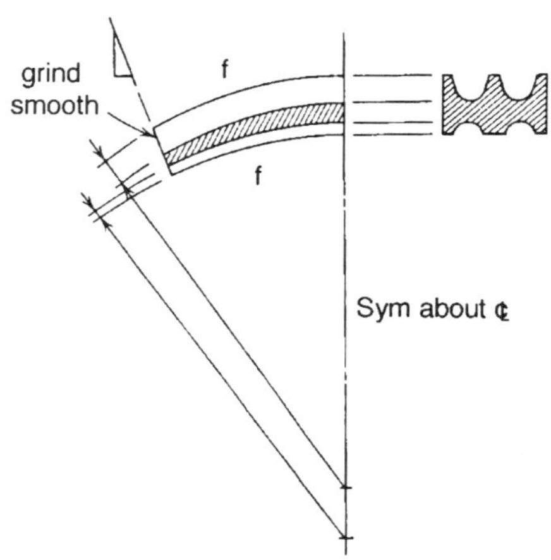

Figure 2-18 Saddle casting for cable-stayed bridge (reproduced from Podolny, W., 1976, *Construction of Cable-Stayed Bridges.* New York: John Wiley and Sons by permission of publisher).

Fittings

Fittings or sockets are the means for connecting cables to the structure itself. There are basically two methods of connecting the end of the cable to the fittings. One is referred to as *zinc poured* and the other *swaged*. In the first case the strands of the cable within the socket are enclosed in zinc or in epoxy that has been poured into the socket and solidified in a monolithic block. In the second case the strands are inside the fittings and this in turn is placed within a die block. A press is applied to the block that squeezes the steel fittings. Under such pressure, the steel fittings are tightly clamped to the strings.

Towers (pylons)

The towers of stayed bridges include typical different types as follows: (a) single tower, (b) portal-type frame, or (c) A-frame. The towers at the base may either be hinged or fixed (Figures 2-19 through 2-21).

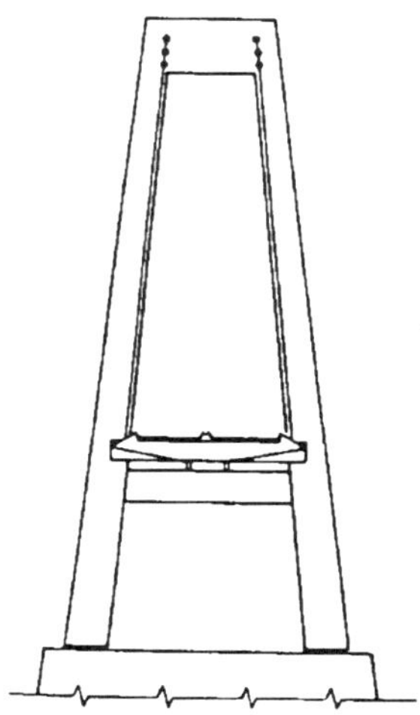

Figure 2-19 Modified A-frame tower for cable-stayed bridge (reproduced from Podolny, W., 1976, *Construction of Cable-Stayed Bridges.* New York: John Wiley and Sons by permission of publisher).

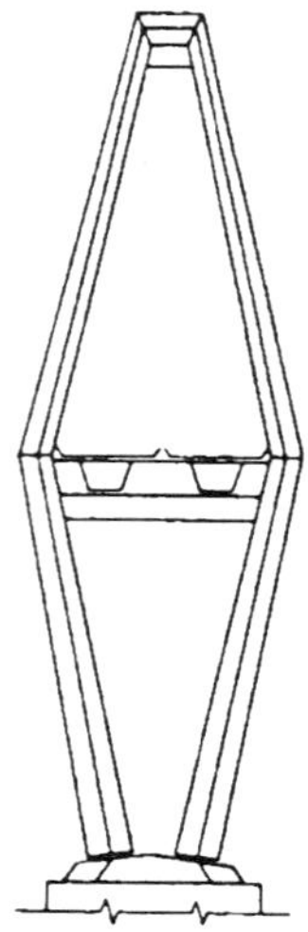

Figure 2-20 Diamond tower for cable-stayed bridge (reproduced from Podolny, W., 1976, *Construction of Cable-Stayed Bridges.* New York: John Wiley and Sons by permission of publisher).

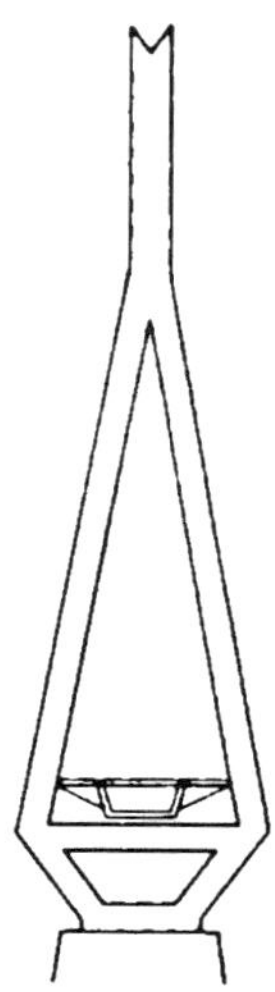

Figure 2-21 Modified diamond (or delta) tower for cable-stayed bridge (reproduced from Podolny, W., 1976, *Construction of Cable-Stayed Bridges.* New York: John Wiley and Sons by permission of publisher).

Span ratios

By surveying a number of existing bridges, it can be noticed that for category one, when one tower is carrying two adjacent spans, the longest of the two is typically about 70 percent of the total length.

In category two, with two towers one span in between and two projecting spans, one on each side, the central span is usually 55 percent of the total and the remaining two will be 22.5 percent each of the total.

In the case of multiple spans with three or more towers, it is typical that all spans are the same, with the exception of the two spans projecting on each side of the outside towers. The stays are usually symmetrical on each side of the towers. It is also typical to insert intermediate spans between two consecutive towers, which facilitate the erection process.

Superstructure types

In theoretical terms there should be no limitations of structural systems to be used for carrying the deck, however, typical solutions for stayed bridges include mostly stiffened trusses, plate-girders, box-girders, single-celled box-girders, multiple-cell box-girders, one box-girder or two box-girders. In practical terms stiffened trusses and single plate-girders are not much use and the plate-girder is preferred. Even more preferred are box-girders of various types. Girder depth to length ratios, of girders typically used in stayed bridges, vary between 1:40 to 1:100. The later is an extreme case.

Clamps

Clamps are fittings used to connect smaller cables to larger ones by friction as in the case of hangers connected to the main cables.

Deck types: orthotropic deck

Deck types mostly used are of the orthotropic box-girder arrangement with box-type cross-sectional configuration, which is usually cantilevered on both sides of the supporting box-girders. More specifically, this type of deck originated in Europe in the 1950s, and has since spread to the rest of the world, including the United States, becoming one of the most popular types of bridge decks (Figure 2.22).

Orthotropic steel plate decks are cross-stiffened steel plate decks consisting of grids of intersecting and mutually supporting floor beams and stringers. They act like a plate having dissimilar elastic properties in the two mutually perpendicular directions.

These decks include two types of longitudinal ribs: open and closed. The open type includes flat bars, bulb sections, angles, and inverted T-sections. The

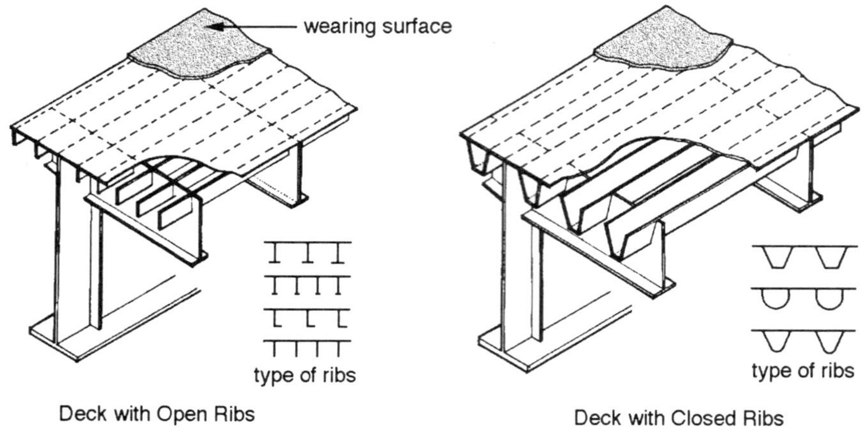

Figure 2-22 Two types of steel orthotropic deck bridges (reproduced courtesy of Best Center, University of Maryland).

ribs are spaced approximately one foot on center. The closed type, consists of trapezoid or rounded cross-sections. Furthermore, the torsional rigidity of the open type is much smaller than that of the closed type.

The deck supported by floor beams, act in a composite action with such beams constituting the top flange of T-beams. The floor beams are usually spaced between 4 ft and 15 ft.

The deck is usually covered with a layer of asphalt from 1 1/2 to 2 1/2 in thick.

More specifically in the open-rib deck system, the flat bars and the bulb sections are the most used. The ribs are arranged in the longitudinal direction of the bridge and run continuously through slots in the transverse floor beams. Their construction is simple to fabricate and these decks can easily be spliced in the field. The deck is open on the bottom part and it can be easily inspected during the life-time of the structure.

The thickness of the deck is not necessarily uniform but it can vary as necessary by reducing the depth of flat bars, bulb sections, etc. Usually, the spacing of the floor beams in the open decks is small. Typically, the open type requires heavier decks than those in the case of closed types. This is due to the fact that the ribs are farther apart so that the deck must be deeper. Further considerations point to the higher cost of welding in the open rib system because the length of the welds is twice as large as that of the closed ribs.

The span for open-rib bridges erected in Europe does not exceed 7 ft, while in the United States, such a span is larger.

In closed-rib bridges, the ribs may have different configurations, but the most efficient one is the trapezoidal, because this type has a higher torsional resistance than the others. The rounded configuration of these closed ribs have a lower torsional rigidity. Between the two types of ribs there is another configuration that is a compromise between the trapezoidal and the round type. Finally, another configuration used is the triangular. For further details, see American Institute of Steel Construction (1964).

Erection Systems

Although every bridge is erected with its own specific methodology, there are three basic types of erection systems that can be identified for cable-stayed structures. These include the following:

- *Staging Method:* this method is usually applied when it is possible to erect temporary supporting piers along the spans of the structure.
- *Push-Out System:* this method is used more often in Europe and is advantageous when no temporary structure can be erected, and when minimum interference is required, with the traffic passing underneath the bridge span. The construction sequence includes the erection of the permanent piers on site and the assembly of the prefabricated parts on an adjacent site. In this area, individual sections of the deck and the upperstructure are built. Each unit includes the full width and extends longitudinally for a relatively short span. The individual units are picked up by a crane, placed over the supports, and made to slide over the supports by means of flat Teflon bearings.
- *Cantilever Method:* this method has been used most typically in the United States and it is used mostly in the same conditions as in the case of the method above.

The staging method usually consists of the following successive operations: the entire suspended structure is first erected on temporary piers, then the towers are erected and the cables are connected. Following this the saddles are jacked up on the piers, stretching the cables until getting the exact configuration, and are stressed to carry the desired tension forces. At this point the temporary piers are removed.

This was the method employed for the Rhine River Bridge at Maxay, Germany, constructed in 1966. The structure consists of one transverse cable plane and two cable spans. The three longitudinal cables configuration are of the fan type. The girder depth/span ration is 1/62. The preassembled individual units were 65 ft long, weighing 27.5 tons. The towers were erected in stages by assembling in succession nine units together, each of which weighed up to 4.4 tons with cross-sectional dimensions of 6.5 to 39.8 ft. The structure includes two spans of 575 and 383 ft supported by a central tower that rises 141 ft above the deck. The deck has a cross-section of 115.8 ft and is supported by a box-girder 39 ft 4 in wide, with an average depth of 9 ft 9 in, creating a satisfactory torsional rigidity.

Box girders for cable-stayed bridges

Box-girders are also applied in the upperstructure of cable-stayed bridges. To obtain a general frame of reference of the cross-sectional dimension of a box-girder applied to a stayed bridge, it is convenient to use an example as follows. The bridge over the Rhine at Maxay, Germany, completed in 1967, includes one central tower and two asymmetrical spans, respectively, of 575 ft. and 385 ft. The cross-section of the superstructure includes a box-girder 10 ft. 6 in. deep, 39 ft. 4 in. wide, with 38 ft. cantilevers on each side supported by inclined struts.

Another example is that of the Railway Bridge in Belgrade, Yugoslavia having a cable-stayed structure with a main span of 833 ft. and two side spans of 164 ft. each. The cross-section includes two stiffening box-girders with respectively a width of 10 ft. 6 in. and a depth of 14 ft. 6 in. The spacing between the two box-girders is 26 ft. 6 in. and the overall width of the bridge is 54 ft.

Steel box-girders are very stable laterally and have a high torsional stiffness. According to the AISC specifications, the buckling of box-girders does not constitute a critical problem when the depth of the box-section is less than six times its width. Beyond such a value of the depth-to-width ratio, the requirements for lateral support have to be determined. Usually steel box-girders are laterally unbraced, have thin webs and are not compact. For allowable bending stresses to be used in the flanges of box-girders, see AISC specifications. In general, for box-shaped sections, if I_{xx} is approximately equal to I_{yy} or if the torsional stiffness is greater than the in-plane bending stiffness, there is no buckling problem.

All the typical box-girder arrangements in common use include the following types:

- Single rectangular box girder
- Central box girder and side single web girders
- Single twin cellular box girder and sloping struts
- Single trapezoidal box girder
- Twin rectangular box girder
- Twin trapezoidal box girder

See Figure 2-2 for cross-sections including the upper deck.

Suspension bridges

This bridge typology is the only structures capable of bridging the longest spans covered by modern bridges. Its economical application, in terms of minimum span length, starts at approximately 984□ftThe spanning capability could go as high as 3,300 m (10,826 ft) as in the case of the Messina bridge, which is still in the planning stage.

The minimum limit of 984 ft previously mentioned does not imply that suspension bridges must only be applied to spans longer than such a limit. In fact, suspension bridges can also be successfully built in shorter lengths without losing their aesthetic appeal. Yet, for spans between 700 and 984 ft, the cable-stayed bridge is considered to be more economical. Other applications of suspended bridges, which can be particularly attractive structures from an architectural point of view, are the pedestrian bridges.

Basic components of suspended bridges are the towers or pylons, the main cables, the vertical hangers, the cable anchorages, the deck and the stiffening girders. From the point of view of suggested proportions, the side spans appear to be more harmonious when their length is less than half the length of the main span. In fact, the smaller the side spans, the more emphasis is placed on the main span, which appears longer giving an impression of higher slenderness for the whole structure. Minimum recommended values for such lateral spans would probably be in the order of 20 percent of the main span.

Besides the relationship of main span to side spans, based purely on aesthetic, there are other considerations based on structural principles that have equal or more meaning. These will be discussed again later when different structural types for suspended bridges are presented.

With regard to overall geometric characteristics of suspension bridges, the height of the deck above the water level, in the case of bridges spanning over water, is related to the length of the main span when the bridge is judged in its visual expression. These two dimensions are fixed, and cannot be altered, nevertheless the longer the span with respect to the height, the more gracious the structure appears.

Usually the stiffening beams or girders under the deck have a great influence on the overall silhouette of the bridge and they definitely improve the aesthetics when they are as shallow as possible. To this effect stiffening trusses are not particularly attractive while box-girders are. Also when considering the cable anchors, it is suggested that appearance should not be too massive. Then considering the towers or pylons, these could be made either of concrete or steel, and although they may consist of an open framework (relatively slender in comparison to old masonry towers), they still retain a sense of main contrast with the cables they support, fulfilling their role without any critical limitation. Their shape can be a main architectural feature that would impress a morphological identity, characterizing each individual bridge. Towers can be in the shape of portals, with or without cross-bracing.

The height of the tower is strictly related to the main span since they support the main cables that need an adequate sag. Under dead load, the bridge upper structure, supported at equal intervals by the hangers, require that the shape of the main cables be parabolic. The shape of the parabola passing between three points (i.e., the two points at the top of the towers and the lower point of the cable at midspan) are determined by the designer and usually imply a sag-to-span ratio of 1:10.

Structural systems for suspension bridges

With reference to "Cable Supported Bridges: Concept and Design" by Niels J. Gimsing, p. 137, the following case is presented (Figure 2-23).

Suspension bridge estimates

The estimates of cable quantity expressed in units of weight can approximately be calculated for preliminary design using the following approach. For the sake of simplicity, a three-span symmetrical suspension bridge is examined. The uniform distributed load on the stiffening girder in the main span is: $w_m + p_m$, while on the side span on the load is: $w_a = p_a$. The step-by-step procedure starts with the determination of the equation of the shape of the cable in the main span:

$$y = 4f_m[x/L_m][(x/L_m)-1]$$

where:

y = vertical coordinate
x = horizontal coordinate
L_m = length of main span
f_m = sag of main span measured from top of tower

The equation of the shape of the cable in the side span is:

$$y = 4f_a(x/L_a)^2 + (4f_a + f_m + b_m)(x/L_a)$$

where:

y = vertical coordinate
x = horizontal coordinate
L_a = length of side span
f_a = sag of side span measured from top of tower

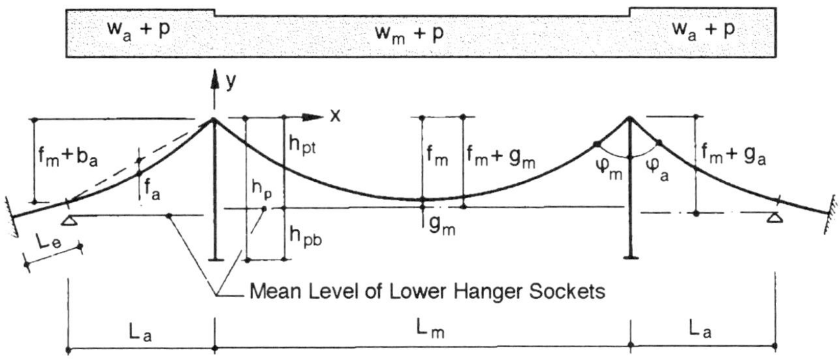

Figure 2-23 Schematic showing geometrical parameters of a symmetrical three-span suspension bridge (reproduced from Grimsing, N.J., *Cable-Supported Bridges: Concept and Design* by permission).

The quantity of steel required for the main cables in the main span is given by:

$$Q_{cm} = \frac{\gamma_c}{\sigma_c}(w_m + p_m) L_m^2 \frac{\sqrt{\left[1 + 16\left(\frac{f_m}{L_m}\right)^2\right]}}{8\frac{f_m}{L_m} - \frac{\gamma_c}{\sigma_c} L_m \sqrt{\left[1+16\left(\frac{f_m}{L_m}\right)^2\right]}}\left[1 + \frac{8}{3}\left(\frac{f_m}{L_m}\right)^2\right]$$

The quantity of steel required for the main cables in both side spans is given by:

$$f_a \cong \frac{w_a + p_a}{w_m + p_m}\left(\frac{L_a}{L_m}\right)^2 f_m$$

where:

γ_c = density of cable material
σ_c = allowable cable stress
$(w_m + p_m)$ = uniform load on main span
L_m = length of main span
f_m = main span cable sag
g_m = vertical distance (no load) from cable low point to bridge deck
$(w_a + p_a)$ = uniform load on side span
f_a = vertical distance from chord and side cable
h_p = height of pylon
h_{pt} = height between top of pylon and main span connector
$h_{pb} + h_{pt}$ = h_p
$f_m + b_a$ = top of pylon to intersection of deck ramp and side cable

The quantity of steel required for the hanger cables in the main span is given by:

$$Q_{hm} = \frac{\gamma_c}{\sigma_c}(w_m + p_m)\left(g_m + \frac{f_m}{3}\right)L_m$$

The quantity of steel required for the hanger cables in both side spans is given by:

$$Q_{ha} = \frac{\gamma_c}{\sigma_c}(w_a + p_a)(f_m - \frac{4}{3}f_a + 2g_a - b_a)L_a$$

Typical geometry of suspended bridges includes one central span and two lateral ones. The ratio between the length of the main span and that of the side spans may vary and, as they do, major changes occur in the structural behavior of the whole bridge. Another factor influencing the behavior of the whole structure is the type of supports chosen for the cables.

In all conditions the ends of the main cables at the anchor boxes at the end of the side spans are considered to be fixed. The supports of the cable on the top of the towers instead can be fixed or the cable could slide on the saddles without any friction, creating two different type of behaviors. Another major structural consideration concerning the way the central spans and the two side spans are supported include the following three cases:

1. Only the central span is supported by the main cables by means of hanger. The other two side spans are not supported by the main side cables but span from the towers to the abutments. In such a condition the main cable over the central span is fixed at the top of both towers. Then the top of the tower is prevented from sliding horizontally by the main side cables, which connect directly atop of the towers with the anchor blocks.
2. In this case the main span and the side spans are carried by the main cables through the action of hangers. The two side main cables are obviously always above the two side decks.
3. This case differs from Case 2 by the fact that the two main side cables sag below the side decks and the two side decks, instead of hanging from the main side cables with hangers, bear over the sagged cables by means of vertical struts.

In Case 1, the system is considered to be very efficient except when the two side spans are very long. In this case the side cables holding the top of the towers are not straight but sag due to their own weight. Therefore side spans should be kept short 30 percent or less of the main span. An example of a bridge with short spans is the George Washington Bridge in New York with side span/main span ratio of 0.17 and 0.19.

Nevertheless, sometimes the side spans need to be long due to site conditions, reaching values up to 0.5. In this latter case, however, even if structurally speaking the system is less efficient, it seems that the bridge is more pleasing to the eye. Another negative effect of long side spans is the making of the main span more flexible.

Case 3 usually occurs when the side spans are very long, such as in the case of the Brooklyn Bridge when the ratio reached 0.59. In the Brooklyn Bridge, the towers had to be designed to be very stiff because the side main cables were not able to give proper support to the top of the towers because of the sag. Furthermore, to add additional support to the towers, stayed cables had to be added.

The stiffening girder extending from one anchor box to the other at both ends of the girder can be hinged on one side and simply supported on the other, while at the midspan the girder may be clamped with a fixed clamp to the main cable, adding substantial rigidity to the assembly.

The clamp is usually connected directly to the top of the stiffening girder; yet in some cases such a central fixed clamp is omitted and the main cable does not touch the stiffening girder at any point. When the central fixed clamp is required, but the main cable does not come down to meet the stiffening girder, the central stiffener is substituted with a rigid member having the same function as the clamp. The hangers in some cases are purposely inclined rather than vertical to increase the rigidity of the structure, especially under wind forces.

Segmental bridges: cast-in-place and posttensioned

Segmental bridges can either be cast-in-place or precast. Their common characteristic is that they are built up in segments which in turn are later connected. In North America, the first cast-in-place segmental-type bridge was the Ste. Adele Bridge, completed in 1964 in Quebec, Canada.

The second one that was built in Vancouver, BC is the Knight Street Bridge, finished in 1973. Many other bridges of the cast-in-place segmental type have since been built utilizing typical construction procedures used for segmental bridges in general. These include the cantilever method of the cast-in-place segmental construction.

This system does not use any falsework, except for that small portion adjacent to the pier. The operation starts by casting on top of the pier a small part of the superstructure. Then the form travelers are attached to this first portion on both sides of the pier so that they can carry the forms and the concrete being cast. Then the form travelers are further extended on each side of the pier after the second segment has set and the third segment is cast. Since the construction continues projecting out from each side of the pier, when the two projections come in contact, the continuity between the two is attained either by casting a connection or by inserting an intermediate span. As the form travelers extend on each side, the weight is balanced. If the form travelers extend only on one side, then the balancing is obtained with fictitious bents on the other side.

Segmental bridges: precast and posttensioned

The literature on precast segmental bridge construction indicates that the first bridge of this type was first erected in New York state in 1952 and consisted of a single span. The concept was also applied in Europe and spread worldwide. This kind of bridge consisted of precasting the girders, posttensioning them, and then lifting them into place by means of cranes. One type of crane is the launching gantry first used in 1964. This operates in such a way that it travels over the spans already completed.

Cable behavior

For understanding the basic principles that made possible the construction of tensile structures, whether applied to suspension bridges, stayed bridges, or suspension roof systems, it is useful to start from a fundamental point of equilibrium of tension members subjected to loads. Fundamental advantages of tensile structures include the uniform distribution of stresses across the cross-section of any tension members, and the fact that tension members (unlike compression members) are not subject to buckling failure. In addition to these basic advantages peculiar to any members in tension, the present-day high-strength steel cable furnishes a material that is extremely efficient cost wise, more so than any other steel members of lower-grade strength.

These previously mentioned advantages are unfortunately counteracted by an inherent instability that forces cables to vibrate under certain dynamic forces such as those generated by wind. Such vibrations under certain conditions may force cables to enter into a state of resonance, which may eventually cause the failure of the cable itself.

Let us consider one of the simplest conditions of static loading for a cable spanning a distance of 300 ft between supports A and B having equal elevation above the ground. The load on it is a uniform distributed load q, which is equal to 200 pounds per foot (dead plus live loads). Let us calculate the tension force T, generated in the cable by the load q, assuming that before any load is applied, including the weight of the cable, the configuration of the cable was such that the cable would sag 20 ft at midspan. Under this condition the tensile force is given by:

$$T = \frac{q\,l^2}{8f}\sqrt{\left[l + 16\left(\frac{f}{l}\right)^2\right]}$$

where:

f = sag (ft)
l = span (ft)
q = load (lb/ft)
T = tension (lb)

Solving for f, one can see that with the values given in the problem the calculated value of f will be slightly larger than the assumed value of $f = 20$ ft and, therefore, no appreciable error will derive by using the previous formula with a value of $f = 20$ ft. In fact, applying the values given in the problem one finds that $T = 116$ kips. At this point let us introduce a cable of adequate cross-section, such as the following: 2-in diameter with cross-sectional area of 2.43 in^2, weighing 8.48 lb/ft. Next, consider the geometry of the cable by finding the angle a, between the horizontal line at the supports and the tangent to the cable at the same point. Therefore:

$$L = l\left[1 + \frac{8}{3}\left(\frac{f}{l}\right)^2\right]$$

Substituting the value for l and f we obtain tan a = 0.267; therefore, $a = 0.2618$ radians. Vertical reaction V is:

$$V = T \sin a$$

Substituting the value for T and for a we obtain $V = 30$ kips. Consequently, the horizontal reaction H is:

$$H = T \cos a$$

Substituting the value for T and for a we obtain $H = 112$ kips. Note that, in this case, the tension in the cable at midspan is equal to the horizontal component of the reaction at the support H. This indicates that the tension in the cable is not uniform throughout its length, but it varies with a maximum the support and a minimum at midspan. However, it is feasible to consider the tension to be uniform if the sag-to-span is small. In such a case such tension force is assumed to be equal to the maximum at the support:

$$T = 116 \text{ kips, in this case}$$

Considering the geometry of the cable, the length of the cable, L, before any load was applied is given by:

$$L = l\left[1 + \frac{8}{3}\left(\frac{f}{l}\right)^2\right]$$

Substituting l is 300, and $f = 20$ we obtain the value $L = 303.5$ ft. Such value does not take into consideration the elastic elongation of the cable due to the tension force produced by the load.

To calculate ΔL, the elongation of the cable due to the tension force T, one finds it from the following expression:

$$\Delta L = \frac{TL}{EA}$$

where:

E = modulus of elasticity (24×10^6 equivalent E for steel)
A = cross-sectional areas of cable (for the cable given above)

Substituting the numerical values, given above, we obtain 0.605 ft (7.25 in). The new length L_1 is therefore:

$$L_1 = L + \Delta L$$

Substituting the values, we obtain $L_1 = 304.105$ ft. The increase in sag Δf, due to the elongation, ΔL, is:

$$\tan \alpha = \frac{4f}{l}$$

which is derived from the known data based on the knowledge that the cable has a parabolic shape. Substituting the numerical values, $\Delta f = 1.75$ ft. This increase in sag added to the previous one makes the total sag equal to 21.75 ft. Therefore, substituting these new values of the sag in the expression for T, calculate the reduced tension force equal to 107 kips. If the tension is 107 kips, the sag will be less than 21.75 ft, then these variations of tension and sag will eventually reach equilibrium, and the tension will between 116 and 107 kips, and closer to the lower value of 107 kips.

Deep-water piers

For obvious reasons, the number of piers in the design of a long span bridge is the first factor to be considered. Naturally the number of piers and their location establish the longest span for a given structure, which automatically reflects the structural type that can be used. Eventually this would be the limiting factor in deciding whether the structure could feasibly be built. Consequently, when an additional pier can be added for reducing the maximum span of a bridge, this may be the key that determines whether a bridge can be built when the extreme limits are reached.

When the site does not allow the erection of a pier under normal engineering practice the problem reaches a dead end. Typical situations of this sort usually occur when a bridge has to be built over water and the depth of the water is so great that any pier cannot be feasible. This was the case for the bridge over the Strait of Messina, where the depth of the water did not allow the erection of any intermediate piers, so the structure had to depend on only two supporting towers sited on dry land, creating the longest span ever built. The possibility of an intermediate support would have greatly reduced the task, but it was considered unfeasible.

A new approach to this problem has now been developed and will shortly be presented herein. When, the depth of the sea bottom is exceptionally great, the concept for building a pier is based on the fact that a floating chamber can be constructed with a buoyancy to carry its own dead load plus an additional load superimposed by the bridge that the pier must carry.

The concept consists of building a floating chamber of sufficient size to develop the required buoyancy force and anchoring the chamber at the desired site and at an established depth. The chamber built of steel or concrete would be multicellular in cross-section, which would guarantee safety and also increase the rigidity of the structure. Any chamber could definitely be built in various components on the ground and be launched and floated to the site individually. Then, on the site, the components can be assembled to form the chamber which eventually will be sunk to the desired depth. To stabilize the chamber there would be a set of vertical cables anchored at the bottom to hold the chamber at the desired depth. Another set of stayed cables placed at an angle will guarantee the horizontal stability of the chamber. To lower the chamber at the desired depth, one of the systems that could be adopted would include the filling up of some of the cells with water and then, after sinking the chamber at the desired depth, the water could be pumped out, achieving the needed buoyancy.

The depth below the water surface where the chamber is designed to rest should be such that any wave action under the water surface would not be felt by the chamber itself. The pier itself, supporting the bridge, would extend from the top of the chamber up to a certain level above water where it would carry the upper structure of the bridge. In this manner the floating chamber will offer a stable bearing capacity unaffected by the wave action. The supporting pier above the chamber extending upward through the water and above it, will be the only portion subjected to the battering of the waves. However, since the pier will be consisting of an open framework, the dynamic action of the waves will exert its force over a minimum surface area.

This concept has been practically accepted now for a new generation of oil drilling facilities in deep waters that substitutes the old-fashioned structures referred to as "Texas towers," which were supported on long structures bearing on the sea bottom. However, it is strange that such a methodology had originally been proposed among the various submissions for the bridge over the Strait of Messina, but at the time it had not been approved, probably because it was considered too avant-garde and not sufficiently tested. Now, however, the use of floating chambers, proven feasible for the new oil rigs, stands out as a new approach to be applied to deep water piers for bridges that hopefully will soon find a viable application.

CRITERIA FOR THE SELECTION OF BRIDGE TYPES

Considering cost as a primary factor for selecting one type of bridge versus another, often only initial costs have been evaluated without considering the total cost over the lifetime of the structure, which is a much more meaningful factor. A change occurred in this respect in 1991, when sections 134 and 135 of the Intermodal Surface Transportation Efficiency Act stated that state and local planning would consider the total life cycle costs of designing and building bridge structures.

Analytical models capable of evaluating total life cycle costs for bridges seem not to be dependable at this time according to recent studies by Nickerson and Veshosky (1992). However, a general scheme upon which total life cycle costs could be formulated according to the same authors should include: initial design; construction, and construction inspection; periodic inspection and maintenance; scheduled maintenance and repair; breakdown maintenance; rehabilitation of the bridge; upgrades to bridge; traffic delay costs; and demolition.

Vehicular clearances on highway bridges

Geometrical parameters for the clearance of vehicular traffic on bridges require standardization both for vertical and horizontal clearances. To this effect, some basic rules have been compiled by the American Concrete Institute and the American Society of Civil Engineers in a report, "ACI-ASCE Committee 343." Specifically, in section 2.2.1.3 of this document, horizontal and vertical clearance data are clearly presented (Figure 2.24).

First, the minimum horizontal clearances required for bridges with low traffic speed and low traffic volumes is 8 ft (2.4 m) larger than in the roadway approaching the bridge. This is an element of safety that reassures the motorist approaching a bridge structure that is perceived as a restriction of the viable space. Secondly, such enlargement should be increased, as the volume and the speed of traffic increases. Such an increase must follow the regulations expressed specifically in AASHTO DS-2, DSOF-3, GD-2, and GU-2.

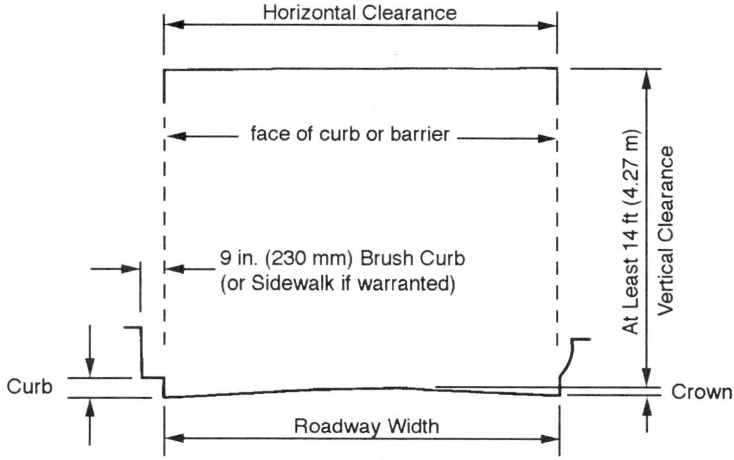

Figure 2-24 Vehicular clearances for highway bridges (reproduced from *Analysis and Design of Reinforced Concrete Bridge Structures,* 1988 by permission of American Concrete Institute).

Vertical clearances on state and interstate highways, in rural and urban areas, should have a minimum of 16 ft (5 m) over the entire width of the roadway, when such values are not unreasonably costly or needed for military traffic. When the last condition does not apply, the minimum vertical clearance could be 14 ft (4.25□m). Allowances should be made for the reduction caused by future resurfacings.

Spanning streams and wetlands

Normative guidelines

Bridging two points sited on both sides of an obstacle is the main function of any bridge. The nature of the obstacle determines many of the parameters that influence the characteristics of bridges from the geometrical point of view, involving the configuration of the structures, as well as from the legal point of view expressed by legal authorities. In the United States, the latter include federal as well as state agencies that have jurisdiction over the structures to be built in certain sites.

The nature of the obstacle that bridges overcome may include urban sites such as fluvial courses or railroad yards. Along roads, bridges are also used to cross over ravines and more often span across water bodies such as streams, lakes, and wetlands. With regard to streams and wetlands in particular, the construction of a bridge is a complex project involving the compliance to legal regulations.

When wetlands are crossed by highway bridges, there is a special sensitivity due to the special status in which wetlands have acquired in the last few decades. Wetland losses are typically occurring under the impact of urbanization and transportation systems which include the construction of bridges. Furthermore, other causes for the potential damage of marshes and swamps comprise the material resulting from excavation for the foundations and other requirements of these structures built in wetland sites. In addition, roads that run close to the wetlands are usually a source of constant runoff pollution.

The wetlands of the United States are presently protected by federal as well as state legislation. To this effect, it is useful to familiarize oneself with the legislative actions that regulate the wetlands before any design project concerning wetlands is begun.

Among such documents in the United States, the most informative is the Clean Water Act of 1997. This document follows the Federal Water Pollution Control Act of 1972 and several amendments such as those of 1987 and 1995, which received congressional approval. The last amendment of 1997 was first introduced by two congressman, Frank Pallone from New Jersey and Peter DeFazio from Oregon. The former congressman's bill is the Federal Clean Water Enforcement Act, while the bill of the latter congressman was the Federal Facilities Clean Water Compliance Act.

The last version of the Clean Water Act includes its basic intent, consisting of the regulations of the treatment and disposal of waste waters. Components include the following: (a) permits for National Pollutant Discharge Elimination System (NPDES); (b) indirect polluted discharges in public treatment works used for waste water treatment; (c) non-point pollution sources; (d) reporting the release of oil and other dangerous substances in the wetlands; (e) disposal of materials that are being dredged; and f) instituting programs by the federal government to enforce the goals of such an act.

Moreover, the act includes the following six titles: Title 1, Research and Related Programs; Title 2, Grants for the Construction of POTWs; Title 3, Standards and Enforcement; Title 4, Permits and Licenses; Title 5, General Provisions; and Title 6, State Water Pollution Control Revolving Funds.

With regard to the wetland definitions, one should note that these have been extended in the 1980s, but constructed wetland systems in upland areas are not regulated by the Clean Water Act.

In order to insure protection of wetlands, the developer should first try to avoid the wetlands altogether. Should avoidance be impossible, the federal act requires the developer to minimize in any way possible the damage that may occur to the wetlands. In mitigating such damage, the developer is required to create new wetlands in substitution of those damaged or restore previous existing wetlands. The site of such new wetlands or recovered wetlands should be adjacent to the project under construction. Moreover, once such wetlands have been created, the contractor or developer should guarantee their functioning capacity for a period of five years.

Case study of wetlands damage mitigation: the Bank System

An illustrative example of wetland damage mitigation programs is included for a better understanding of the practical applications of viable programs guiding similar projects. The case study in question considers specifically the Ohio Department of Transportation (ODOT), which is required to dispose of fill into the wetlands due to 100 projects in its jurisdiction.

Of these, small wetland areas are affected, but for 28 projects permits are required and eventually the ODOT is required to compensate for the damages being imposed. In attempting to mitigate the damage, one should consider that not all lands can be converted into efficient wetlands. Moreover, the presence of other projects next to the site in question may interfere, adding further complications. Based on cost analysis of this type of work of mitigation, the department indicates that the costs ranges from $66,000 to $306,000 per acre.

An efficient system to mitigate such costs is referred to as "wetland mitigation banking." The reason for such terminology lies in the similitude of this program with that of a real bank. Its application is currently underway by the ODOT which in turn has inspired a nonprofit organization called Ohio Wetland Foundation Bank. The concept is based on the fact that when a contractor cannot build a new wetlands area on or near a project site, he can create such wetlands areas at locations which do not relate to the project in question. In so doing, the developer would receive from the bank, a credit which in turn becomes a commodity owned in the bank. The credit can be used toward other future projects where required mitigation is necessary or it can be sold to other developers who find themselves in the same situation.

The Ohio Department Of Transportation has currently mitigated the loss of three acres of wetlands by purchasing credits from such a bank and is ready to extend the operation to another project consisting of 38 acres of wetlands. Special attention should be paid to the fact that under the Clean Water Act, the bank system discussed cannot be applied in all cases, but only when the restoration of wetlands on or near the site is impossible. In any other case where the developer can mitigate damage on or adjacent to the project site, the bank cannot be used.

Stream crossing

Besides the regulatory norms protecting the wetlands from the impact of negative factors expressed in the Clean Water Act, further regulations concerning bridge crossings over streams and other waterbodies are under the jurisdiction of the U.S. Army Corps of Engineers. Bridge builders are required to obtain permits under the Nation-wide Permit program. Among such authorities are the U.S. Coast Guard, which this has been given the task of reviewing such bridge construction pursuant to section 9 of the Rivers and Harbors Act of 1899 by the U.S. Army Corps of Engineers. Section 404 of the Clean Water Act defines this delegation of authority. However, such a decision involving the U. S. Coast Guard has been criticized as being incomplete.

However, a U.S. Army Corps of Engineers permit, pursuant to Section 404 of the Clean Water Act, remains mandatory. The permit controls the deposit of dredged or fill material into waters of the United States in connection with the construction of bridges. This includes the excavations for the foundation of bridge piers in water areas such as streams and wetlands.

More specifically, the U. S. Coast Guard permits for bridges regulate the discharge of dredged material as well as fill that derive from the construction of bridges within United States navigable waters. Such materials can derive from various sources such as cofferdams, abutments, foundation seals, piers, and temporary construction and access fills. However, causeway and approach fills are specifically omitted and require a regional permit of Section 404 of the Clean Water Act.

The U.S. Army Corps of Engineers' jurisdiction over this topic involves the enforcement of a regulatory program, within which is the authority of issuing permits according to Section 320.2 of the 33 CFR. Relevant to issuing bridge permits is the implementation of Section 9 of the Rivers and Harbors Act of 1899. Under such an act, the construction of dams and dikes across navigable waters in the United States is strictly prohibited if it is does not have approval by Congress and the approval by the Chief of Engineers and the Secretary of the Army. Moreover, if the water body through which work is performed lies solely within one state, it is possible to attain permits from the state itself, providing that the project has been approved by the Chief of Engineers and the Secretary of the Army.

In terms of the construction of bridges and causeways, Section 9 of the Rivers and Harbors Act states that the authority for the approval of the plans is no longer within the jurisdiction of the Chief of Engineers and the Secretary of the Army, but it is transferred to the Secretary of Transportation under the Department of Transportation Act of October 15,1966 [49 U.S.C. 1155g(B)(A)].

A particular case concerning the approval of structures to be built within the navigable waters of the United States covers the situation in which a federal project could be impacted by the construction of the structures for which the permit is requested. In such cases the permit applicant should be informed in writing of the situation and be advised that there could be a possibility that the desired project may have to be removed or reconstructed. Furthermore, the applicant for the permit should be well informed that the United States government does not assume any responsibility for damages or injuries to those structures that have been authorized under Section 9 of the Rivers and Harbors Act of 1899 or by section 404 of the Clean Water Act. Similarly, the government is also not assuming any responsibility for damages and injuries caused by future work aimed at the conservation or improvement of navigation. The government also declines any claim to compensation caused by federal projects.

From the U. S. Code (Title 33—Navigation and Navigable Waters), Chapter 11, "Bridges over Navigable Waters" is relevant in the present context. Chapter 11 includes the following specific parts: Subchapter I, General Provisions; Subchapter II, Alteration of Bridges; Subchapter III, General Bridge Authority; and Subchapter IV, International Bridges. The first subchapter dealing with the general provisions is further subdivided into sections 491 to 508: 491 Approval of and Deviation from Plans, Exemptions; 492 Bridge as Post Route, Limitation to Charges Against Government, Telegraph and Telephone lines; 493 Use of Railroad Bridges by other Railroad Companies; 494 Obstruction of Navigation, Alterations and Removals, Lights and Signals, Draws; 495 Violations of Orders Respecting Bridges and Accessory Works; 496 Time for Commencement and Completion of Bridge; 497 "Persons" Defined; 498 Reservation of Right to Alter or Repeal; 498a, 498b, repealed; 499 Regulations for Drawbridges; 500 Deflection of Current, Liability to Riparian Owners; 501 Omitted; 502 Alteration, Removal, or Repair of Bridge or Accessory Obstructions to Navigation; 503 to 507, repealed; and 508 Amount of Tolls.

The second subchapter has the following divisions: 511 Definitions; 512 Obstruction of Navigation; 513 Notice, Hearings, and Findings; 514 Submission and Approval of General Plans and Specifications; 515 Contracts for Project, Guarantee of Cost; 516, Apportionment of Cost; 517, Payment of share of United States; 518, Authorization of appropriations; 519, Noncompliance with Orders, Penalties, Removal of Bridge; 520, Review of Findings and Orders; 521, Regulations and Orders; 522, Existing Provisions of Law; 523, Relocation of Bridges; 524, Applicability of Administrative Procedure Provisions.

Subchapter III has ten divisions: 525 Construction and Operation of Bridges; 526 and 526a, repealed; 527 Acquisition of Interstate bridges by Public Agencies, Amount of Damages; 528 Statement of Construction Costs of Privately Owned Interstate Bridges, Investigation of Costs, Conclusiveness of Findings, Review; 529 repealed; 530 Bridges Included and Excluded; 531 International Bridges; 532 Eminent Eomain; 533 Penalties for Violations; 534 Conveyance of Right, Title, and Interest of United States in Bridges Transferred to States or Political Subdivisions, Terms and Conditions.

Finally, Subchapter IV has one primary division: 535 Congressional Consent to Construction, Maintenance, and Operation of International Bridges, Conditions of Consent; 535a Congressional Consent to State Agreements with Canada and Mexico, Secretary of State's Approval of Agreements; 535b Presidential Approval, Recommendations of Federal Officials; 535c Secretary of Transportation's Approval, Commencement and Completion Requirements, Extension of Time Limits; 535d repealed; 535e Ownership; 535f Applicability of Provisions; 535g Federal Navigable Waters and Commerce Jurisdiction Unaffected; 535h Report of Secretary of Transportation's Approval during Fiscal Year; 535i Reservation of Right to Alter or Repeal.

AASHTO (1996, Section 1.3.1), refers to requirements for the design of bridges crossing waterways. In the preliminary design of bridges crossing waterways, provisions must include the selection of special sites favorable for the bridge crossing. This involves the study of stream meanders and verification of the need for channel changes, river training works, and other methods. The main goals of such required studies is the reduction of erosion problems and prevention of any possible damage to the bridge structure being built.

With reference to the foundations of bridges, this same section states that when a channel has been realigned, such foundations should be designed for the possibility of deepening and widening of the channel. Furthermore, in order to prevent the loss of the structure, it is necessary to lower the approach embankments so that unusual floods will pass over the highway. Considering the case in which bridge reliefs are needed, the code calls for caution for proportioning the size and selecting the site of the structure in order to avoid scouring and changing the course of the main channel of the stream being crossed.

The code further states that waterways flowing in the channel at bridge crossings are usual designed to accommodate a maximum flood expected in accordance to the class and type of the highway. More precisely, when designing the waterway opening the upper stream, pounding must be considered as well as the passage of ice and debris. Considerations should be paid also to the possibility of scouring of the foundation of the bridge. However, larger openings of the channel at bridge crossings are pointed out to be necessary in order to allow the flow of exceptional floods of extreme magnitude.

In order to minimize the effects of adverse flood flow conditions, relief openings, *spur dikes* (a barrier built out from the river bank to prevent erosion and sand movements), debris deflectors, and channel training works need to be included in the project. When designing bridge piers and abutments in water, consideration should be given to the use of protective measures to prevent scouring. The code requires that rip-rap, mattresses, retards, spur-dikes, etc. be used to protect embankment slopes adjacent to the structure. Considering the bridge opening over the waterway, The code also requires also that the natural vegetation such as brush and trees be left undisturbed around the bridge opening to prevent high flow velocity which in turn could cause scouring. Also, borrow pits must be specifically sited to minimize scouring in the stream.

Hydraulic studies

In the AASHTO *Standard Specifications for Highway Bridges*, Section 1.3.2 addresses specifically the hydraulic studies of bridge sites. These studies are an essential part in the preliminary design of bridges which is specifically related to the main topical issues of the book. These studies should be conducted according to the following requirements consisting of three major points as follows: site data, hydrologic analysis, and hydraulic analysis.

Site data sources include the following:

- maps, stream cross sections, and aerial photographs
- complete data of bridge construction dates, performance during previous floods, etc.
- available highwater marks and their dates
- information on channel stability, debris, and ice
- effects on water stages from other flood control projects, reservoirs, streams, and tides, and
- geomorphic changes in channel flow

Hydrologic analysis contains:

- data for the estimation of floods at the site
- flood-frequency curve
- design distribution of flow and velocities of flood discharges at the site
- stage-discharge curve for the site
- backwater lines and velocity at the bridge opening must be calculated for special discharges and for different proposed bridge lengths
- estimated scour at piers and abutments
- effects from natural stream pattern changes, and
- effects on nearby structures from geomorphic changes of the stream

Computer software programs available for discharge calculations required in (a) include Water Surface Profile (WSPRO) available from the Federal Highway Administration (FHWA) and HEC-2 available from U.S. Army Corps of Engineers.

The AASHTO specifications Section 7.3.2 deals specifically with the protection of bridge piers. Starting from the potential of collision of piers from river traffic, a risk analysis is required in order to provide sufficient barriers to prevent any damage due to impact. With regard to protection measures against collisions for railroad bridges, the code requires collision walls to be placed between columns extending 6 ft above the top of the rails. Particular attention is commonly given to the effect of scouring on various types of foundations.

To this effect, the depth of scour should be determined by subsurface exploration and by hydraulic studies. Further precautions for the protection of bridge piers in waterways consist of shaping the nose of the pier in such a manner that it would deflect or even break ice or drift material. Moreover, further defensive systems for piers is to provide steel plates and angles over the noses or facing the piers with granite.

Earthquake considerations

Particularly sensitive to ground motions, bridges in seismic-prone areas respond to particular criteria promoted by various codes in the United States and in other countries worldwide. Reference material along the topic concerned with design specifications and construction of bridges is easily found through selected bibliographies. One of the latest references of major interest is the *Workshop on Seismic Design Methodologies for the Next Generation of Codes* held in Bled, Slovenia, in June 1997. Forty experts from different countries were invited to compare ideas and determine a consensus for the improvement and the expansion of present codes for seismic design. One specific objective was design and performance evaluation approaches for buildings and bridges. The reader is referred to the proceedings of that workshop for the most update concepts in the design of bridges in seismic zones.

Another significant reference on this topic is the Proceedings of the Fifth SECED Conference on European Seismic Design Practice, held in Chester, UK, on October 26-27, 1995. Several topical issues concerning the seismic design analysis in different countries of Europe are grouped together with a special regard to bridge behavior given in one of the sections.

3

COMPUTER SOFTWARE FOR BRIDGE DESIGN

INTRODUCTION

In the organization of the office it is essential to set up a work sequence, following guidelines, prior to engaging in the preparation of the preliminary design of bridges for the submission of proposals. At a certain phase during this period of organization, it is strongly recommended that a structural engineer, together with his/her partners or consultants, check the market of software dealing with bridge design from references concerning commercially available software and in-house programs. Some of the data are presented in this section. In the pages that follow, major programs and their sources are listed as a reference guide for the benefit of the reader, who should investigate the content of each program and decide on which should be part of the office library. Appendix 5 includes sources of such programs.

SOFTWARE SOURCES

American Iron and Steel Institute

Available at the American Iron and Steel Institute are plans and software for short-span steel bridges. Prepared by the Transportation Structures Committee of The American Iron and Steel Institute is a comprehensive set of predesigned bridge superstructures (plans) covering a variety of roadway widths and beam-girder types and spacing.

Also included is a computer program for generating customized designs and appropriate supporting calculations. The software can also be used to reproduce the design, except end-bolted beams.

Such a package can be used for generating preliminary designs and cost-effective final designs. For more information see the Appendix.

California Department of Transportation

CALATRANS (California Department of Transportation) uses a variety of bridge engineering software that include three programs of public domains, and other public domain software, as well as proprietary software. The following is a specific list:

- COMP — Composite Steel Girder/Concrete Deck design and analysis, Simple Spans; will develop LL shear and moments, flange areas and stresses, deflections.
- PROP— calculates section properties by coordinates; plane geometric figures described by line segments; maximum 200 nodes; also calculates area, centroids, moments of inertia, and products of inertia determined.
- PSGIRDER— prestressed girder design; simple beams only; gross and net transformed sections; pretensioning or posttensioning; elastic theory.

Portland Cement Association

Several computer programs are available through the Portland Cement Association, including:

- BRADD — analysis & design by AASHTOWARE PRODUCT
- ADAPT - ABI Analysis & Design of Concrete Bridges and Frames (developed by Bijan, Florian & Associates, Inc., Redwood City, California).

American Institute of Steel Construction

The following are available from the American Institute of Steel Construction:

- AISC database V 2.0 — to be incorporated with in-house routines; gives electronic access to the dimensions and properties of the more than 1,200 shapes in AISC's Manual of Steel Construction- WMS and HP shapes, channels (C and MC), angles (L and 2L), tees (WT, ST, and MT), structural tubing (TS), and steel pipe (P).
- AISC for AutoCAD — draw AISC shapes, including end, elevation, and plan views W, S, M, and HP shapes, American Standard Channels (C), Miscellaneous Channels (MC), Structural Tees cut from W, M, and S shapes (WT, #MT, #ST), Single and Double Angles, Structural Tubing and Pipe, to full scale.
- CONXPRT — designs of appropriate steel connections including shear, moment connections, column stiffeners and doublers. The software comes with present guidelines, but can be modified to meet individual shop standards.
- WEBOPEN — designs steel webs in which openings have been cut off, including round and rectangular openings with or without reinforced openings.
- CD-ROM/LRFD Manuals (second edition) — includes two volumes of the LRFD Manual of Steel Construction, plus the 1992 and 1993 AISC's Engineering Journal; allows quick cross-referencing throughout the manuals; includes introduction to LRFD electronically linked to the specification and commentary.

SIMON Systems

These programs design straight steel, plate-girder bridges in accordance with the Standard Specifications for Highway Bridges of the American Association of State Highway and Transportation Officials (AASHTO), and will design I-shaped or multiple single-celled box-shaped girders with up to 12 continuous spans that may contain hinges.

- SIMPLE — optimizes the vertical web depth of an I- or box-girder.
- PONTIS — network level Bridge Management System (BMG), which incorporates dynamic, probabilistic models, and a detailed bridge data base to predict maintenance and improvement needs.
- BRIDGIT — Bridge Management System developed in 1985 directed by the National Cooperative Highway Research Program

(NCHRP). Its intent is to provide a microcomputer-based bridge management system software package to be used by transportation agencies. The system is operable through an IBM PC or compatible hardware environment.

ADAPT

Structural Concrete Software System includes a variety of software:

- ADAPT-TS - designs posttensioned beams, slabs, and floor systems.
- ADAPT-PT — advanced posttensioning software for design of beams, slabs, and floor systems.
- ADAPT-Floor — three-dimensional finite element software for analysis and design of posttensioned floor systems and mat foundations.
- ADAPT-RC — for design and analysis of non-prestressed concrete beams, slabs, and floor systems.
- FELT — calculates friction, elongation, and long-term stress losses in prestressing.
- PULT — ultimate strength computation of prestressed sections.
- ADAPT-ABI — time-dependent analysis of concrete bridges and frames, with or without posttensioning.
- ADAPT-ABS — three-dimensional analysis of posttensioned box-girder bridges.
- ADAPT Posttensioning Details — structural details for building construction.

Georgia Institute of Technology

- GT STRUDL — presently available from the Georgia Institute of Technology was originally prepared at MIT under contract by IBM. In recent years the program has been expanded by the Georgia Institute of Technology and is currently available under the new acronym GT STRUDL (Georgia Tech Structural Design Language). Each scope is extremely large. It includes structure in two-dimensional and three-dimensional space, it analyzes hinged as well as rigid frames, and cables under static and dynamic loads. It also covers thin shells. Its wide range of action applies to bridge design as well.

BEST Center (University of Maryland)

The following is an excerpt from the introduction describing the BEST Center at the University of Maryland: "The Bridge Engineering Software and Technology (BEST) Center of the University of Maryland was established in 1986 to provide bridge related technology development to other states' design agencies and to private consultants. Also its creation provides a mechanism whereby the bridge oriented software which was developed for the Maryland State Highway Administration is made available on a commercial basis to other state design agencies and private consultants." The following programs are available from the BEST Center:

- MERLIN-DASH — for use by bridge design engineers who function in a production environment. Its scope includes a sequence of operations for work stress design method or load factor design method. These include the analysis of dead load and live load conditions in accordance with the American Association of State Highway and Transportation Officials (AASHTO); design of the size of structural members; complete and detailed code check of all steel or reinforced concrete beam components; detailed inventory and operating rating of all beam components.
- PBEAM — a program for the design and analysis of prestressed concrete bridge girders.
- TRAP-jr — a program for the design for continuous trusses up to six spans, including prestressed trusses.
- RETWALL — a program for the analysis design of cantilevered retaining walls.
- PILEGP — a program for the design of pile systems capable of analyzing groups of piles, up to 200 randomly placed.
- SABRE — a system for the design of highway signs.
- BDGM — a program for the geometrics that are required for the design, detailing, and construction of highway bridges.

Engineering Design Programmers, Inc.

- MDX Software Available Program: curved and straight steel bridge design.

North Carolina Department of Transportation

The following list includes a number of in-house computer programs in current use by the North Carolina Department of Transportation. The topics covered in these programs include the designs of straight as well as curved steel bridges.

- LAYOUT AND ELEVATIONS (E7000M) — performs various tasks associated with the layout and elevations of a bridge; also used for designing retaining walls.
- APPROACH SLAB LAYOUT (E7005M) — computes the layout of an approach slab when given fill face stations and skew angles. This program will also determine required reinforcement and show a Bar Schedule of that reinforcement.
- COMPOSITE I-BEAM (E7020M) — a program that will design one or more rolled steel I-beams, with or without cover plates, suitable for a given span, slab, beam spacing and live load using either Working Stress Design or Load Factor Design.
- BRIDGE LENGTH/HORIZONTAL CLEARANCE (E7030M) — calculates the required bridge length when given the horizontal and vertical alignment, the roadway geometry, and the roadway slopes.
- CULVERT LENGTH (E7032M) — designs a plate girder using either load factor or working stress design, with or without composite action, with any standard AASHTO live loading.
- BRIDGE SLAB A-BARS (E7240M) — designs the main bars and cut-off bars for any number of spans, and for a tangent or circular alignment. Weights of bars are given, together with subtotals, totals, and grand totals.
- COMPOSITE PLATE GIRDER DESIGN (E7290M) — designs a plate girder using either Load Factor or Working Stress Design, with or without composite action, with any standard AASHTO live loading (or none).
- COMPOSITE PRESTRESSED GIRDER DESIGN (E7300M) — designs or analyzes a composite pretensioned prestressed concrete girder for a simple span.
- BOLTED FIELD SPLICE ANALYSIS (E7340M) — analyzes a given bolted field splice, with or without composite action, for a plate girder or an I-beam without a cover plate.
- END BENT DESIGN (E7490M) — analyzes and designs an end bent pile supported substructure for bridges on the North Carolina Highway System.

- TRUCK LOADS (E7510M) — calculates the beamline reactions from the superstructure due to the positioning of AASHTO truck loadings within the design lanes by simple beam distribution of the truck wheel loads.
- GEORGIA PIER (E7570M) — aids in the design of substructures for highway, railroad, and pedestrian bridges.
- PRESTRESSED CORED SLAB DESIGN (E7580M) — FORTRAN computer program to design or analyze a pretensioned, prestressed concrete cored slab unit to be used on bridge structures on the North Carolina Highway System.
- CULVERT DESIGN PROGRAM (E7670M) — analyzes and designs a one-, two-, three-, or four-cell reinforced concrete box culvert with or without bottom slab.

Skilling Ward Magnusson Barkshire, Inc.

This software was developed for the Washington Department of Transportation, and includes the following moduli:

- Concrete reinforced box-girders, slabs, and T-beam modules.
- Prestress, pretension, and posttension for box-girder and beams modules.
- Steel plate girder, standard beam sections, and riveted girder modules.
- Three-dimensional version for trusses and box-girders for both concrete and steel modules.

Florida Department of Transportation

The following programs are in use by the Florida Department of Transportation.

- MERLIN DASH — used for steel girder design.
- SIMON — used for steel girder design.
- DESCUS — for curved steel box and girder design.
- TANGO — for posttensioned concrete bridges.
- ADAPT — for posttensioned concrete bridges.
- GTSTRUDL — for structural analysis.

A list of in-house software can be found at the Florida Department of Transportation web site: http://www.dot.state.fl.us/. To see or download these programs look under Doing Business at FOOT and then go to the Structures Design page.

LEAP Software, Inc.

Programs include:

- CONSPAN — for continuous precast/prestressed concrete bridge girder design software.
- CONSYS 2000 — for LRFD and AASHTO Standard Live Load and Static Load Analysis Software.
- RC-PIER Metric — for reinforced concrete bridge pier design and analysis.
- GEOMATH Version 3 — for bridge layout and geometrics.

EXAMPLE PROBLEMS

CONSPAN

The following tutorial illustrates the practical applications of the software that they produced in the field of bridge design. For an understanding of the basic criteria for the bridge beam design theory used in this context, see Appendix 10.

Problem 3.1: Continuous 3-Span Bridge with Bulb Tee Girder and Continuity Design

The basic features of CONSPAN are demonstrated in this tutorial. We will be designing beam 2 of span 1 of a three-span continuous bridge. The main objective of this tutorial is to demonstrate the step-by-step process of designing a precast bridge beam (See Figure 3-1).
The information for Tutorial 1 is contained in a CONSPAN input file called "tutor1.cpf." As an alternative to inputting all the information from scratch, as directed in this chapter, you can load this file and only work on the sections that you wish. To load this file, select File I Open. (Select File from the pull-down main menu, and then select Open.) Move to the CONSPAN subdirectory, if it is not showing already, and select file "tutor1.cpf."

- Pier-to-Pier: Distance between center lines of piers. The pier-to-pier span dimensions of the continuous beam model are used for the dead load acting on the composite section and for live load analyses. Precast Length: The overall length of the precast beam.

- Pier CL-Precast: Horizontal distance from the left pier center line to the left precast beam end. This value is positive if the CL pier is to the left of the precast end; otherwise, it is negative.
- Bearing-to-Bearing, ft [m]: The design or bearing-to-bearing span length.
- Clear Span: The span length measured between faces of bearing. It is also adequate to input it as equal to the bearing-to-bearing span.
- Release Span: The span length used to compute moments at release for the release stress check. Release span is typically set equal to the precast length.

PROBLEM 3-1 DATA

Concrete properties:

Strength:
Precast, release: 5000 psi
Precast, 28 days: 6000 psi
Cast-in-place topping, 28 days: 3500

Density, all concrete: 150 pcf

Strand Properties:
Strand type: 1/2"-27OK-low relaxation
Two-point depressed pattern
Depression point at 0.4L

Rebar Properties:
Grade 60 rebar

Figure 3-1 Example precast-concrete bridge design using LEAP software (reproduced courtesy Leap Software, Inc.).

Prestress Loss Information:

Calculation method: AASHTO
Hours to release: 18
Relative humidity: 75%

Live Load:

AASHTO HS-25 truck

Superimposed Dead Loads:

On precast: 180 plf, due to stay-in-place forms
On composite: 30.50 psf, due to future wearing surface

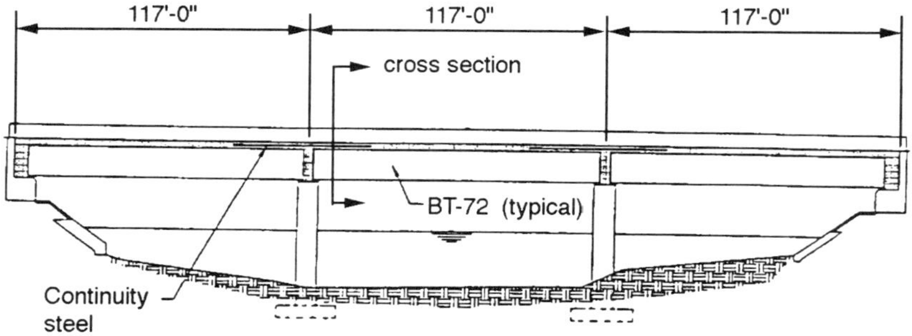

Elevation View

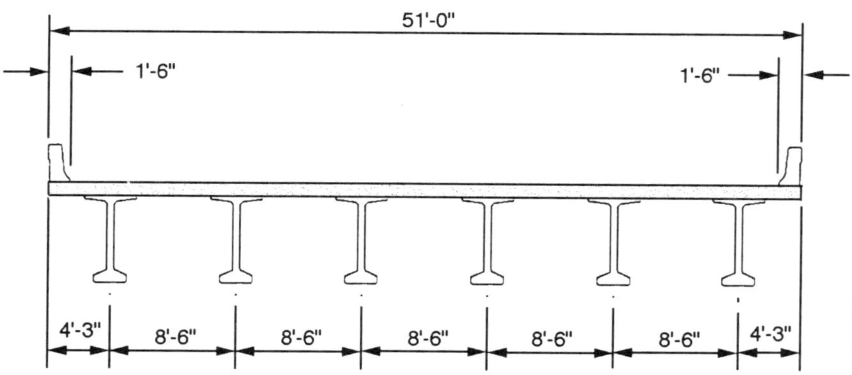

Cross-section View

Figure 3-1 (continued)

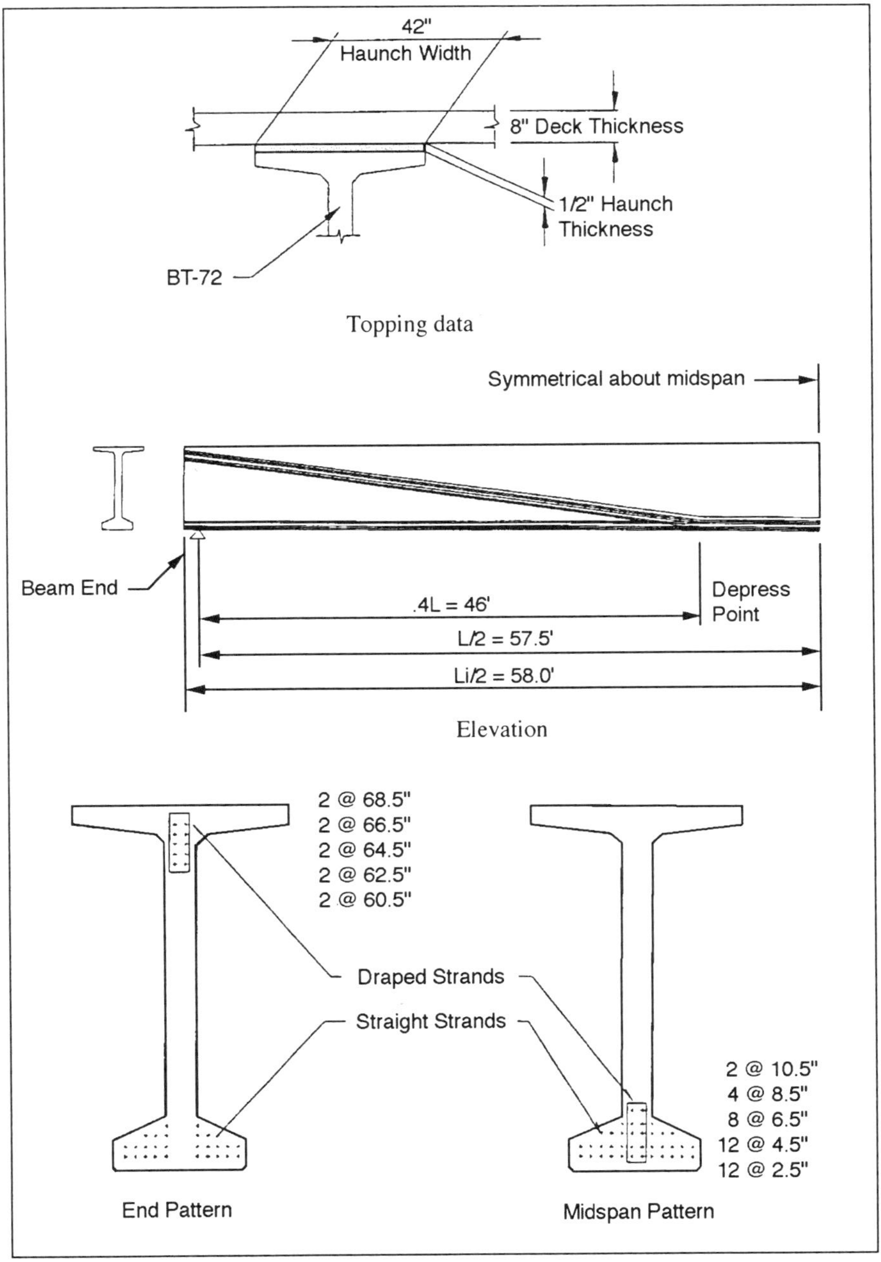
42"
Haunch Width
8" Deck Thickness
1/2" Haunch
Thickness
BT-72
Topping data
Symmetrical about midspan
Beam End
.4L = 46'
Depress
Point
L/2 = 57.5'
Li/2 = 58.0'
Elevation
2 @ 68.5"
2 @ 66.5"
2 @ 64.5"
2 @ 62.5"
2 @ 60.5"
Draped Strands
Straight Strands
2 @ 10.5"
4 @ 8.5"
8 @ 6.5"
12 @ 4.5"
12 @ 2.5"
End Pattern
Midspan Pattern

- ❑ Start *CONSPAN* by double-clicking the *CONSPAN* icon, in the LEAP Applications Group, while in the Windows Program Manager.

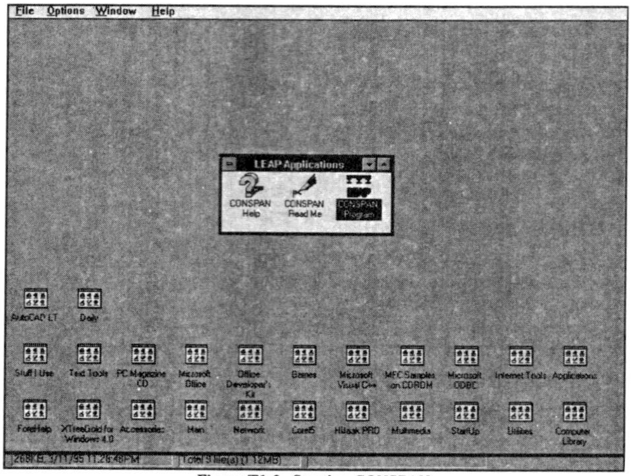

Figure T1-2. Starting *CONSPAN*

- ❑ The introductory screen will appear. Click the **OK** button. You will be presented with the main screen.

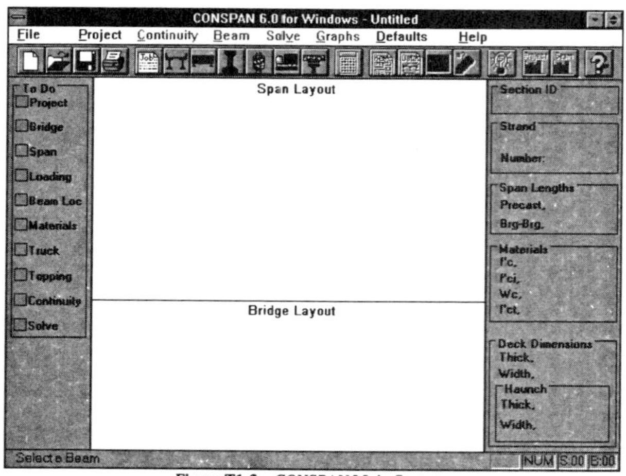

Figure T1-3 . *CONSPAN* Main Screen

Figure 3-1 (continued)

- ☐ Select **Project | Project Description** from the main menu. The *Project Description - Project* screen will appear.

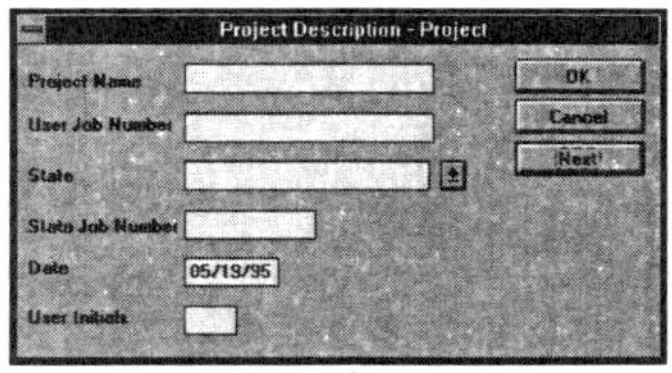

Figure T1-4. Project Description
(Project | Project Description)

- ☐ Fill in each field as shown. Move to each field by clicking on that particular field. You can also use the **Tab** key to move to the next field. Pressing the **Shift** and **Tab** keys together will move the cursor to the previous field. The "State" field has a pull-down menu. If you click on the down arrow symbol, a list of available states will appear. You may choose the appropriate state from the list by clicking on it. You may also type in the state's name manually.

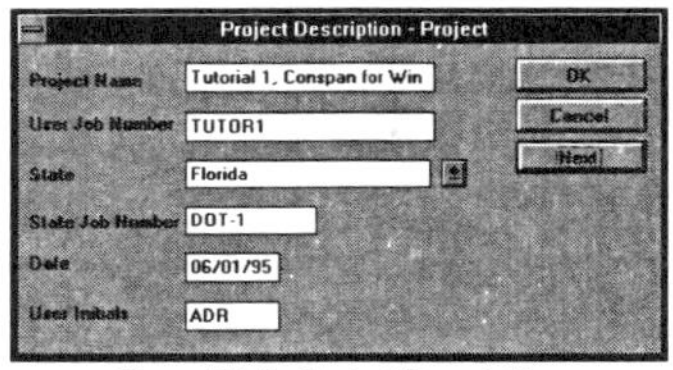

Figure T1-5. Project Description
(Project | Project Description)

- ☐ Click the **Next** button to go to the next screen, *Bridge Layout - Project*. Fill in the information shown in Figure T1-1b. Click the **Recalc** button to calculate the Curb-to-Curb dimension. This will also be calculated when you exit this screen.

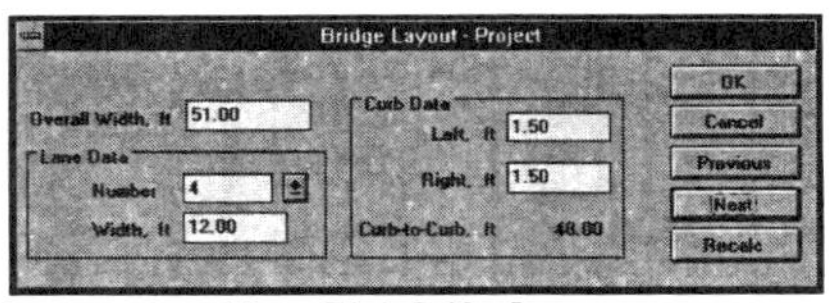

Figure T1-6. Bridge Layout
(Project | Bridge Layout)

- ☐ We will now illustrate the use of *CONSPAN*'s Help feature. You can bring up Help at any time by clicking the **F1** key. *CONSPAN*'s Help is context-sensitive; that is, it will show information about the particular screen that is active at the moment. Click **F1** to see the screen shown. Use the scroll bar on the right to view more information.

  If you want to see more information about any screen during this Tutorial, press the **F1** key, and you will be provided with help about the active screen. Exit *CONSPAN Help* by double-clicking the control box at the top left corner of the Help window, or by pressing **Alt-F4**.

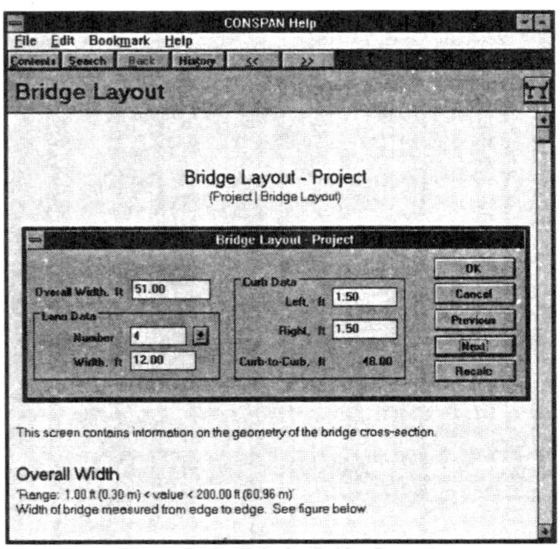

Figure T1-7. Help for Bridge Layout

- Now that we have exited Help, we can continue with the input for Tutorial 1. Click **Next** to advance to the next screen, *Span Data - Project*. Enter the values shown in Figure T1-1c. To edit the data, click on the span you want to edit. The values will appear on the input boxes at the top, where you can modify them. When you are done, click the **Modify** button. To insert an additional span above the highlighted one, click the **Insert** button. To insert an additional span after the last one, click the **Add** button.

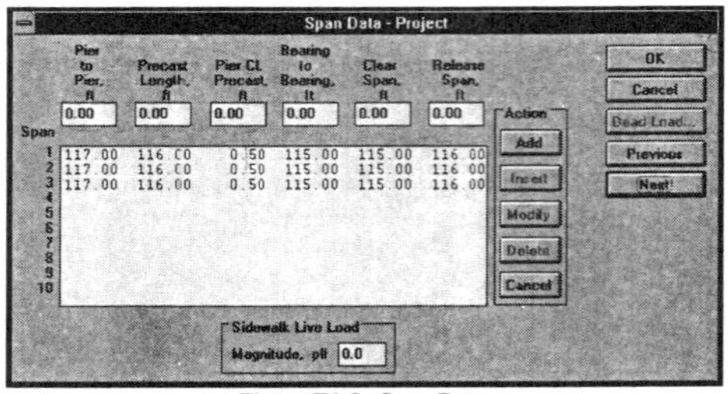

Figure T1-8. Span Data
(Project | Span Data)

- Highlight Span 1 by clicking on it, and then click the **Dead Load** button. You will see the *Superimposed Dead Loads - Project-Span:1* screen. Fill in this screen as shown. Click the **Copy to ALL Spans** buttons, under "SDL on Precast" and under "SDL on Composite" since this load is applied to all three spans.

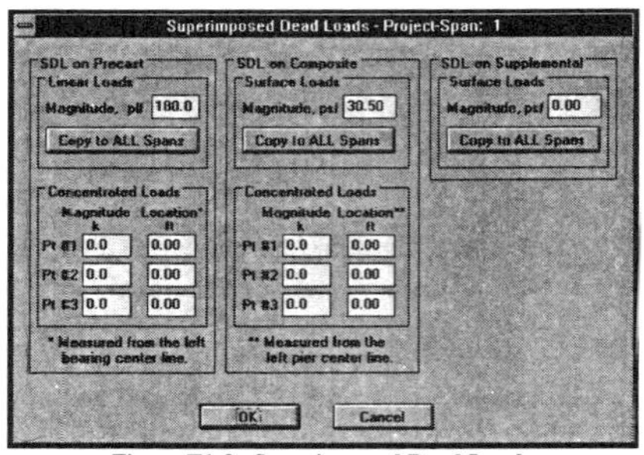

Figure T1-9. Superimposed Dead Loads
(Project | Span Data | Dead Loads)

- Click the **OK** button to return to the *Span Data* screen. Then, click **Next** to advance to the *Beam Location Data - Project* screen, and fill in the data according to Figure T1-1b.

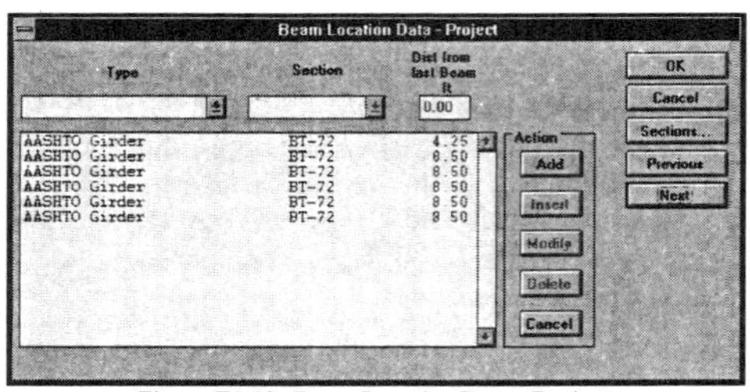

Figure T1-10. Beam Location Data - Project
(Project | Beam Location Data)

❑ Click the **Next** button to advance to the *Materials - Project* screen. Click the **Concrete** button, and you will see the *Concrete Data - Project* screen. Modify the concrete strength values as shown. You can select the values for Unit Weights and Strength by typing them in or by selecting them from a drop-down list, which appears when you click on the arrow.

The "Auto Recalculation" box is checked by default. This means that whenever you click **OK** or **Update**, the modulus of elasticities will be recalculated, according to AASHTO Eq. 9-8. If you want to use a modulus of elasticity different from the ones calculated, uncheck the "Auto Recalculation" box by clicking on it, and type in the desired values.

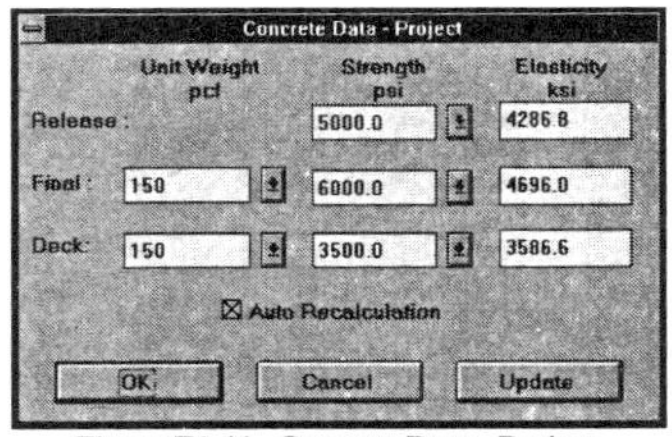

Figure T1-11. Concrete Data - Project
(Project | Materials | Concrete)

❑ Click the **OK** button to go back to the *Materials - Project* screen, and then click **Strand**. You will see the *Strand Data - Project* screen. Fill it in as shown. This information corresponds to strands type 1/2"-270K-Low-relaxation and a two-point depressed pattern. The depression points are located at distances of 0.4L and 0.6L from the left bearing, where L is the bearing to bearing distance, 115 ft in this case.

Select a Midspan Increment of 2.00 inches. This value determines the vertical distance between strands at midspan when the program generates the draped pattern automatically. The first row is determined by the strand template In this case, the first row will be placed at 2.50". Successive rows are located at a distance from the previous one equal to the Midspan Increment. For this case, the locations will be: second row at 4.50", third row at 6.50", etc.

The Shielding Length Increment of 3 ft means that if you have the program automatically generate the shielding pattern, it will generate shielding length intervals of 3 ft (i.e. 3 ft, 6 ft, 9 ft, etc.) Please notice that the program will let you combine draping and shielding in the same beam; only straight strands may be shielded. We will not use any debond constraints, so leave this box unchecked. In Tutorial 2, we will illustrate the use of this feature.

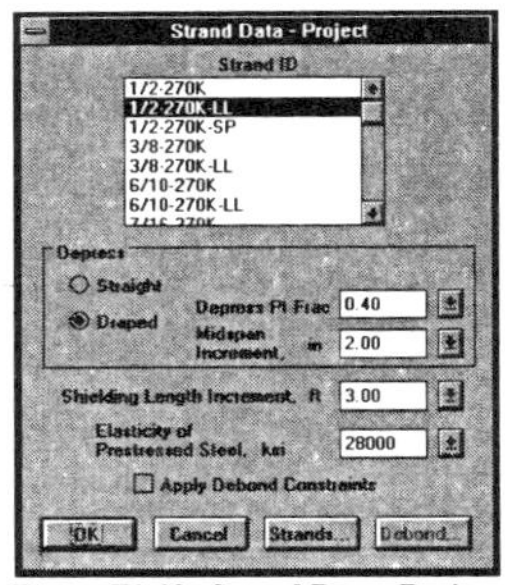

Figure T1-12. Strand Data - Project
(Project | Materials | Strands)

❑ Click the **OK** button to go back the *Materials - Project* screen. Click **Rebar**, and you will see the *Rebar - Project* screen. Click **OK** to accept the default values, since we are using grade 60 steel for both longitudinal and stirrup rebars.

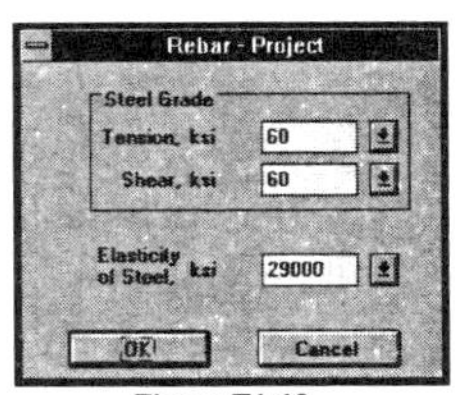

Figure T1-13.
Rebar Data -Project
(Project | Materials | Rebar)

❑ Click **OK** to go back to the *Materials - Project* screen Click **Next**, and you will see the *Truck Data - Project* screen. To select an HS-25 truck, highlight it by clicking on it and then click **Add-->**.

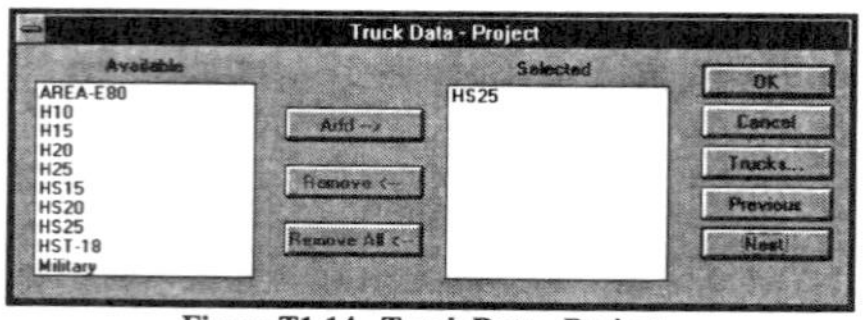

Figure T1-14. Truck Data - Project
(Project | Truck Data)

❑ Click **Next** to go to the *Topping Data - Project* screen. Fill in the information shown in Figure T1-1d. The supplemental thickness will always be zero unless you are using a supplemental layer. See "Supplemental Layer" in the "Theory" section.

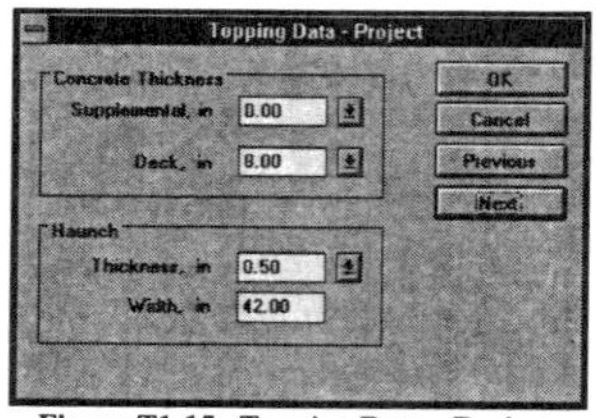

Figure T1-15. Topping Data - Project
(Project | Topping Data)

❑ Click **Next** to go to the *Loss Data - Project* screen. Fill it in as shown.

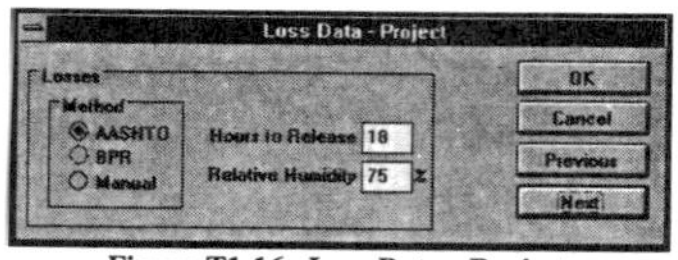

Figure T1-16. Loss Data - Project
(Project | Misc. | Loss Data)

❑ Click **Next** to go to the *Miscellaneous Multipliers - Project* screen. Click the **Next** button to accept the default values and move to the next screen.

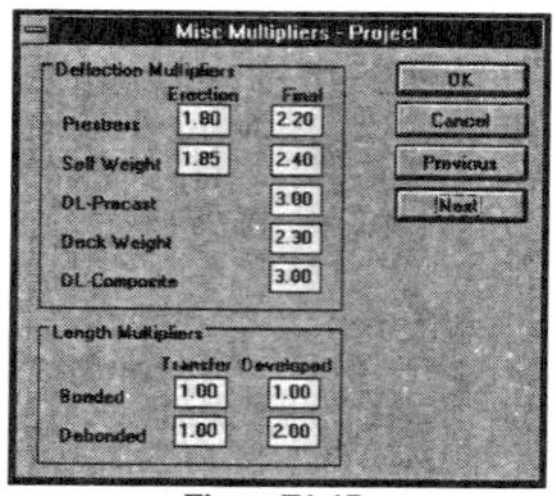

Figure T1-17.
Miscellaneous Multipliers - Project
(Project | Misc | Misc Multipliers)

- ❑ On the *Allowable Stresses -Project* screen, click the **Factors** button to go to the *Allowable Stress Factors - Project* screen. Shown are the default values, according to Art. 9.15 and 9.7.2.4. For the purpose of illustrating how to make changes to this screen, change the values and input the numbers as shown in Figure T1-18. These factors are multiplied by the concrete strengths to obtain the allowable stresses shown in the next screen. For example, the allowable compression at bottom at release is 3000 psi, which is obtained by multiplying the concrete strengths at release, 5000 psi, by the corresponding factor, 0.6. Click the **OK** button to go back to the *Allowable Stresses - Project* screen.

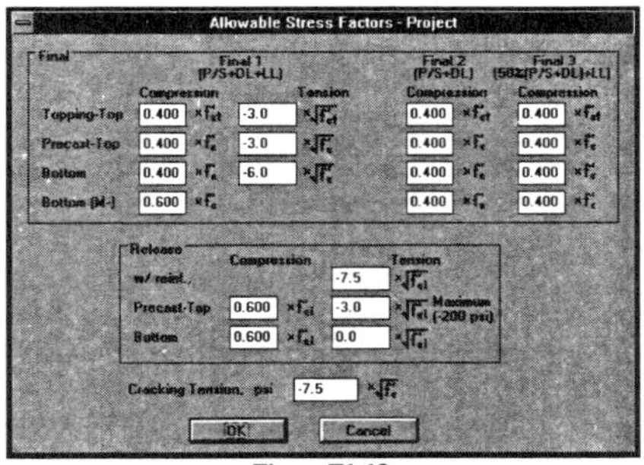

Figure T1-18.
Allowable Stresses Factors - Project
(Project | Misc | Allowable Stresses | Factors)

- ❑ The values in the *Allowable Stresses - Project* screen are computed using the factors specified in the *Allowable Stress Factors - Project* screen, unless the "Compute using Factors" box is unchecked.

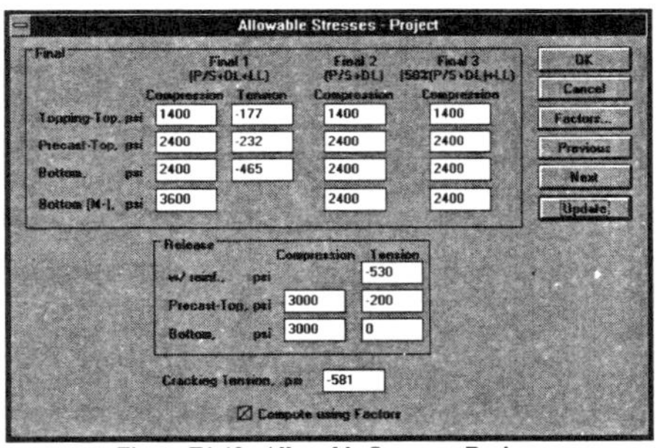

Figure T1-19. Allowable Stresses - Project
(Project | Misc | Allowable Stresses)

- ❑ Click **Next** to go to the *Miscellaneous Factors - Project* screen. We will use the default values as shown. We want the program to calculate the moment impact factor and use this same value for the shear impact factor. We also want to have the program calculate the axle distribution factor. We will also use the Group I load factors specified in AASHTO table 3.22.1A.

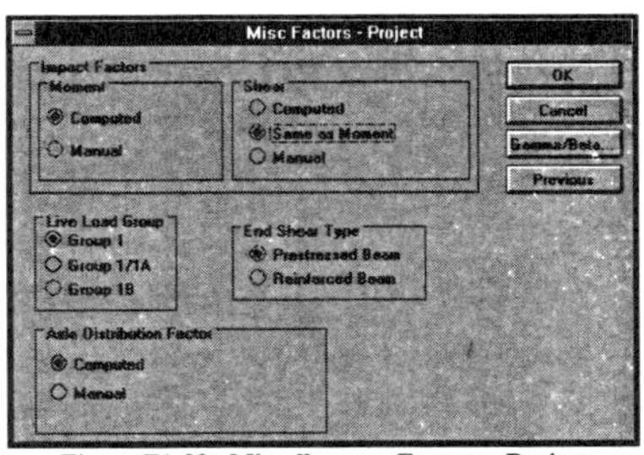

Figure T1-20. Miscellaneous Factors - Project
(Project | Misc | Misc Factors)

❑ Click the **Gamma/Beta** button to go to the *Gamma/Beta Factors - Project* screen. Shown are the default values, according to AASHTO table 3.22.1A.

This is the last screen of Project information. As explained under "Project and Beam Information," in the "General Operation" section, all project information is accessed under **Project** from the main screen. Beam information is accessed under **Beam** from the main menu and is valid only for the selected beam.

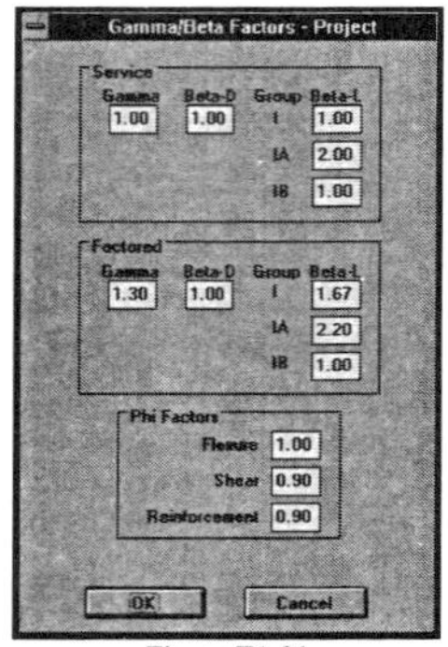

Figure T1-21.
Gamma/Beta Factors - Project
(Project | Misc | Misc Factors | Gamma/Beta)

❑ Click **OK** to accept the default Gamma and Beta Factors. Click **OK** again to return to the main screen. Now that all the Project information has been entered, we can perform the "Continuity Analysis." In the "Continuity Analysis," the program calculates moments and shears due to live loads and dead loads that are applied after the topping has hardened, and are, therefore, resisted by the continuous composite section. (See "Structural Modeling," in the "Theory" section.) Select **Continuity | Continuity Analysis** from the main menu to perform the analysis. When the analysis is complete, you will notice that the "Continuity" box in the "To Do" list of the main screen is now checked.

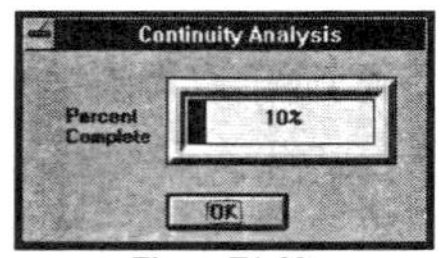

Figure T1-22.
Continuity Analysis Being Performed

- ❑ The next step is to select the beam that will be designed. Since we are going to design beam 2 of span 1, select the beam by clicking it with the mouse. It will be enclosed by a rectangle to indicate that it is the selected beam. You can also see at the bottom right hand corner of the screen "S:01 B:02," indicating that beam 2 of span 1 is selected. After doing this, you will notice that the **Beam** selection from the main menu becomes enabled.

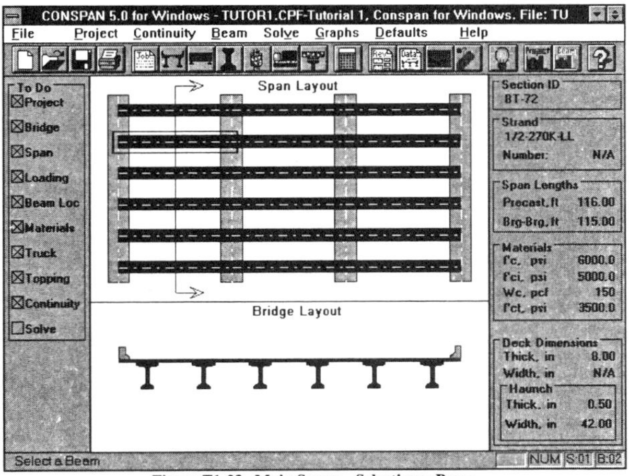

Figure T1-23. Main Screen. Selecting a Beam

- ❑ Select **Beam | Review Project**, and the *Review Project* screen will appear. This is a review of the project information that has been entered. If you want to modify any of these values, you have to go to the appropriate screen under the **Project** pull-down menu on the main screen.

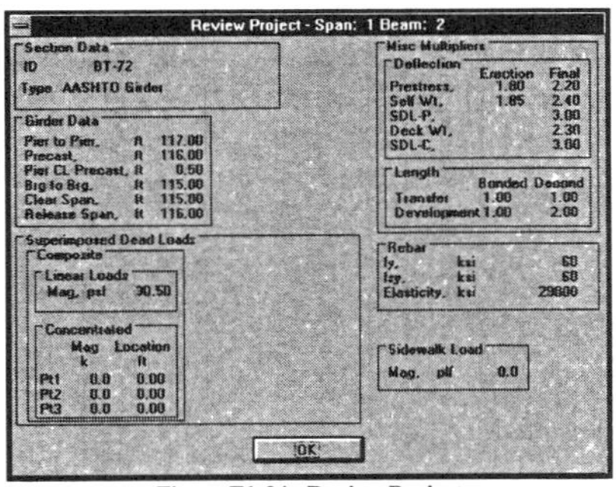

Figure T1-24. Review Project
(Beam | Review Project)

❑ Click **Design Status**. A dialog box will appear, asking you "About to regenerate groups. Continue? Yes/No." Click Yes. It means that it is about to regenerate strand groups, as explained under "Strand groups" in the "Theory" section. While doing so, it will set all debonding lengths equal to zero. The only time you will want to answer "No" to this question is when you have manually input the shielding pattern and you want to keep it.

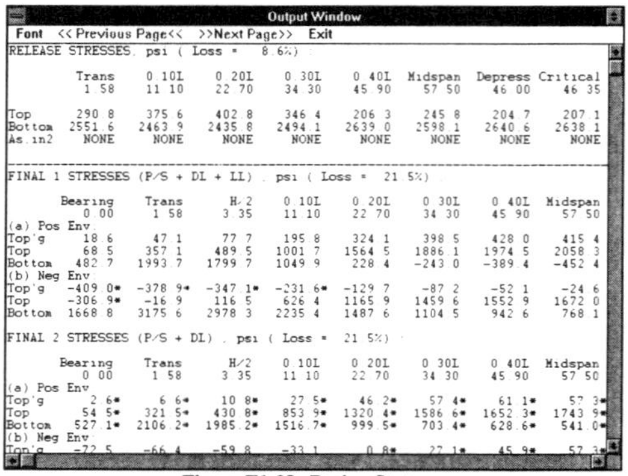

Output Window

Font << Previous Page<< >>Next Page>> Exit

RELEASE STRESSES. psi (Loss = 8.6%)

	Trans	0.10L	0.20L	0.30L	0.40L	Midspan	Depress	Critical
	1.58	11.10	22.70	34.30	45.90	57.50	46.00	46.35
Top	290.8	375.6	402.8	346.4	206.3	245.8	204.7	207.1
Bottom	2551.6	2463.9	2435.8	2494.1	2639.0	2598.1	2640.6	2638.1
As.in2	NONE	NONE	NONE	NONE	NONE	NONE	NONE	NONE

FINAL 1 STRESSES (P/S + DL + LL). psi (Loss = 21.5%)

	Bearing	Trans	H/2	0.10L	0.20L	0.30L	0.40L	Midspan
	0.00	1.58	3.35	11.10	22.70	34.30	45.90	57.50
(a) Pos Env.								
Top'g	18.6	47.1	77.7	195.8	324.1	398.5	428.0	415.4
Top	68.5	357.1	489.5	1001.7	1564.5	1886.1	1974.5	2058.3
Bottom	482.7	1993.7	1799.7	1049.9	228.4	-243.0	-389.4	-452.4
(b) Neg Env.								
Top'g	-409.0*	-378.9*	-347.1*	-231.6*	-129.7	-87.2	-52.1	-24.6
Top	-306.9*	-16.9	116.5	626.4	1165.9	1459.6	1552.9	1672.0
Bottom	1668.8	3175.6	2978.3	2235.4	1487.6	1104.5	942.6	768.1

FINAL 2 STRESSES (P/S + DL). psi (Loss = 21.5%)

	Bearing	Trans	H/2	0.10L	0.20L	0.30L	0.40L	Midspan
	0.00	1.58	3.35	11.10	22.70	34.30	45.90	57.50
(a) Pos Env.								
Top'g	2.6*	6.6*	10.8*	27.5*	46.2*	57.4*	61.1*	57.3*
Top	54.5*	321.5*	430.8*	853.9*	1320.4*	1586.6*	1652.3*	1743.9*
Bottom	527.1*	2106.2*	1985.2*	1516.7*	999.5*	703.4*	628.6*	541.0*
(b) Neg Env.								
Top'g	-72.5	-66.4	-59.8	-33.1	0.8*	27.1*	45.9*	57.3*

Figure T1-28. Design Status
(Beam | Pattern | Design Status)

A screen showing Release Stresses, Final Stresses and Ultimate Capacity will appear. Values exceeding the allowables are marked with an (*). Under Final Stresses, there are several locations at the top of the cast-in-place topping and one location at precast top that exceed the allowable stresses.

The over-stresses at the top of the topping are not influenced by the pattern. They will be taken care of later by the placement of rebars in the slab, which will be shown under "Reinforced Design" in the output.

The stress of -306.9 psi at the bearing under negative envelope, exceeds the allowable stress of-232 psi. It could possibly be corrected by increasing draping. Click **Exit** to go back to the *Pattern* screen. Then, click **Increment Draping**. You will see that it has no effect. The pattern does not change because the strands are already draped as much as possible. In fact, if they could be draped more, the automatic generation would have already done it to try to lower the stress of -306.9 psi.

To bring this stress within the allowable we could try choosing a smaller beam spacing, or a higher concrete strength, or a different beam. We can also decide to accept the overstress if it is not too high, or because it occurs at the bearing, where there is confinement steel present. For the purpose of continuing with the tutorial we will accept the overstress.

Click **Exit** to go back to the *Pattern* screen. A dialog box will appear, asking "Change in pattern requires regeneration of groups. Continue? Yes/No." Click **Yes**.

Click **OK** to accept the pattern. Now that all the required information has been entered, we can proceed to calculate the information required to create a printout, which includes stresses, shear reinforcement, camber, detensioning and slab reinforcement. To do this, select **Solve | Solve**. After a few seconds, the calculations will be complete. The "Solve" box at the left of the main screen, which was unchecked previously, will now be checked.

❑ Next, we will show how to display graphs. *CONSPAN* can display and print graphs of moments and shears along the entire structure (under **Graphs | Project Level**), and also graphs of stresses, for the selected beam (under **Graphs | Beam Level**).

To display the graph for final stresses under positive moment envelope for the beam being designed, select **Graphs | Beam Level**. From the pull-down menu under "Graph Type," select "Final Stresses - Pos. Moment." Click **Draw**. The stresses that are graphed are the same that appear in the output, under "FINAL STRESSES - POSITIVE MOMENT ENVELOPE". You may send this graph to the printer by clicking **Print**. Now click **OK** to return to the main screen.

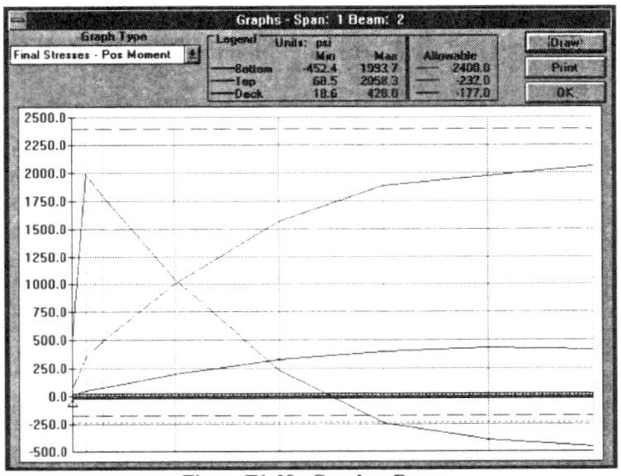

Figure T1-29. Graphs - Beam
(Graphs | Beam Level)

❑ The next step is to create a printout. Select **File | Print**. The *Output Options* screen will appear. To print all the sections of the output, check the boxes as shown, and click **OK**.

Since we requested the output to go to the screen, the first page of the output will appear. To view different pages click **Next Page** and **Previous Page**. To navigate through a page, use the scroll bars at the right. When you are finished viewing the printout, click **Exit**, and you will return to the main screen. A copy of this printout is reproduced starting on the next page.

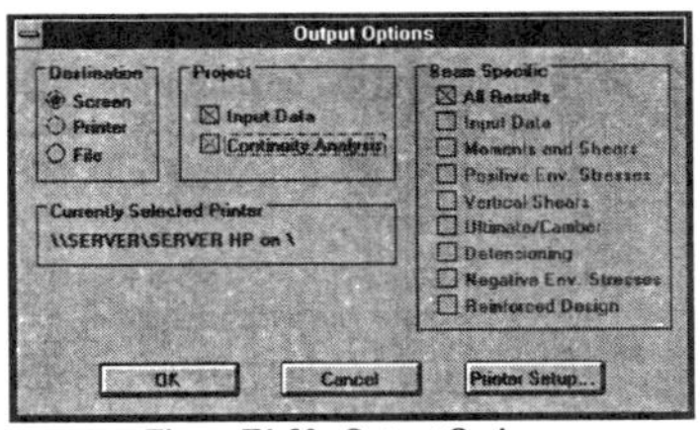

Figure T1-30. Output Options
(File | Print)

❑ The last step is to exit the program. Select **File** from the main menu, then select **Exit**. Since we have not saved it, and we have not named the file yet, the program will ask you "Save Changes to Untitled?". Click **Yes**. Enter "TEMP_TUT" without an extension. The program will automatically append an extension of ".CPF." Then click **OK** to confirm, and the program will end.

As with any other program, it is recommended that you periodically save your work. This way, if your work gets lost or corrupted, due to, say, a power outage, you can go back and retrieve the last file that you saved. (To save a file, select **File | Save** from the main screen.)

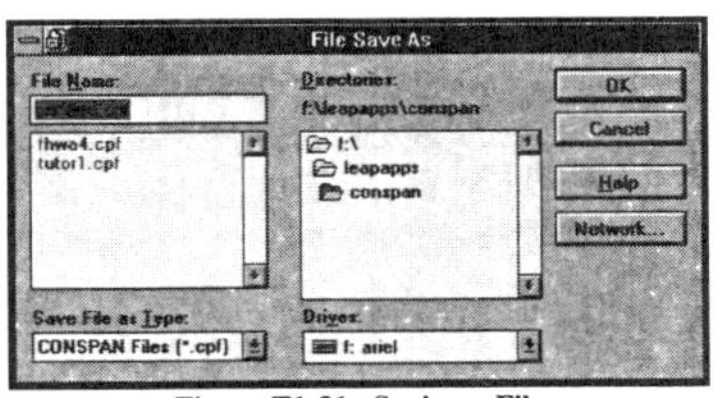

Figure T1-31. Saving a File

```
LEAP Software, Inc.                      PHONE: (813) 985-9170  | SHEET   OF
11602 N. 51st Street, Suite 200          Tampa, FL       33617  | JOB NO TUTOR1
PROGRAM: CONSPAN -V6.0  by LEAP Software Inc., Tampa, Florida   | SK   10/18/96
PHONE  : TOLL-FREE 1-800-451-5327  (TAMPA AREA: 813-985-9170)   | TUTOR1.CPF
```

PROJECT DATA

Project Tutorial 1, Conspan for Windows. File: TUTOR1.CPF
User job number: TUTOR1
State : Florida   State Job #: DOT-1

BRIDGE LAYOUT

Overall width ,ft	=	51.00
Left curb ,ft	=	1.50
Right curb ,ft	=	1.50
Curb-to-curb width,ft	=	48.00
Number of Spans	=	3
Beam Lines	=	6
Number of Lanes	=	4
Topping Thickness,in	=	8.00
Haunch Thickness,in	=	0.50
Moment of Inertia,in4	=	6609699

CONCRETE PROPERTIES

	Precast	C.I.P.
f'c, psi =	6000.0	3500.0
Wc,pcf =	150	150
Ec,ksi =	4696.0	3586.6
f'ci,psi =	5000.0	

SPAN DATA

Span	Pier-to-pier ft	Precast ft	Release ft	Brg-to-brg ft	Clear ft	End-offset ft
1	117.00	116.00	116.00	115.00	115.00	0.50
2	117.00	116.00	116.00	115.00	115.00	0.50
3	117.00	116.00	116.00	115.00	115.00	0.50

BEAM LINE DATA

Line	Beam Id	Area in2	M.I. in4	Height in	Yb in	Loc-prev ft	B-Topg in	B-Trib ft
1	BT-72	767	545894	72.00	36.60	4.25	42.00	8.50
2	BT-72	767	545894	72.00	36.60	8.50	42.00	8.50
3	BT-72	767	545894	72.00	36.60	8.50	42.00	8.50
4	BT-72	767	545894	72.00	36.60	8.50	42.00	8.50
5	BT-72	767	545894	72.00	36.60	8.50	42.00	8.50
6	BT-72	767	545894	72.00	36.60	8.50	42.00	8.50

Figure 3-1 (continued)

```
LEAP Software, Inc.                       PHONE: (813) 985-9170   | SHEET   OF
11602 N. 51st Street, Suite 200           Tampa, FL       33617   | JOB NO TUTOR1
PROGRAM: CONSPAN -V6.0  by LEAP Software Inc., Tampa, Florida     | SK    10/18/96
PHONE   : TOLL-FREE 1-800-451-5327  (TAMPA AREA: 813-985-9170)    | TUTOR1.CPF
--------------------------------------------------------------------------------
Project : Tutorial 1, Conspan ver 6.0

LOADS ON COMPOSITE
------------------
Superimposed Dead Load:
                    ---CL #1--          ---CL #2--          ---CL #3--
 Span     UDL       Mag     Loc         Mag     Loc         Mag     Loc
          psf        k       ft          k       ft          k       ft

   1    30.50       0.0    0.00         0.0    0.00         0.0    0.00
   2    30.50       0.0    0.00         0.0    0.00         0.0    0.00
   3    30.50       0.0    0.00         0.0    0.00         0.0    0.00

LIVE LOADS
----------
ID: HS25    Width, ft:  10.00         Wheel Spg., ft:   6.00

  Lane Load (adjustable)
        UDL, klf :  0.800          Conc. Load, k :
                                         Mom.:     22.5
                                         Shear:    32.5

  Truck load:     (Spacing increment=  14.00 ft)
        Mag, k          Min. spac., ft           Max. spac., ft
         10.0               -----                    -----
         40.0               14.00                    14.00
         40.0               14.00                    30.00

        Preceding uniform load :
                Mag., klf : 0.000   Dist, ft:  0.00   Len, ft:  0.00
        Trailing uniform load  :
                Mag., klf : 0.000   Dist, ft:  0.00   Len, ft:  0.00

Sidewalk Live Load:     0.0 plf
```

LEAP Software, Inc. PHONE: (813) 985-9170 | SHEET OF
11602 N. 51st Street, Suite 200 Tampa, FL 33617 | JOB NO TUTOR1
PROGRAM: CONSPAN -V6.0 by LEAP Software Inc., Tampa, Florida | SK 10/18/96
PHONE : TOLL-FREE 1-800-451-5327 (TAMPA AREA: 813-985-9170) | TUTOR1.CPF

Project : Tutorial 1, Conspan ver 6.0
Spans 01-03

STATIC ANALYSIS

Loads: SDL on composite

Loc. (ft)	Mom (kft)	Shear (k)	Reac (k)
Span 1			-72.8
0.00	0.0	72.8	
12.10	767.0	54.0	
23.70	1288.4	35.9	
35.30	1600.6	17.9	
46.90	1703.5	-0.2	
58.50	1597.0	-18.2	
70.10	1281.2	-36.2	
81.70	756.1	-54.3	
93.30	21.8	-72.3	
104.90	-921.9	-90.4	
117.00	-2129.3	-109.2	
Span 2			-200.2
0.00	-2129.3	91.0	
12.10	-1142.1	72.2	
23.70	-409.6	54.1	
35.30	113.7	36.1	
46.90	427.7	18.0	
58.50	532.3	-0.0	
70.10	427.7	-18.0	
81.70	113.7	-36.1	
93.30	-409.6	-54.1	
104.90	-1142.1	-72.2	
117.00	-2129.3	-91.0	
Span 3			-200.2
0.00	-2129.3	109.2	
12.10	-921.9	90.4	
23.70	21.8	72.3	
35.30	756.2	54.3	
46.90	1281.2	36.2	
58.50	1597.0	18.2	
70.10	1703.5	0.2	
81.70	1600.6	-17.9	
93.30	1288.4	-35.9	
104.90	767.0	-54.0	
117.00	0.0	-72.8	
			-72.8

Figure 3-1 (continued)

LEAP Software, Inc. PHONE: (813) 985-9170 | SHEET OF
11602 N. 51st Street, Suite 200 Tampa, FL 33617 | JOB NO TUTOR1
PROGRAM: CONSPAN -V6.0 by LEAP Software Inc., Tampa, Florida | SK 10/18/96
PHONE : TOLL-FREE 1-800-451-5327 (TAMPA AREA: 813-985-9170) | TUTOR1.CPF

Project : Tutorial 1, Conspan ver 6.0
Spans 01-03
LIVE LOAD ENVELOPE

Results are for axle load, unfactored, no-impact.
Trucks:
1) HS25

Loc. (ft)	MaxM(+) (kft)	-----V (k)	Trk	MaxM(-) (kft)	----V (k)	Trk	MaxV (k)	-----M (kft)	Trk
Span 1		(Support 1 R-max, k			7.1/	-81.0)			
0.00	0.0	81.0	01	0.0	73.9	01	81.0	0.0	01
12.10	839.2	69.4	01	-84.4	7.0	01	69.4	839.2	01
23.70	1386.2	58.5	01	-165.4	7.0	01	58.5	1386.2	01
35.30	1701.4	42.2	01	-246.3	7.0	01	48.0	1695.3	01
46.90	1830.1	32.0	01	-327.3	7.0	01	38.1	1786.9	01
58.50	1786.0	39.9	01	-408.2	7.0	01	45.9	1737.6	01
70.10	1592.4	49.3	01	-489.1	7.0	01	55.6	1570.4	01
81.70	1245.5	57.9	01	-570.1	7.0	01	64.6	1238.9	01
93.30	786.0	65.6	01	-651.0	7.0	01	72.6	781.6	01
104.90	357.9	27.9	01	-990.1	52.7	01	79.7	243.7	01
117.00	262.5	2.2	01	-1757.8	75.3	01	90.2	-1277.6	01
Span 2		(Support 2 R-max, k			13.5/	-144.9)			
0.00	262.5	11.2	01	-1757.8	71.3	01	87.1	-1277.6	01
12.10	323.0	69.3	01	-1054.3	41.5	01	76.0	321.1	01
23.70	827.8	60.7	01	-784.2	11.2	01	67.0	826.9	01
35.30	1216.3	51.2	01	-749.8	0.0	01	57.2	1203.6	01
46.90	1452.9	41.1	01	-749.8	0.0	01	47.0	1420.5	01
58.50	1517.9	30.8	01	-749.8	0.0	01	36.8	1463.0	01
70.10	1452.9	41.1	01	-749.8	0.0	01	47.0	1420.5	01
81.70	1216.3	51.2	01	-749.8	0.0	01	57.2	1203.6	01
93.30	827.8	60.7	01	-784.2	11.2	01	67.0	826.9	01
104.90	323.0	69.3	01	-1054.3	41.5	01	76.0	321.1	01
117.00	262.5	11.2	01	-1757.8	71.3	01	87.1	-1277.6	01
Span 3		(Support 3 R-max, k			13.5/	-144.9)			
0.00	262.5	2.2	01	-1757.8	75.3	01	90.2	-1277.6	01
12.10	357.9	27.9	01	-990.0	52.7	01	79.7	243.7	01
23.70	786.0	65.6	01	-651.0	7.0	01	72.6	781.6	01
35.30	1245.5	57.9	01	-570.1	7.0	01	64.6	1238.9	01
46.90	1592.4	49.3	01	-489.1	7.0	01	55.6	1570.4	01
58.50	1786.0	39.9	01	-408.2	7.0	01	45.9	1737.6	01
70.10	1830.1	32.0	01	-327.2	7.0	01	38.1	1786.9	01
81.70	1701.4	42.2	01	-246.3	7.0	01	48.0	1695.3	01
93.30	1386.2	58.5	01	-165.4	7.0	01	58.5	1386.2	01
104.90	839.2	69.4	01	-84.4	7.0	01	69.4	839.2	01
117.00	0.0	81.0	01	0.0	81.0	01	81.0	0.0	01
		(Support 4 R-max, k			7.1/	-81.0)			

Moment causing bot tens and Support Reac downwards are positive

```
LEAP Software, Inc.                       PHONE: (813) 985-9170   | SHEET    OF
11602 N. 51st Street, Suite 200           Tampa, FL         33617  | JOB NO TUTOR1
PROGRAM: CONSPAN -V6.0  by LEAP Software Inc., Tampa, Florida | SK    10/18/96
PHONE   : TOLL-FREE 1-800-451-5327  (TAMPA AREA: 813-985-9170) | TUTOR1.CPF
--------------------------------------------------------------------------------
Project : Tutorial 1, Conspan ver 6.0
Span    :  1            Beam   : 2

PRECAST DATA:                                GENERAL BRIDGE DATA:
-------------                                --------------------
Section Id: BT-72                            System: AASHTO Girder
Type       : AASHTO Girder                   Bridge width     , ft =  51.00
Flng wid: Top= 42.00 in Bot=  26.00 in       Curb-to-curb     , ft =  48.00
     thk: Top=  2.00 in Bot=   6.00 in       Beam Spac. Lt/Rt, ft =   8.50/  8.50
Stems:  No.   =   1                          Lane width       , ft =  12.00
        Top   =  6.00 in                     No. of lanes          =      4
        Bot   =   N/A                        No. of beam lines     =      6
Shear width   =  6.00 in                     Int./Ext.             = Interior
Wide top flange : No

GENERAL LOAD DATA                            TOPPING DATA
-----------------                            ------------
Dead Loads on Precast:                       Concrete thickness:
  Linear loads:                                Deck          , in =    8.00
                plf                            Supplemental, in =    0.00
      Mag.  180.0                            Haunch:
  Concentrated loads:                          Thickness    , in =    0.50
                 k       ft                    Width        , in =   42.00
     Pt # 1     0.0     0.00                 Effective width:
     Pt # 2     0.0     0.00                   Composite    , in = 102.00
     Pt # 3     0.0     0.00

Dead Loads on Composite
---------------------------
See Project info for loads.
Tributary fraction: 0.167

Impact (Moment)   =  1.208 (computed)
Impact (Shear)    = same as for moment

LOAD EQUATIONS:  Gamma  Beta    Beta   D.F.  ( Art. 3.22.2 )
---------------
Service  Load   =  1.00*(1.00*DL+1.00*0.773*(LL+I))  Group I
Factored Load   =  1.30*(1.00*DL+1.67*0.773*(LL+I))  Group I

GENERAL SPAN DATA:                      MISCELLANEOUS
------------------                      -------------
Overall Length,ft = 116.00              Kern pts:        Upper,in = 56.05
Release Span   ,ft = 116.00                              Lower,in = 16.49
Design Span    ,ft = 115.00             Trans len mult,Bonded    =  1.00
Clear Span     ,ft = 115.00                              Debonded =  1.00
                                        Dev len mult,  Bonded    =  1.00
                                                         Debonded =  2.00
                                        Sidewalk Tributary frac = 0.167
                                        End shear design  : Prestressed
```

Figure 3-1 (continued)

```
LEAP Software, Inc.                        PHONE: (813) 985-9170   | SHEET   OF
11602 N. 51st Street, Suite 200            Tampa, FL        33617  | JOB NO TUTOR1
PROGRAM: CONSPAN -V6.0  by LEAP Software Inc., Tampa, Florida      | SK    10/18/96
PHONE  : TOLL-FREE 1-800-451-5327  (TAMPA AREA: 813-985-9170)      | TUTOR1.CPF
------------------------------------------------------------------------------------
Project : Tutorial 1, Conspan ver 6.0
Span    :  1              Beam   : 2

SECTION PROPERTIES:
-------------------
                         PRECAST        COMPOSITE
                         -------        ---------
Area             =       767 in2         1406 in2*
Total Height     =     72.00 in         80.50 in *
Mom of Inertia   =    545894 in4      1101617 in4*
Ht. of c.g.      =     36.60 in         54.69 in *
Density          =       150 pcf          150 pcf
Self-weight      =     799.0 plf       1670.8 plf
At Final 1 (P/S + DL + LL):
  Strength       =    6000.0 psi       3500.0 psi
  Max comp: Top=      3600 psi         2100 psi
    Pos Mom,Bot=      3600 psi
    Neg Mom,Bot=      3600 psi
  Max tens: Top=      -232 psi         -177 psi
            Bot=      -465 psi
  Crack. tens. =      -581 psi
  Elasticity   =    4696.0 ksi       3586.6 ksi
At Final 2 (P/S + DL):
  Max comp: Top=      2400 psi         1400 psi
    Pos Mom,Bot=      2400 psi
    Neg Mom,Bot=      3600 psi
At Final 3 (50% P/S + 50% DL + LL):
  Max comp: Top=      2400 psi         1400 psi
    Pos Mom,Bot=      2400 psi
    Neg Mom,Bot=      3600 psi
At Release:
  Strength       =    5000.0 psi
  Max comp: Bot=      3000 psi
            Top=      3000 psi
  Max tens: Bot=         0 psi
            Top=      -200 psi
        w/reinf=      -530 psi
  Elasticity   =    4286.8 ksi

* Of total section using Ect/Ec = 0.7638

  Use transformed strands and rebar: No
```

```
LEAP Software, Inc.                       PHONE: (813) 985-9170  | SHEET   OF
11602 N. 51st Street, Suite 200           Tampa, FL       33617  | JOB NO TUTOR1
PROGRAM: CONSPAN -V6.0  by LEAP Software Inc., Tampa, Florida | SK    10/18/96
PHONE  : TOLL-FREE 1-800-451-5327  (TAMPA AREA: 813-985-9170) | TUTOR1.CPF
------------------------------------------------------------------------------
Project : Tutorial 1, Conspan ver 6.0
Span    :  1             Beam  : 2

PRESTRESSED STEEL:  38  strands, 1/2-270K-LL, Low relaxation strands
                      Depressed at 0.40L ( 46.50 ft from member end).
END PATTERN (Ycg= 20.29 in):
     10 @    2.50 in  10 @    4.50 in    6 @    6.50 in    2 @    8.50 in
      2 @   60.50 in   2 @   62.50 in    2 @   64.50 in    2 @   66.50 in
      2 @   68.50 in
MID PATTERN (Ycg=  5.03 in):
(A) DRAPED    :       2 @    2.50 in    2 @    4.50 in    2 @    6.50 in
      2 @    8.50 in   2 @   10.50 in
(B) STRAIGHT :       10 @    2.50 in  10 @    4.50 in    6 @    6.50 in
      2 @    8.50 in

Strand Dia          =  0.5000 in      Ult strength(f's) =  270000 psi
Strand Area         =  0.1530 in2     Initial prestress =   0.75 f's= 202.5 ksi
Total strand area   =  5.8140 in2     Initial pull      = 1177.3 k
Trans len,basic     =    2.08 ft      Dev len,  basic   =   6.62 ft
Trans len,bonded    =    2.08 ft      Dev len,  bonded  =   6.62 ft
Trans len,debonded  =    2.08 ft      Dev len,  debonded=  13.24 ft
------------------------------------------------------------------------------
REINFORCING STEEL:  Tension steel: fy =    60 ksi    fs =    24 ksi
   Shear steel: fsy =    60 ksi   Modulus of elasticity= 29000 ksi
------------------------------------------------------------------------------
Midspan: Str. area=   5.814 in2 Ycg=  5.03 in P-init= 1177.3 k  Ecc= 31.57 in
Hours to release = 18 Rel. Humid. (RH) = 75%  Es= 28000 ksi  Eci =  4287 ksi

AASHTO LOSSES:              Release                    Final  (Art. 9.16.2)
Steel relaxation RET*     1600 psi   CRs(Eq 9-10a)   2135 psi
Elastic shortening ES    15897 psi   ES   (Eq 9-6)  15897 psi (Fcir=   2434 psi)
Concrete shrinkage                   SH   (Eq 9-4)    5750 psi
Concrete creep                       CRc  (Eq 9-9)  19748 psi (Fcds=  -1351 psi)
                          -----                     -----
Total                    17498 psi (  8.6 %)        43531 psi ( 21.5 %)

* Steel relax. before release - Ref: PCI Journal Vol 20, No.4, Jul-Aug 1975
------------------------------------------------------------------------------
SHIELDING AND REDUCED INITIAL PULLS:
  NONE
```

Figure 3-1 (continued)

```
LEAP Software, Inc.                        PHONE: (813) 985-9170  | SHEET   OF
11602 N. 51st Street, Suite 200            Tampa, FL        33617  | JOB NO TUTOR1
PROGRAM: CONSPAN -V6.0  by LEAP Software Inc., Tampa, Florida  | SK    10/18/96
PHONE   : TOLL-FREE 1-800-451-5327  (TAMPA AREA: 813-985-9170)  | TUTOR1.CPF
```

Project : Tutorial 1, Conspan ver 6.0
Span : 1 Beam : 2

SERVICE LOAD MOMENTS AND SHEARS (k ,ft):

		Bearing	Trans	H/2	0.10L	0.20L	0.30L	0.40L	Midspan
Location, ft		0.00	1.58	3.35	11.10	22.70	34.30	45.90	57.50
Self wt.	: M	0.0	71.7	149.6	460.7	837.0	1105.8	1267.0	1320.8
	V	45.9	44.7	43.3	37.1	27.8	18.5	9.3	0.0
DL-Prec.	: M	0.0	16.2	33.7	103.8	188.6	249.1	285.5	297.6
	V	10.4	10.1	9.7	8.4	6.3	4.2	2.1	0.0
Topping	: M	0.0	78.3	163.2	502.8	913.4	1206.7	1382.7	1441.3
	V	50.1	48.8	47.2	40.5	30.3	20.2	10.1	0.0
DL-Comp	: M	12.0	30.5	50.5	128.1	215.2	267.3	284.5	266.7
	V	11.9	11.5	11.0	9.0	6.0	3.0	-0.0	-3.0
LL+I	: M+	74.5	188.8	311.5	783.6	1294.3	1588.6	1708.8	1667.6
	V	74.7	73.3	71.7	64.8	54.6	39.4	29.9	37.3
LL+I	: M-	-6.5	-16.8	-28.4	-78.8	-154.4	-230.0	-305.6	-381.1
	V	63.9	55.7	46.5	6.5	6.5	6.5	6.5	6.5
LL+I	:Vmx	74.7	73.3	71.7	64.8	54.6	44.8	35.6	42.9
	M	74.5	188.8	311.5	783.6	1294.3	1582.9	1668.4	1622.4

		0.60L	0.70L	0.80L	0.90L	H/2	Trans	Bearing
Location, ft		69.10	80.70	92.30	103.90	111.65	113.42	115.00
Self wt.	: M	1267.0	1105.8	837.0	460.7	149.6	71.7	0.0
	V	-9.3	-18.5	-27.8	-37.1	-43.3	-44.7	-45.9
DL-Prec.	: M	285.5	249.1	188.6	103.8	33.7	16.2	0.0
	V	-2.1	-4.2	-6.3	-8.4	-9.7	-10.1	-10.4
Topping	: M	1382.7	1206.7	913.4	502.8	163.2	78.3	0.0
	V	-10.1	-20.2	-30.3	-40.5	-47.2	-48.8	-50.1
DL-Comp	: M	214.0	126.3	3.6	-154.0	-278.7	-309.4	-337.5
	V	-6.1	-9.1	-12.1	-15.1	-17.1	-17.6	-18.0
LL+I	: M+	1486.9	1163.0	733.9	334.2	238.6	236.0	239.8
	V	46.0	54.1	61.2	26.0	10.7	7.2	4.1
LL+I	: M-	-456.7	-532.3	-607.9	-924.4	-1337.9	-1455.1	-1567.1
	V	6.5	6.5	6.5	49.2	62.7	65.8	68.6
LL+I	:Vmx	51.9	60.3	67.8	74.4	80.7	82.1	83.4
	M	1466.3	1156.7	729.8	227.5	-576.3	-812.8	-1040.8

```
LEAP Software, Inc.                         PHONE: (813) 985-9170  | SHEET   OF
11602 N. 51st Street, Suite 200             Tampa, FL        33617  | JOB NO TUTOR1
PROGRAM: CONSPAN -V6.0  by LEAP Software Inc., Tampa, Florida | SK    10/16/96
PHONE   : TOLL-FREE 1-800-451-5327  (TAMPA AREA: 813-985-9170) | TUTOR1.CPF
```

Project : Tutorial 1, Conspan ver 6.0
Span : 1 Beam : 2

RELEASE STRESSES, psi (Loss = 8.6%)

	Trans	0.10L	0.20L	0.30L	0.40L	Midspan	Depress	Crit
Location, ft	1.58	11.10	22.70	34.30	45.90	57.50	46.00	46.35
Prest.								
Precast-top	217.0	-0.9	-266.5	-532.1	-797.6	-799.9	-799.9	-799.9
Bottom	2627.9	2853.2	3127.7	3402.3	3676.9	3679.3	3679.3	3679.3
Self wt.								
Precast-top	73.8	376.5	669.3	878.4	1003.9	1045.7	1004.6	1007.1
Bottom	-76.3	-389.2	-692.0	-908.2	-1037.9	-1081.2	-1038.7	-1041.2
Total								
Precast-top	290.8	375.6	402.8	346.4	206.3	245.8	204.7	207.1
Bottom	2551.6	2463.9	2435.8	2494.1	2639.0	2598.1	2640.6	2638.1
As-top, in2	NONE	NONE	NONE	NONE	NONE	NONE	NONE	NONE

Figure 3-1 (continued)

```
LEAP Software, Inc.                      PHONE: (813) 985-9170  | SHEET   OF
11602 N. 51st Street, Suite 200          Tampa, FL       33617  | JOB NO TUTOR1
PROGRAM: CONSPAN -V6.0  by LEAP Software Inc., Tampa, Florida  | SK   10/18/96
PHONE  : TOLL-FREE 1-800-451-5327  (TAMPA AREA: 813-985-9170)  | TUTOR1.CPF
```

Project : Tutorial 1, Conspan ver 6.0
Span : 1 Beam : 2

FINAL STRESSES, psi (Loss = 21.5%) - POSITIVE MOMENT ENVELOPE :

	Bearing	Trans	H/2	0.10L	0.20L	0.30L	0.40L	Midspan
Location, ft	0.00	1.58	3.35	11.10	22.70	34.30	45.90	57.50
Prest.								
Precast-top	52.2	186.5	151.6	-0.8	-229.0	-457.2	-685.4	-687.4
Bottom	534.2	2258.1	2294.1	2451.7	2687.6	2923.6	3159.5	3161.5
Self wt.								
Precast-top	0.0	55.8	116.4	358.5	651.3	860.5	986.0	1027.8
Bottom	0.0	-57.7	-120.4	-370.7	-673.4	-889.6	-1019.4	-1062.6
DL-Prec.								
Precast-top	0.0	12.6	26.2	80.8	146.7	193.9	222.1	231.6
Bottom	0.0	-13.0	-27.1	-83.5	-151.7	-200.4	-229.7	-239.4
Topping								
Precast-top	0.0	60.9	127.0	391.2	710.8	939.0	1075.9	1121.6
Bottom	0.0	-63.0	-131.3	-404.5	-734.9	-970.8	-1112.4	-1159.6
DL-Comp								
Topping-top	2.6	6.6	10.8	27.5	46.2	57.4	61.1	57.3
Precast-top	2.3	5.8	9.5	24.2	40.6	50.4	53.6	50.3
Bottom	-7.2	-18.2	-30.1	-76.3	-128.2	-159.2	-169.5	-158.9
LL+I(+)								
Topping-top	16.0	40.5	66.9	168.3	277.9	341.1	366.9	358.1
Precast-top	14.0	35.6	58.7	147.8	244.1	299.6	322.2	314.4
Bottom	-44.4	-112.5	-185.5	-466.8	-771.1	-946.4	-1018.0	-993.4
Total, Final 1 (P/S + DL + LL)								
Topping-top	18.6	47.1	77.7	195.8	324.1	398.5	428.0	415.4
Precast-top	68.5	357.1	489.5	1001.7	1564.5	1886.1	1974.5	2058.3
Bottom	482.7	1993.7	1799.7	1049.9	228.4	-243.0	-389.4	-452.4
Total, Final 2 (P/S + DL)								
Topping-top	2.6	6.6	10.8	27.5	46.2	57.4	61.1	57.3
Precast-top	54.5	321.5	430.8	853.9	1320.4	1586.6	1652.3	1743.9
Bottom	527.1	2106.2	1985.2	1516.7	999.5	703.4	628.6	541.0
Total, Final 3 (50% P/S + 50% DL + LL)								
Topping-top	17.3	43.8	72.3	182.0	301.0	369.8	397.5	386.7
Precast-top	41.3	196.4	274.1	574.7	904.3	1092.8	1148.4	1186.4
Bottom	219.2	940.6	807.1	291.6	-271.3	-594.7	-703.7	-722.9

```
LEAP Software, Inc.                        PHONE: (813) 985-9170  | SHEET   OF
11602 N. 51st Street, Suite 200            Tampa, FL       33617  | JOB NO TUTOR1
PROGRAM: CONSPAN -V6.0  by LEAP Software Inc., Tampa, Florida | SK   10/18/96
PHONE  : TOLL-FREE 1-800-451-5327  (TAMPA AREA: 813-985-9170) | TUTOR1.CPF

Project : Tutorial 1, Conspan ver 6.0
Span    :  1              Beam  : 2
VERTICAL SHEAR :
```

	Bearing	Trans	H/2	0.10L	0.20L	0.30L	0.40L
Location, ft	0.00	1.58	3.35	11.10	22.70	34.30	45.90
Vd,k	118.3	-121.1	-117.3	-101.0	70.4	-52.0	-27.5
Md,kft	12.0	-143.2	67.9	913.3	2154.1	2687.8	3149.1
Ml,kft	74.5	-1455.1	-1337.9	334.2	1294.3	1163.0	1486.9
Vu,k	300.3	327.7	327.7	292.8	210.1	198.5	148.5
Mu,kft	177.3	-3345.1	-2816.2	1912.8	5610.3	6019.0	7321.8
Vmu,k	316.0	300.2	288.7	187.7	210.1	185.0	135.7
Mmax,kft	165.3	-3202.0	-2884.1	999.5	3456.2	3331.2	4172.7
Vi,k	197.7	179.2	171.3	86.8	139.7	133.0	108.2
fpe,psi	534.2	0.0	0.0	2451.7	2687.6	2923.6	3159.5
fd,psi	-7.2	-66.4	-59.8	-767.0	-1688.2	-2136.1	-2488.9
Mcr,kft	1664.8	-1343.7	-1374.4	3608.1	2457.8	2101.9	1905.8
d,in	64.40	76.25	76.25	64.40	67.83	71.63	75.44
Vci-com,k	2127.5	217.5	220.2	432.1	188.7	155.9	98.0
Vci-min,k	50.9	60.2	60.2	50.9	53.6	56.6	59.6
Vci,k	2127.5	217.5	220.2	432.1	188.7	155.9	98.0
fpc,psi	168.1	750.6	804.5	1013.3	1243.3	1374.2	1406.2
Vp,k	6.1	25.5	25.5	25.5	25.5	25.5	25.5
Vcw,k	130.4	252.6	260.0	247.8	287.7	319.3	339.2
Vc,k	130.4	217.5	220.2	247.8	188.7	155.9	98.0
Vs-reqd,k	203.2	146.6	143.9	77.5	44.8	64.6	67.0
Vs-max,k	239.4	283.5	283.5	239.4	252.2	266.3	280.5
Av-com,in2/ft	0.63	0.38	0.38	0.24	0.13	0.18	0.18
Av-min,in2/ft	0.06	0.06	0.06	0.06	0.06	0.06	0.06
Av,in2/ft	0.63	0.38	0.38	0.24	0.13	0.18	0.18
Vs-crit,k	119.7	141.8	141.8	119.7	126.1	133.2	140.2
Max spc,in	12.00	12.00	12.00	24.00	24.00	24.00	24.00

Figure 3-1 (continued)

```
LEAP Software, Inc.                          PHONE: (813) 985-9170  | SHEET   OF
11602 N. 51st Street, Suite 200         Tampa, FL        33617  | JOB NO TUTOR1
PROGRAM: CONSPAN -V6.0  by LEAP Software Inc., Tampa, Florida | SK   10/18/96
PHONE  : TOLL-FREE 1-800-451-5327  (TAMPA AREA: 813-985-9170) | TUTOR1.CPF
--------------------------------------------------------------------------------
Project : Tutorial 1, Conspan ver 6.0
Span    :  1              Beam   : 2
VERTICAL SHEAR (continued)

                      Midspan   Depress
Location, ft           57.50     46.00

Vd,k                    -3.0     -27.3
Md,kft                3326.4    3151.8
Ml,kft                1667.6    1489.0
Vu,k                    97.1     148.1
Mu,kft                7944.5    7330.1
Vmu,k                   84.8     135.3
Mmax,kft              4618.2    4178.2
Vi,k                    81.8     108.0
fpe,psi               3161.5    3161.5
fd,psi               -2620.5   -2491.0
Mcr,kft               1688.3    1905.7
d,in                   75.47     75.47
Vci-com,k               54.0      97.6
Vci-min,k               59.6      59.6
Vci,k                   59.6      97.6
fpc,psi               1454.7    1406.0
Vp,k                     0.0      25.5
Vcw,k                  320.4     339.3
Vc,k                    59.6      97.6
Vs-reqd,k               48.2      66.9
Vs-max,k               280.6     280.6
Av-com,in2/ft           0.13      0.18
Av-min,in2/ft           0.06      0.06
Av,in2/ft               0.13      0.18
Vs-crit,k              140.3     140.3
Max spc,in             24.00     24.00
```

```
LEAP Software, Inc.                      PHONE: (813) 985-9170  | SHEET   OF
11602 N. 51st Street, Suite 200          Tampa, FL       33617  | JOB NO TUTOR1
PROGRAM: CONSPAN -V6.0  by LEAP Software Inc., Tampa, Florida   | SK    10/18/96
PHONE   : TOLL-FREE 1-800-451-5327  (TAMPA AREA: 813-985-9170)  | TUTOR1.CPF
```

Project : Tutorial 1, Conspan ver 6.0
Span : 1 Beam : 2

HORIZONTAL SHEAR : (Fsy = 60 ksi bv= 42.00 in)

	Bearing	Trans	H/2	0.10L	0.20L	0.30L	0.40L
Location, ft	0.00	1.58	3.35	11.10	22.70	34.30	45.90
Vu,k	342.8	335.7	327.7	292.8	246.6	198.5	148.5
d,in	76.50	76.50	76.50	76.50	67.83	71.63	75.44
Vnh-req,psi	118.6	116.1	113.3	101.2	96.2	73.3	52.1
Surf,in2/ft	504.00	504.00	504.00	504.00	504.00	504.00	504.00
Avhmin,in2/ft	0.42	0.42	0.42	0.42	0.42	0.42	0.42
Max spc,in	24.00	24.00	24.00	24.00	24.00	24.00	24.00
Avh-sm,in2/ft	1.23	1.18	1.12	0.87	0.76	0.42	0.42
Avh-rg,in2/ft	0.42	0.42	0.42	0.42	0.42	0.00	0.00

	Midspan	Depress
Location, ft	57.50	46.00
Vu,k	97.1	148.1
d,in	75.47	75.47
Vnh-req,psi	34.0	51.9
Surf,in2/ft	504.00	504.00
Avhmin,in2/ft	0.42	0.42
Max spc,in	24.00	24.00
Avh-sm,in2/ft	0.42	0.42
Avh-rg,in2/ft	0.00	0.00

ULTIMATE CAPACITY (f'c-eff = 3500.0 psi Beta-1 = 0.850) :

	Trans	H/2	0.10L	0.20L	0.30L	0.40L	Midspan	Depress
Location, ft	1.58	3.35	11.10	22.70	34.30	45.90	57.50	46.00
A*s,in2	1.830	3.386	5.814	5.814	5.814	5.814	5.814	5.814
Ycg,in	19.61	19.02	16.48	12.67	8.87	5.06	5.03	5.03
p* (A*s/bd)	0.00029	0.00054	0.00089	0.00084	0.00080	0.00076	0.00076	0.00076
f*su,ksi	268.0	266.3	263.9	264.2	264.5	264.8	264.8	264.8
a,in	1.62	2.97	5.06	5.06	5.07	5.07	5.07	5.07
Mu-prv'd,kft	2455.1	4505.1	7855.4	8352.6	8849.9	9347.3	9351.6	9351.6
Mu-req'd,kft	665.6	1192.3	3255.1	5610.3	7126.5	7895.3	7944.5	7898.7
Crkg Mom,kft	2110.2	3096.4	4716.5	4607.1	4984.9	5250.0	5209.7	5252.6
Crkg Ratio				1.74	1.78	1.78	1.80	1.78

CAMBER AND DEFLECTIONS (Ref: PCI Design Handbook - 3rd Ed., Table 4.6.3) :
Units: in

	Release	Mult	Erection	Mult	Final
Prestress :	3.149	1.80	5.669	2.20	6.929
Self weight :	-1.391	1.85	-2.573	2.40	-3.338
SDL-Precast :			-0.282	3.00	-0.845
Topping weight :			-1.338	2.30	-3.078
SDL-Composite :			-0.110	3.00	-0.330
Live load :					-0.523
Total :	1.759		1.366		-1.185

Positive value indicates upward deflection

Figure 3-1 (continued)

```
LEAP Software, Inc.                              PHONE: (813) 985-9170  | SHEET    OF
11602 N. 51st Street, Suite 200                  Tampa, FL        33617  | JOB NO TUTOR1
PROGRAM: CONSPAN -V6.0  by LEAP Software Inc., Tampa, Florida  | SK    10/18/96
PHONE  : TOLL-FREE 1-800-451-5327  (TAMPA AREA: 813-985-9170)  | TUTOR1.CPF
-----------------------------------------------------------------------------------
Project : Tutorial 1, Conspan ver 6.0
Span    :  1              Beam  : 2

DETENSIONING ( Groups 01-10  Loss=  8.6%  Units:psi) :
Grp  Str  Ys,in        2.08ft
 01   2 E   2.50 Ft       22
        M   2.50 Fb      127

 02   2 E  68.50 Ft      204
        M  10.50 Fb       90

 03   2 E  66.50 Ft      378
        M   8.50 Fb       60

 04   2 E  64.50 Ft      545
        M   6.50 Fb       38

 05   2 E  62.50 Ft      704
        M   4.50 Fb       23

 06   2 E  60.50 Ft      856
        M   2.50 Fb       16

 07   2 E   8.50 Ft      827
        M   8.50 Fb      196

 08   2 E   6.50 Ft      790
        M   6.50 Fb      384

 09   2 E   6.50 Ft      753
        M   6.50 Fb      572

 10   2 E   6.50 Ft      716
        M   6.50 Fb      760
```

```
LEAP Software, Inc.                     PHONE: (813) 985-9170  | SHEET    OF
11602 N. 51st Street, Suite 200         Tampa, FL        33617  | JOB NO TUTOR1
PROGRAM: CONSPAN -V6.0  by LEAP Software Inc., Tampa, Florida | SK    10/18/96
PHONE  : TOLL-FREE 1-800-451-5327  (TAMPA AREA: 813-985-9170) | TUTOR1.CPF
------------------------------------------------------------------------------
Project : Tutorial 1, Conspan ver 6.0
Span    : 1               Beam  : 2

DETENSIONING ( Groups 01-10  Loss=  8.6%  Units:psi) :
Grp  Str  Ys,in       2.08ft
 01   2 E   2.50 Ft      22
        M   2.50 Fb     127

 02   2 E  68.50 Ft     204
        M  10.50 Fb      90

 03   2 E  66.50 Ft     378
        M   8.50 Fb      60

 04   2 E  64.50 Ft     545
        M   6.50 Fb      38

 05   2 E  62.50 Ft     704
        M   4.50 Fb      23

 06   2 E  60.50 Ft     856
        M   2.50 Fb      16

 07   2 E   8.50 Ft     827
        M   8.50 Fb     196

 08   2 E   6.50 Ft     790
        M   6.50 Fb     384

 09   2 E   6.50 Ft     753
        M   6.50 Fb     572

 10   2 E   6.50 Ft     716
        M   6.50 Fb     760
```

Figure 3-1 (continued)

```
LEAP Software, Inc.                       PHONE: (813) 985-9170  | SHEET   OF
11602 N. 51st Street, Suite 200           Tampa, FL       33617  | JOB NO TUTOR1
PROGRAM: CONSPAN -V6.0  by LEAP Software Inc., Tampa, Florida    | SK   10/18/96
PHONE  : TOLL-FREE 1-800-451-5327  (TAMPA AREA: 813-985-9170)    | TUTOR1.CPF
```

Project : Tutorial 1, Conspan ver 6.0
Span : 1 Beam : 2

FINAL STRESSES, psi (Loss = 21.5%) - NEGATIVE MOMENT ENVELOPE :

	Bearing	Trans	H/2	0.10L	0.20L	0.30L	0.40L	Midspan
Location, ft	0.00	1.58	3.35	11.10	22.70	34.30	45.90	57.50
Prest.								
Precast-top	52.2	186.5	151.6	-0.8	-229.0	-457.2	-685.4	-687.4
Bottom	534.2	2258.1	2294.1	2451.7	2687.6	2923.6	3159.5	3161.5
Self wt.								
Precast-top	0.0	55.8	116.4	358.5	651.3	860.5	986.0	1027.8
Bottom	0.0	-57.7	-120.4	-370.7	-673.4	-889.6	-1019.4	-1062.6
DL-Prec.								
Precast-top	0.0	12.6	26.2	80.8	146.7	193.9	222.1	231.6
Bottom	0.0	-13.0	-27.1	-83.5	-151.7	-200.4	-229.7	-239.4
Topping								
Precast-top	0.0	60.9	127.0	391.2	710.8	939.0	1075.9	1121.6
Bottom	0.0	-63.0	-131.3	-404.5	-734.9	-970.8	-1112.4	-1159.6
DL-Comp								
Topping-top	-72.5	-66.4	-59.8	-33.1	0.8	27.1	45.9	57.3
Precast-top	-63.6	-58.3	-52.5	-29.0	0.7	23.8	40.3	50.3
Bottom	201.1	184.3	166.0	91.7	-2.2	-75.2	-127.5	-158.9
LL+I(-)								
Topping-top	-336.5	-312.5	-287.3	-198.5	-130.5	-114.3	-98.1	-81.8
Precast-top	-295.5	-274.4	-252.3	-174.3	-114.6	-100.4	-86.1	-71.9
Bottom	933.6	866.9	797.0	550.7	362.1	317.1	272.1	227.1
Total, Final 1 (P/S + DL + LL)								
Topping-top	-409.0*	-378.9*	-347.1*	-231.6*	-129.7	-87.2	-52.1	-24.6
Precast-top	-306.9*	-16.9	116.5	626.4	1165.9	1459.6	1552.9	1672.0
Bottom	1668.8	3175.6	2978.3	2235.4	1487.6	1104.5	942.6	768.1
Total, Final 2 (P/S + DL)								
Topping-top	-72.5	-66.4	-59.8	-33.1	0.8	27.1	45.9	57.3
Precast-top	-11.4	257.4	368.7	800.7	1280.5	1560.0	1639.0	1743.9
Bottom	735.3	2308.7	2181.3	1684.7	1125.5	787.4	670.6	541.0
Total, Final 3 (50% P/S + 50% DL + LL)								
Topping-top	-372.7	-345.7	-317.2	-215.0	-130.1	-100.7	-75.1	-53.2
Precast-top	-301.2	-145.7	-67.9	226.0	525.6	679.6	733.4	800.1
Bottom	1301.2	2021.2	1887.7	1393.1	924.9	710.8	607.4	497.6

LEAP Software, Inc. PHONE: (813) 985-9170 | SHEET OF
11602 N. 51st Street, Suite 200 Tampa, FL 33617 | JOB NO TUTOR1
PROGRAM: CONSPAN -V6.0 by LEAP Software Inc., Tampa, Florida | SK 10/18/96
PHONE : TOLL-FREE 1-800-451-5327 (TAMPA AREA: 813-985-9170) | TUTOR1.CPF

Project : Tutorial 1, Conspan ver 6.0
Span : 1 Beam : 2

REINFORCED DESIGN
(fy = 60 ksi f'c= 6000.0 psi f'ct= 3500.0 psi Phi = 0.90):
Note: End sections are designed using f'c (Art. 9.7.2.3.3)

Sec	Dist ft	Mu kft	b in	hf in	bw in	d in	d' in	1.2*Mcr kft	Asb in2	Ast in2
(a) NEGATIVE MOMENTS ALONG SPAN (f'cEff = 6000.0 psi)										
1	0.00	0.0	42.00	10.00	6.00	78.50	4.00	-1170.2	0.00	0.00
2	12.10	-4.6	26.00	6.00	6.00	76.50	2.00	-2479.5	0.00	0.02
3	23.70	-55.5	26.00	6.00	6.00	76.50	2.00	-2479.5	0.00	0.21
4	35.30	-151.8	26.00	6.00	6.00	76.50	2.00	-2479.5	0.00	0.59
5	46.90	-293.5	26.00	6.00	6.00	76.50	2.00	-2479.5	0.00	1.14
6	58.50	-480.7	26.00	6.00	6.00	76.50	2.00	-2479.5	0.00	1.87
7	70.10	-713.4	26.00	6.00	6.00	76.50	2.00	-2479.5	0.00	2.78
8	81.70	-991.4	26.00	6.00	6.00	76.50	2.00	-2479.5	0.00	3.87
9	93.30	-1314.9	26.00	6.00	6.00	76.50	2.00	-2479.5	0.00	5.16
10	104.90	-2207.1	26.00	6.00	6.00	76.50	2.00	-2479.5	0.00	7.36
11	117.00	-4025.5	26.00	6.00	6.00	76.50	2.00	-2479.5	0.00	12.13

(b) POSITIVE MOMENTS AT PIERS
None

Figure 3-1 (continued)

EXAMPLE PROBLEMS FROM THE BEST CENTER

The following examples furnished by the BEST Center (Department of Civil Engineering, University of Maryland) are hereby reproduced with the permission of the copyright holders. Specifically, they are included to give a clear illustration on the practical applications of the software available at the BEST Center (Figures 3-2 through 3-4).

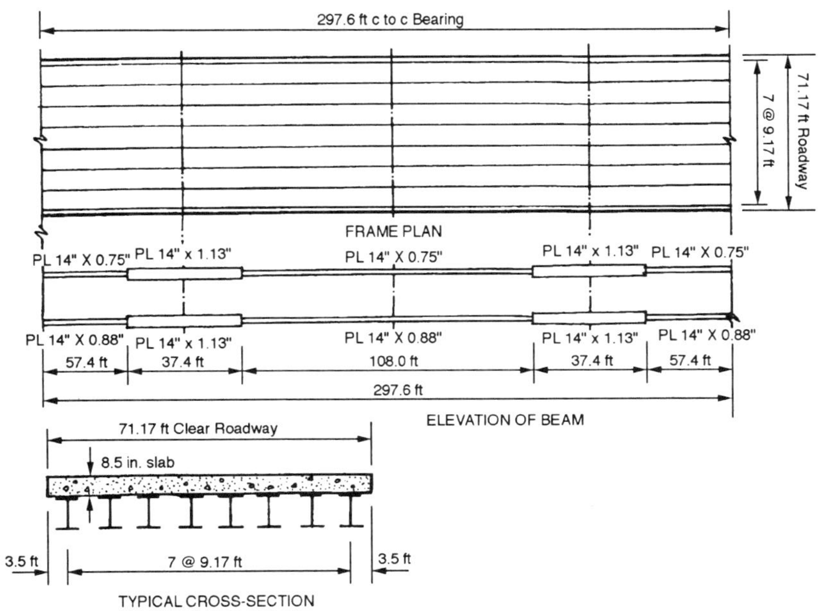

Figure 3-2 Example composite steel bridge design using MERLIN-DASH software (reproduced courtesy BEST Center, University of Maryland).

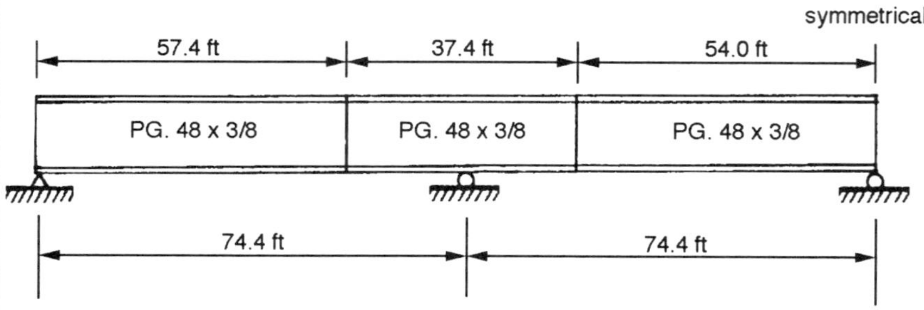

Effective Flange Width (AASHTO ART. 10.38.3.1. pp.147)

= Smallest Value of
- 1/4 Span Length
- Girder Spacing = 9.17 ft = 110 in.
- $12 \times T_{slab} = 12 \times 8.5 = 102$ in.

= 102 in.

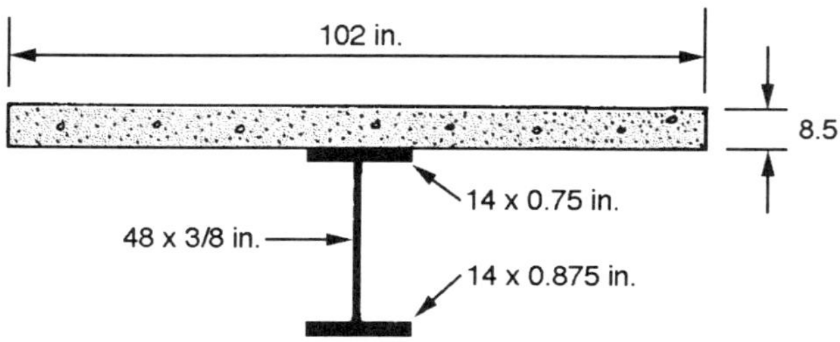

for N = 24.0	for N = 8.0
$I_x = 33579.6\ in^4$	$I_x = 43045.8\ in^4$
$S_t = 2773.7\ in^3$	$S_t = 9465.8\ in^3$
$S_b = 895.0\ in^3$	$S_b = 954.9\ in^3$
$S_c = 1691.1\ in^3$	$S_c = 3500.4\ in^3$

For Positive Moment Region

Applicable Output Tables: 1.2.4.2 and 1.2.4.3

Figure 3-2 (continued)

Input Data

```
01011
01012  CHOPTANK RIVER BRIDGE
01021
01022 DESIGN EXAMPLE 3
01031
01032     1          10       1                      1          3
03011
03012   8    1    71.167     3.5                                      1.1
03021
03022 74.4  74.4  74.4  74.4
03041
03041  9.17  9.17  9.17  9.17
04011
04012  1PG            48.0  0.375 14.00 0.75 14.00  0.875
04012  2PG            48.0  0.375 14.0  1.12514.0   1.125
05011
05012  1  1  1    57.4
05012  2  2  2    37.4
05012  3  1  1   108.0
05012  4  2  2    37.4
05012  5  1  1    57.4
06011
06012  HS 20                       1
10011
10012  1    SLAB LOAD     9.0  8.5     24.  8.       1.192    0.0    297.6
11011
11012  1       BARRIERS           0.118    0.0 297.6
11012  2       MEDIAN             0.056    0.0 297.6
11012  3       WEARING SUR        0.229    0.0 297.6
13011
13012    0.  297.60 50.00
20012
```

```
                    BRIDGE ENGINEERING SOFTWARE CENTER                    MERLIN V 6.4
                     DEPARTMENT OF CIVIL ENGINEERING                      COMPOSITE
                         UNIVERSITY OF MARYLAND                           LFD -- 1995
                                                                          INPUT

TABLE 0.0.1.1 PROJECT DATA
              ************
                  DESCRIPTION                                              DATE
                  -----------                                              ----
 CHOPTANK RIVER BRIDGE

CONTRACT NUMBER      STR NO     STR UNIT  DES  CHK  SPECS. USED
---------------      ------     --------  -BY- -BY- ------------
 DESIGN EXAMPLE 3

TABLE 0.0.1.2 GENERAL PROGRAM OPTIONS
              ***********************
 OUTPUT    SPAN    CONSTRUCTION         ANALYSIS CODE                PROGRAM
  LEVEL  INTERVAL  ------------  --------------------------           FLOW
  (0,1)  (MAX=20)  1= COMPOSITE   CODE    YEAR  UNIT DESIGN          CONTROL
                   2= NONCOMP.     ID           TYPE OPTION
 ------  --------  ------------  -------------------------- -------------

   1        10          1        AASHTO  1995     0     1              3
-------------------------------------------------------------------------
 * output level     : 0 = basic output
                      1 = detailed output
 * span interval    : maximum = 20
                      default = 10
 * structural type  : 1 = composite (default)
                      2 = noncomposite
 * type of unit     : 0 = English (default)
                      1 = Metric
                      2 = Metric input English output
                      3 = English input Metric output
 * design option    : 0 = WSD (default)
                      1 = LFD
 * program flow     : 0 = DL analysis only
                      1 = DL + LL analysis
                      2 = code check
                      3 = rating
                      4 = design
                      5 = design + code check
                      6 = design + recycle + code check
                      7 = DL stage only
                      8 = DL stage + LL

TABLE 0.0.3.1 STRUCTURAL DETAILS
              ******************


          BEAM      WIDTH               EDGE OF              COMPOSITE  STEEL
        POSITION   BETWEEN   OVERHANG  SLAB TO    HAUNCH     PERCENTAGE  LOAD
NUMBER  --------   CURBS OR    WIDTH    CURB   ------------  AT NEG MOM DETAIL
  OF     1=INT.    BARRIER   --------  ------- DEPTH WIDTH.    REGION   FACTOR
GIRDERS  2=EXT.      (ft)       (ft)     (ft)   (in)  (in)      (%)     >= 1.0
------------------------------------------------------------------------------

   8       1        71.17       3.50     0.00   0.00  0.00      0.00     1.10

------------------------------------------------------------------------------
* WIDTH BETWEEN CURBS OR BARRIERS (ROAD WIDTH) is used for the
  determination of traffic lanes
```

Figure 3-2 (continued)

TABLE 0.0.3.2 SPAN LENGTHS --- in feet

SPAN-1	SPAN-2	SPAN-3	SPAN-4	SPAN-5	SPAN-6	SPAN-7	SPAN-8	SPAN-9	SPAN-10
74.40	74.40	74.40	74.40	0.00	0.00	0.00	0.00	0.00	0.00

TABLE 0.0.3.4 BEAM SPACING --- in feet

SPAN-1	SPAN-2	SPAN-3	SPAN-4	SPAN-5	SPAN-6	SPAN-7	SPAN-8	SPAN-9	SPAN-10
9.17	9.17	9.17	9.17	0.00	0.00	0.00	0.00	0.00	0.00

TABLE 0.0.4.1 DEFINITION OF SECTIONS

SECTION		STANDARD SECTN		PLATE GIRDER		ROLLED SECTIONS WITH COVER PLATES OR PLATE GIRDERS ... (in)				REINFORCED CONCRETE SECTION	
		NOMINAL DEPTH	WEIGHT	WEB DEPTH	WEB THICK.	TOP PLATE		BOT. PLATE		AREA	Ix
NO.	ID.	(in)	(lb/ft)	(in)		WIDTH	THICK.	WIDTH	THICK.	(in**2)	(in**4)
1	PG			48.0	0.3750	14.00	0.7500	14.00	0.8750		
2	PG			48.0	0.3750	14.00	1.1250	14.00	1.1250		

TABLE 0.0.5.1 DEFINITION OF MEMBERS

MEMBER NUMBER (IN ORDER)	MEMBER END SECTION ID LEFT	RIGHT	MEMBER DESCRIPTION LENGTH (ft)	-->TYPE<-- 0=PRISMATIC	PARAMETERS FOR NONPRISMATIC MEMBER S(0)	S(1)
1	1	1	57.40			
2	2	2	37.40			
3	1	1	108.00			
4	2	2	37.40			
5	1	1	57.40			

NOTE: [1] maximum allowable member number is 70.

BRIDGE ENGINEERING SOFTWARE CENTER
DEPARTMENT OF CIVIL ENGINEERING
UNIVERSITY OF MARYLAND

MERLIN V 6.4
COMPOSITE
LFD -- 1995
ANALYSIS

TABLE 1.2.5.4=MOMENT SUMMARY FOR COMPOSITE CONSTRUCTION (FACTORED)

			DEAD LOAD		LL+I:N= 8.0 LOAD TYPE=H,HS,L..				TOTAL MAXIMUM (L F D)	
SP NO	IN NO	D FROM L SUPT (ft)	NON COMP. N=Infin. (k-ft)	COMP. N=24.0 (k-ft)	MAXIMUM POSITIVE (k-ft)	GOVERN LOAD TYPE	MAXIMUM NEGATIVE (k-ft)	GOVERN LOAD TYPE	POSITIVE (k-ft)	NEGATIVE (k-ft)
1	0	0.00	0.0	0.0	0.0	L	0.0	HS	0.0	0.0
1	1	7.44	323.4	102.5	890.7	HS	-77.4	HS	1316.7	348.5
1	2	14.88	549.6	176.0	1504.7	HS	-154.8	HS	2230.3	570.7
1	3	22.32	678.4	220.5	1853.8	HS	-232.3	HS	2752.7	666.7
1	4	29.76	710.0	236.0	2021.6	HS	-309.7	HS	2967.6	636.3
1	5	37.20	644.3	222.5	1999.3	HS	-387.1	HS	2866.0	479.7
1	6	44.64	481.2	180.0	1830.5	HS	-464.5	HS	2491.7	196.7
1	7	52.08	220.9	108.5	1470.8	HS	-541.9	HS	1800.3	-212.5
1	8	59.52	-136.7	8.0	945.9	HS	-619.4	HS	817.2	-748.0
1	9	66.96	-591.6	-121.5	278.4	L	-722.5	HS	-434.7	-1435.5
1	10	74.40	-1143.8	-280.0	146.5	HS	-1240.9	L	-1277.2	-2664.7
2	0	0.00	-1143.8	-280.0	146.5	HS	-1240.9	L	-1277.2	-2664.7
2	1	7.44	-656.6	-139.7	280.9	HS	-759.9	HS	-515.5	-1556.2
2	2	14.88	-266.9	-28.4	894.6	HS	-656.2	HS	599.3	-951.5
2	3	22.32	25.5	53.8	1345.5	HS	-552.5	HS	1424.8	-473.2
2	4	29.76	220.5	107.1	1630.8	HS	-473.9	L	1958.4	-146.3
2	5	37.20	318.0	131.4	1702.7	HS	-458.9	L	2152.0	-9.6
2	6	44.64	318.2	126.6	1652.7	HS	-444.0	L	2097.5	0.7
2	7	52.08	220.9	92.9	1393.4	HS	-433.8	HS	1707.2	-120.0
2	8	59.52	26.3	30.2	959.9	HS	-516.7	HS	1016.4	-460.3
2	9	66.96	-265.8	-61.6	342.4	HS	-620.4	HS	15.1	-947.7
2	10	74.40	-655.2	-182.3	211.5	L	-1027.4	L	-626.0	-1864.9
3	0	0.00	-655.2	-182.3	211.5	L	-1027.4	L	-626.0	-1864.9
3	1	7.44	-265.8	-61.6	342.4	HS	-620.4	HS	15.1	-947.7
3	2	14.88	26.3	30.2	959.9	HS	-516.7	HS	1016.4	-460.3
3	3	22.32	220.9	92.9	1393.4	HS	-433.8	HS	1707.2	-120.0
3	4	29.76	318.2	126.6	1652.7	HS	-444.0	L	2097.5	0.7
3	5	37.20	318.0	131.4	1702.7	HS	-458.9	L	2152.0	-9.6
3	6	44.64	220.4	107.1	1630.8	HS	-473.9	L	1958.4	-146.3
3	7	52.08	25.5	53.8	1345.5	HS	-552.5	HS	1424.8	-473.2
3	8	59.52	-266.9	-28.4	894.6	HS	-656.2	HS	599.3	-951.5
3	9	66.96	-656.6	-139.7	280.9	HS	-759.9	HS	-515.5	-1556.2
3	10	74.40	-1143.8	-280.0	146.5	HS	-1240.9	L	-1277.2	-2664.7
4	0	0.00	-1143.8	-280.0	146.5	HS	-1240.9	L	-1277.2	-2664.7
4	1	7.44	-591.6	-121.5	278.4	L	-722.5	HS	-434.7	-1435.5
4	2	14.88	-136.7	8.0	945.9	HS	-619.4	HS	817.2	-748.0
4	3	22.32	220.9	108.5	1470.8	HS	-541.9	HS	1800.3	-212.5
4	4	29.76	481.2	180.0	1830.5	HS	-464.5	HS	2491.7	196.7
4	5	37.20	644.3	222.5	1999.3	HS	-387.1	HS	2866.0	479.7
4	6	44.64	710.0	236.0	2021.6	HS	-309.7	HS	2967.6	636.3
4	7	52.08	678.4	220.5	1853.8	HS	-232.3	HS	2752.7	666.7
4	8	59.52	549.6	176.0	1504.7	HS	-154.8	HS	2230.3	570.7
4	9	66.96	323.4	102.5	890.7	HS	-77.4	HS	1316.6	348.5
4	10	74.40	0.0	0.0	0.0	L	0.0	L	0.0	0.0

Figure 3-2 (continued)

BRIDGE ENGINEERING SOFTWARE CENTER
DEPARTMENT OF CIVIL ENGINEERING
UNIVERSITY OF MARYLAND

MERLIN V 6.4
COMPOSITE
LFD -- 1995
ANALYSIS

TABLE 1.2.6.4=SHEAR SUMMARY FOR COMPOSITE CONSTRUCTION (FACTORED)

SP NO	IN NO	D FROM L SUPT (ft)	TOTAL DEAD LOADS (k)	TOTAL L+I , LOAD TYPE = HS,L,MIL.. POSITIVE (k)	GOVERN. LOAD TYPE	NEGATIVE (k)	GOVERN. LOAD TYPE	L F D MAX. SHEAR (kips) POS.	NEG.
1	0	0.00	65.7	149.2	HS	-10.2	HS	214.9	55.6
1	1	7.44	48.8	120.7	HS	-13.2	L	169.5	35.5
1	2	14.88	31.8	102.8	HS	-22.4	L	134.5	9.4
1	3	22.32	14.8	85.1	HS	-36.4	HS	99.9	-21.6
1	4	29.76	-2.2	68.0	HS	-54.3	HS	65.8	-56.4
1	5	37.20	-19.1	51.2	HS	-73.3	HS	32.0	-92.5
1	6	44.64	-36.1	35.3	HS	-91.5	HS	-0.8	-127.6
1	7	52.08	-53.1	21.3	HS	-107.9	HS	-31.8	-161.0
1	8	59.52	-70.1	10.6	L	-123.3	HS	-59.5	-193.3
1	9	66.96	-87.0	4.9	L	-136.9	HS	-82.2	-223.9
1	10	74.40	-104.0	1.8	L	-158.8	HS	-102.2	-262.8
2	0	0.00	92.8	156.0	HS	-10.9	HS	248.8	81.9
2	1	7.44	75.8	131.9	HS	-10.9	HS	207.8	64.9
2	2	14.88	58.8	116.0	HS	-17.0	L	174.9	41.8
2	3	22.32	41.9	98.7	HS	-25.8	L	140.6	16.1
2	4	29.76	24.9	80.7	HS	-41.7	HS	105.5	-16.8
2	5	37.20	7.9	61.9	HS	-59.6	HS	69.8	-51.8
2	6	44.64	-9.1	43.6	HS	-78.2	HS	34.5	-87.3
2	7	52.08	-26.1	29.6	L	-96.3	HS	3.5	-122.4
2	8	59.52	-43.1	20.5	L	-114.0	HS	-22.6	-157.0
2	9	66.96	-60.1	13.9	HS	-130.4	HS	-46.2	-190.5
2	10	74.40	-77.1	13.9	L	-155.1	HS	-63.1	-232.2
3	0	0.00	77.1	155.1	HS	-13.9	L	232.2	63.1
3	1	7.44	60.1	130.4	HS	-13.9	HS	190.5	46.2
3	2	14.88	43.1	114.0	HS	-20.5	L	157.0	22.6
3	3	22.32	26.1	96.3	HS	-29.6	L	122.4	-3.5
3	4	29.76	9.1	78.2	HS	-43.6	HS	87.3	-34.5
3	5	37.20	-7.9	59.6	HS	-61.9	HS	51.8	-69.8
3	6	44.64	-24.9	41.7	HS	-80.7	HS	16.8	-105.5
3	7	52.08	-41.9	25.8	L	-98.7	HS	-16.1	-140.6
3	8	59.52	-58.8	17.0	L	-116.0	HS	-41.8	-174.9
3	9	66.96	-75.8	10.9	HS	-131.9	HS	-64.9	-207.8
3	10	74.40	-92.8	10.9	HS	-156.0	HS	-81.9	-248.8
4	0	0.00	104.0	158.8	HS	-1.8	L	262.8	102.2
4	1	7.44	87.0	136.9	HS	-4.9	L	223.9	82.2
4	2	14.88	70.1	123.3	HS	-10.6	L	193.3	59.5
4	3	22.32	53.1	107.9	HS	-21.3	HS	161.0	31.8
4	4	29.76	36.1	91.5	HS	-35.3	HS	127.6	0.8
4	5	37.20	19.1	73.3	HS	-51.2	HS	92.5	-32.0
4	6	44.64	2.2	54.2	HS	-68.0	HS	56.4	-65.8
4	7	52.08	-14.8	36.4	HS	-85.1	HS	21.6	-99.9
4	8	59.52	-31.8	22.4	L	-102.8	HS	-9.4	-134.5
4	9	66.96	-48.8	13.2	L	-120.7	HS	-35.5	-169.5
4	10	74.40	-65.7	10.2	HS	-149.2	HS	-55.6	-214.9

BRIDGE ENGINEERING SOFTWARE CENTER
DEPARTMENT OF CIVIL ENGINEERING
UNIVERSITY OF MARYLAND

MERLIN V 6.4
COMPOSITE
LFD -- 1995
ANALYSIS

TABLE 1.2.9.5=FACTORED TOTAL(DL+LL+I) STRESS SUMMARY (FACTORED)

SP NO	IN NO	D FROM L SUPT (ft)	YIELD STRESS Fy (ksi)	TOTAL POSITIVE ,(ksi) CONCR. TOP	STEEL BEAM TOP	STEEL BEAM BOT	TOTAL NEGATIVE ,(ksi) CONCR. TOP	STEEL BEAM TOP	STEEL BEAM BOT
1	0	0.00	50.	0.00	0.00	0.00	0.00	0.00	0.00
1	1	7.44	50.	-0.41	-7.48	18.01	-0.03	-6.25	5.85
1	2	14.88	50.	-0.70	-12.70	30.52	-0.05	-10.60	9.67
1	3	22.32	50.	-0.86	-15.69	37.68	-0.07	-13.05	11.46
1	4	29.76	50.	-0.94	-16.55	40.53	-0.07	-13.59	11.23
1	5	37.20	50.	-0.92	-15.26	38.96	-0.07	-12.24	8.97
1	6	44.64	50.	-0.84	-11.89	33.52	-0.05	-8.98	4.68
1	7	52.08	50.	-0.66	-6.37	23.66	-0.03	-3.82	-1.63
1	8	59.52	50.	0.00	-10.97	10.97	0.00	10.04	-10.04
1	9	66.96	50.	0.00	5.83	-5.83	0.00	19.27	-19.27
1	10	74.40	50.	0.00	17.14	-17.14	0.00	35.76	-35.76
2	0	0.00	50.	0.00	17.14	-17.14	0.00	35.76	-35.76
2	1	7.44	50.	0.00	6.92	-6.92	0.00	20.89	-20.89
2	2	14.88	50.	0.00	-8.04	8.04	0.00	12.77	-12.77
2	3	22.32	50.	-0.59	-2.40	18.06	-0.02	0.00	-5.79
2	4	29.76	50.	-0.73	-6.56	25.64	-0.03	-3.89	-0.81
2	5	37.20	50.	-0.77	-8.53	28.51	-0.04	-5.79	1.35
2	6	44.64	50.	-0.75	-8.45	27.82	-0.04	-5.79	1.48
2	7	52.08	50.	-0.62	-6.20	22.48	-0.03	-3.89	-0.49
2	8	59.52	50.	-0.42	-1.83	12.91	-0.01	0.04	-5.65
2	9	66.96	50.	0.00	-0.28	0.25	0.00	17.31	-15.96
2	10	74.40	50.	0.00	11.43	-10.54	0.00	34.05	-31.41
3	0	0.00	50.	0.00	11.43	-10.54	0.00	34.05	-31.41
3	1	7.44	50.	0.00	-0.28	0.25	0.00	17.31	-15.96
3	2	14.88	50.	-0.42	-1.83	12.91	-0.01	0.04	-5.65
3	3	22.32	50.	-0.62	-6.20	22.48	-0.03	-3.89	-0.49
3	4	29.76	50.	-0.75	-8.45	27.82	-0.04	-5.79	1.48
3	5	37.20	50.	-0.77	-8.53	28.51	-0.04	-5.79	1.35
3	6	44.64	50.	-0.73	-6.56	25.64	-0.03	-3.89	-0.81
3	7	52.08	50.	-0.59	-2.40	18.06	-0.02	0.00	-5.79
3	8	59.52	50.	0.00	-8.04	8.04	0.00	12.77	-12.77
3	9	66.96	50.	0.00	6.92	-6.92	0.00	20.89	-20.89
3	10	74.40	50.	0.00	17.14	-17.14	0.00	35.76	-35.76
4	0	0.00	50.	0.00	17.14	-17.14	0.00	35.76	-35.76
4	1	7.44	50.	0.00	5.83	-5.83	0.00	19.27	-19.27
4	2	14.88	50.	0.00	-10.97	10.97	0.00	10.04	-10.04
4	3	22.32	50.	-0.66	-6.37	23.66	-0.03	-3.82	-1.63
4	4	29.76	50.	-0.84	-11.89	33.52	-0.05	-8.98	4.68
4	5	37.20	50.	-0.92	-15.26	38.96	-0.07	-12.24	8.97
4	6	44.64	50.	-0.94	-16.55	40.53	-0.07	-13.59	11.23
4	7	52.08	50.	-0.86	-15.69	37.68	-0.07	-13.05	11.46
4	8	59.52	50.	-0.70	-12.70	30.52	-0.05	-10.60	9.67
4	9	66.96	50.	-0.41	-7.48	18.01	-0.03	-6.25	5.85
4	10	74.40	50.	0.00	0.00	0.00	0.00	0.00	0.00

Figure 3-2 (continued)

PRECAST PRESTRESS BRIDGE DESIGN

Example No. 1 – Single Span I-Girder Bridge

1.0 DESIGN INPUT

1.1 Design Conditions

(a) Simple span of 75' long and 30' wide
(b) HS20 live load – 2 lanes
(c) 4 PCI Standard I-Girders at 8' spacing
(d) Composite Construction with 7½" deck slab
(e) 2" future wearing surface

1.2 Materials

(a) Precast concrete: normal weight
fc' = 5000 psi
fci' = 4000 psi

(b) Cast-in-place concrete: normal weight
fc' = 4000 psi

(c) Prestressing steel: ½" diameter 270 ksi stress-relieved strand
strand area = 0.153 in^2
Es = 28×10^6 psi

(d) Reinforcing bars: Fy = 60,000 psi

1.3 Composite Section properties

(a) Modulus ratio:
Cast-in-place $E_c = 150^{1.5} \times 33\sqrt{fi'} = 3.834 \times 10^6$ psi
Precast $E_c = 4.286 \times 10^6$ psi
n = 3.6/4.03 = 0.89

(b) Effective flange width:
1/4 span = $(75 \times 12)/4$ = 225"
Distance Center-to-Center of girders = 96" (governs)
$12 \times$ slab thickness + flange width = $12 \times 7.5 + 20 = 110$"

Figure 3-3 Example composite prestressed-concrete bridge design using MERLIN-PBEAM software (reproduced courtesy BEST Center).

Preliminary Girder Selection

Using span table in "Short Span Bridges Manual" as a guide, both Type III and Type IV I-Girders are possibilities at a 75 ft span. Considering the wide girder spacing and provision for future wearing surface, select a Type IV section. Bridge layout and girder section properties are as follows.

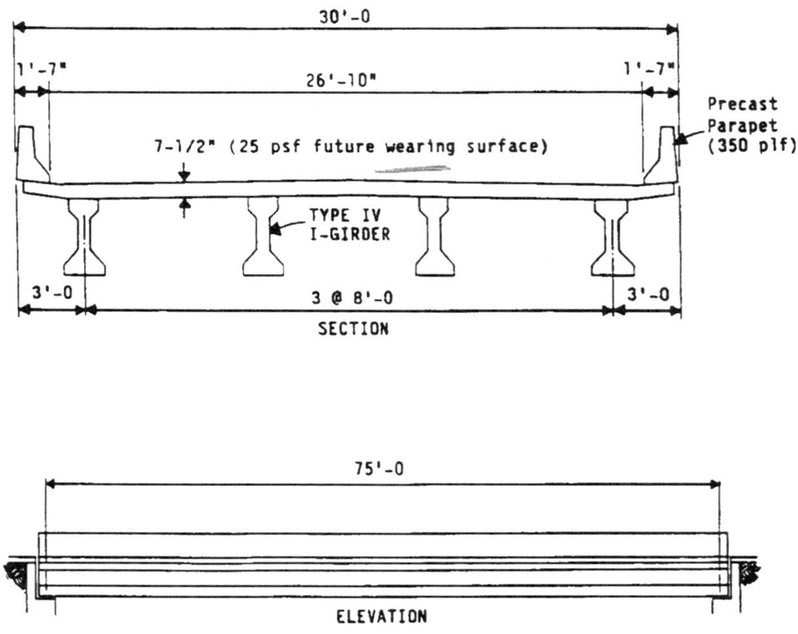

Figure 3-3 (continued)

INPUT DATA

```
101 PBEAM EXAMPLE 1 (PCI EXAMPLE 3)
102 1-SPAN, 4GIRDER BRIDGE
103 000001
104 1   0   1    0.0        0.0       0
105  4  1     26.83        3.0 0.0
201      70.     150.    5000.    4000.      150.    4000.
206   1       0.5    0.153     270.      230.    28000. 0.0
301      7.5    10 0.0
302  1 11
304   1  10   1
306   1      96.       7.5      96.      7.5        0.       0.    0.343
311   8.0 0.0
312   75. 0.0
401  20    1   1 0
402  0.727 0.0
501  1  0  4 0.0
511  1  1 12  1   1      3.  11     3. 0   0.0    0  0.0   0  0.0   0  0.0
511  1  2 12  1   1      5.  11     5. 0
999
```

OUTPUT

```
                CONTINUOUS PRESTRESS BEAM ANALYSIS SYSTEM

PBEAM EXAMPLE 1 (PCI EXAMPLE 3)
1-SPAN, 4GIRDER BRIDGE

TABLE 1.0 DEFINITION OF GENERAL INFORMATION
*******************************************

OUTPUT LEVEL...........................=  1   (1-Basic, 2-Detail)
BOND TYPE..............................=  0   (0-Bonded, 1-Unboded)
ANALYSIS/DESIGN/RATING CONTROL.........=  1   (1-Analysis, 2-Design, 3-Rating)
BETA FACTOR............................= 1.67 (default = 5/3)
GAMMA FACTOR...........................= 1.30 (default = 1.3)
SI/ENGLISH SYSTEM CONTROL..............=  0   (0-English, 1-SI)


TABLE 1.1 DEFINITION OF GENERAL PARAMETERS
******************************************

         BEAM       WIDTH              EDGE OF
       POSITION    BETWEEN   OVERHANG  SLAB TO
NUMBER --------    CURBS OR   WIDTH     CURB
  OF    1=INT.     BARRIER   --------  -------
GIRDERS 2=EXT.      (Ft)       (Ft)     (Ft)
----------------------------------------------
   4      1        26.00      3.00      .00


TABLE 1.2 DEFINITION OF MATERIAL PROPERTIES
*******************************************

RELATIVE HUMIDITY (%)..........................=   70.00
PRESTREEESED GIRDER DENSITY (PCF)..............=  150.00
PRESTRESSED GIRDER ULTIMATE STRENGTH (PSI).....= 5000.00
PRESTRESSED GIRDER INITIAL STRENGTH (PSI)......= 4000.00
SLAB DENSITY (PCF).............................=  150.00
SLAB ULTIMATE STRENGTH (PSI)...................= 4000.00
TABLE 1.3 DEFINITION OF GIRDER PROPERTIES AT ULTIMATE STRENGTH
**************************************************************

COMPRESSION ALLOWABLE STRESSES (PSI)...........=-2000.00
TENSION ALLOWABLE STRESSES (PSI)...............=  424.26
CRACK ALLOWABLE STRESSES (PSI).................=  530.33
MODULUS OF ELASTICITY (KSI)....................= 4286.83
```

INPUT DATA

```
101 PBEAM EXAMPLE 1 (PCI EXAMPLE 3)
102 1-SPAN, 4GIRDER BRIDGE
103 000001
104 1    0    1     0.0        0.0        0
105  4   1      26.83        3.0 0.0
201       70.     150.    5000.    4000.      150.    4000.
206   1       0.5    0.153     270.      230.   28000. 0.0
301       7.5   10 0.0
302  1 11
304   1  10   1
306   1       96.       7.5      96.      7.5       0.       0.    0.343
311   8.0 0.0
312   75. 0.0
401  20    1    1 0
402  0.727 0.0
501  1  0  4 0.0
511  1  1 12  1   1     3.  11     3. 0   0.0    0   0.0    0   0.0    0   0.0
511  1  2 12  1   1     5.  11     5. 0
999
```

OUTPUT

```
                 CONTINUOUS PRESTRESS BEAM ANALYSIS SYSTEM

 PBEAM EXAMPLE 1 (PCI EXAMPLE 3)
 1-SPAN, 4GIRDER BRIDGE

 TABLE 1.0 DEFINITION OF GENERAL INFORMATION
 *****************************************

 OUTPUT LEVEL...........................=  1   (1-Basic, 2-Detail)
 BOND TYPE..............................=  0   (0-Bonded, 1-Unboded)
 ANALYSIS/DESIGN/RATING CONTROL.........=  1   (1-Analysis, 2-Design, 3-Rating)
 BETA FACTOR............................= 1.67 (default = 5/3)
 GAMMA FACTOR...........................= 1.30 (default = 1.3)
 SI/ENGLISH SYSTEM CONTROL..............=  0   (0-English, 1-SI)


 TABLE 1.1 DEFINITION OF GENERAL PARAMETERS
 *****************************************

            BEAM       WIDTH               EDGE OF
          POSITION    BETWEEN   OVERHANG  SLAB TO
 NUMBER   --------    CURBS OR   WIDTH     CURB
   OF      1=INT.     BARRIER   --------  -------
 GIRDERS   2=EXT.      (Ft)      (Ft)      (Ft)
 ---------------------------------------------------
    4       1         26.00      3.00      .00


 TABLE 1.2 DEFINITION OF MATERIAL PROPERTIES
 *****************************************

 RELATIVE HUMIDITY (%)............................=   70.00
 PRESTREEESED GIRDER DENSITY (PCF)................=  150.00
 PRESTRESSED GIRDER ULTIMATE STRENGTH (PSI).......= 5000.00
 PRESTRESSED GIRDER INITIAL STRENGTH (PSI)........= 4000.00
 SLAB DENSITY (PCF)...............................=  150.00
 SLAB ULTIMATE STRENGTH (PSI).....................= 4000.00
 TABLE 1.3 DEFINITION OF GIRDER PROPERTIES AT ULTIMATE STRENGTH
 **********************************************************

 COMPRESSION ALLOWABLE STRESSES (PSI)...........=-2000.00
 TENSION ALLOWABLE STRESSES (PSI)...............=  424.26
 CRACK ALLOWABLE STRESSES (PSI).................=  530.33
 MODULUS OF ELASTICITY (KSI)....................= 4286.83
```

Figure 3-3 (continued)

1 CONTINUOUS PRESTRESS BEAM ANALYSIS SYSTEM

PBEAM EXAMPLE 1 (PCI EXAMPLE 3)

TABLE 4.1 SUMMARY OF DEAD LOAD MOMENTS (KIP-FT)
****

SPAN NO	INCR. NO	DIST. (FEET)	GIRDER (1)	SLAB (2)	ADD.SLAB (3)	SDL (4)	ADD.SDL (5)
1	0	.00	.00	.00	.00	.00	.00
1	1	7.50	208.04	189.85	.00	86.82	.00
1	2	15.00	369.86	337.51	.00	154.35	.00
1	3	22.50	485.44	442.98	.00	202.59	.00
1	4	30.00	554.80	506.26	.00	231.53	.00
1	5	37.50	577.92	527.35	.00	241.18	.00
1	6	45.00	554.80	506.26	.00	231.53	.00
1	7	52.50	485.45	442.98	.00	202.59	.00
1	8	60.00	369.86	337.51	.00	154.35	.00
1	9	67.50	208.05	189.85	.00	86.82	.00
1	10	75.00	.00	.00	.00	.00	.00

1 CONTINUOUS PRESTRESS BEAM ANALYSIS SYSTEM

PBEAM EXAMPLE 1 (PCI EXAMPLE 3)

TABLE 4.2 SUMMARY OF LIVE LOAD MOMENTS (KIP-FT)
****

SPAN NO	INCR. NO	DIST. (FEET)	ADD.LL (6)	+(LL+I) (7)	-(LL+I) (8)	PS-INT. (9)	PS-ULT. (10)	TOT.MAX (11)	TOT.MIN (12)
1	0	.00	.00	.00	.00	-1357.79	-1096.23	.00	.00
1	1	7.50	.00	380.59	.00	-1365.58	-1120.30	1454.73	630.13
1	2	15.00	.00	663.03	.00	-1372.42	-1140.58	2556.80	1120.23
1	3	22.50	.00	847.33	.00	-1376.78	-1153.99	3306.18	1470.31
1	4	30.00	.00	953.84	.00	-1379.49	-1162.24	3747.00	1680.36
1	5	37.50	.00	972.39	.00	-1380.40	-1165.00	3857.22	1750.38
1	6	45.00	.00	953.86	.00	-1379.49	-1162.25	3747.07	1680.38
1	7	52.50	.00	847.34	.00	-1376.78	-1153.99	3306.24	1470.33
1	8	60.00	.00	663.04	.00	-1372.42	-1140.58	2556.83	1120.24
1	9	67.50	.00	380.59	.00	-1365.58	-1120.30	1454.75	630.14
1	10	75.00	.00	.00	.00	-1357.79	-1096.23	.00	.00

NOTE: FACTORED TOTAL MAXIMUM AND MINIMUM WITHOUT PRESTRESSING EFFECTS

1 CONTINUOUS PRESTRESS BEAM ANALYSIS SYSTEM

PBEAM EXAMPLE 1 (PCI EXAMPLE 3)

TABLE 4.4 SUMMARY OF LIVE LOAD SHEARS (KIP)
****

SPAN NO	INCR. NO	DIST. (FEET)	ADD.LL (6)	+(LL+I) (7)	-(LL+I) (8)	PS-INT. (9)	PS-ULT. (10)	TOT.MAX (11)	TOT.MIN (12)
1	0	.00	.00	57.29	.00	.00	.00	217.47	93.35
1	1	7.50	.00	51.14	-3.02	.00	.00	185.49	68.13
1	2	15.00	.00	44.92	-6.45	.00	.00	153.34	42.03
1	3	22.50	.00	38.61	-12.50	.00	.00	121.00	10.26
1	4	30.00	.00	32.21	-18.75	.00	.00	88.47	-21.95
1	5	37.50	.00	25.56	-25.55	.00	.00	55.38	-55.36
1	6	45.00	.00	18.75	-32.21	.00	.00	21.96	-88.46
1	7	52.50	.00	12.50	-38.61	.00	.00	-10.26	-121.01
1	8	60.00	.00	6.45	-44.92	.00	.00	-42.03	-153.34
1	9	67.50	.00	3.02	-51.14	.00	.00	-68.13	-185.49
1	10	75.00	.00	.00	-57.29	.00	.00	-93.35	-217.48

NOTE: FACTORED TOTAL MAXIMUM AND MINIMUM WITHOUT PRESTRESSING EFFECTS

CONTINUOUS PRESTRESS BEAM ANALYSIS SYSTEM

PBEAM EXAMPLE 1 (PCI EXAMPLE 3)
TABLE 6.11 SUMMARY OF ULTIMATE MOMENT (KIP-FT)
****

SPAN NO	MEM NO	DIST. (FEET)	TOTAL MOMENT (1)	MAX ULT MOMENT (2)	CRACK MOMENT (3)	CHECK (4)	TOTAL MOMENT (5)	MIN ULT MOMENT (6)	CRACK MOMENT (7)	CHECK (8)
1	1	7.50	.0	4527.6	871.5		.0	.0		
			1454.7	4527.6	4050.7		.0	.0		
1	2	15.00	1454.7	4527.6	4050.7		.0	.0		
			2556.8	4527.6	3905.6		.0	.0		
1	3	22.50	2556.8	4527.6	3905.6		.0	.0		
			3306.2	4527.6	3798.6		.0	.0		
1	4	30.00	3306.2	4527.6	3798.6		.0	.0		
			3747.0	4527.6	3735.0		.0	.0		
1	5	37.50	3747.0	4527.6	3735.0		.0	.0		
			3857.2	4527.6	3713.8		.0	.0		
1	6	45.00	3857.2	4527.6	3713.8		.0	.0		
			3747.1	4527.6	3735.0		.0	.0		
1	7	52.50	3747.1	4527.6	3735.0		.0	.0		
			3306.2	4527.6	3798.6		.0	.0		
1	8	60.00	3306.2	4527.6	3798.6		.0	.0		
			2556.8	4527.6	3905.6		.0	.0		
1	9	67.50	2556.8	4527.6	3905.6		.0	.0		
			1454.7	4527.6	4050.7		.0	.0		
1	10	75.00	1454.7	4527.6	4050.7		.0	.0		
			.0	4527.6	871.5		.0	.0		NOT OK

1 CONTINUOUS PRESTRESS BEAM ANALYSIS SYSTEM

PBEAM EXAMPLE 1 (PCI EXAMPLE 3)

TABLE 6.12 SUMMARY OF WEB SHEAR REINFORCEMENT
****

SPAN NO	INCR. NO	DIST. (feet)	Vu (kip)	Vc (kip)	# 3 Bar Vs (kip)	# 3 Bar Spacing (inch)	# 3 Bar Note	# 4 Bar Vs (kip)	# 4 Bar Spacing (inch)	# 4 Bar Note
1	0	.00	217.47	174.94	69.00	11.00	2	69.00	20.00	2
1	1	7.50	185.49	176.29	36.14	21.00	2	57.50	24.00	2
				176.29	36.14	21.00	2	57.50	24.00	2
1	2	15.00	153.34	177.42	31.63	24.00	2	57.50	24.00	2
				177.42	31.63	24.00	2	57.50	24.00	2
1	3	22.50	121.00	131.07	31.63	24.00	2	57.50	24.00	2
				131.07	31.63	24.00	2	57.50	24.00	2
1	4	30.00	88.47	55.30	42.17	18.00	2	57.50	24.00	2
				55.30	42.17	18.00	2	57.50	24.00	2
1	5	37.50	55.38	55.30	31.63	24.00	2	57.50	24.00	2
				55.30	31.63	24.00	2	57.50	24.00	2
1	6	45.00	88.46	82.69	31.63	24.00	2	57.50	24.00	2
				82.69	31.63	24.00	2	57.50	24.00	2
1	7	52.50	121.01	113.49	31.63	24.00	2	57.50	24.00	2
				113.49	31.63	24.00	2	57.50	24.00	2
1	8	60.00	153.34	177.42	31.63	24.00	2	57.50	24.00	2
				177.42	31.63	24.00	2	57.50	24.00	2
1	9	67.50	185.49	176.29	36.14	21.00	2	57.50	24.00	2
				176.29	36.14	21.00	2	57.50	24.00	2
1	10	75.00	217.48	78.46	151.80	5.00	2	153.33	9.00	2

Figure 3-3 (continued)

CONTINUOUS PRESTRESS BEAM ANALYSIS SYSTEM

PBEAM EXAMPLE 1 (PCI EXAMPLE 3)

TABLE 6.8A SUMMARY OF BOTTOM STRESSES AT SERVICE LOAD (KSI)
****

SPAN NO	INCR. NO	DIST. (FEET)	PRESTRESS (1)	NON-COMP. DEAD LD (2)	COMPOSITE DEAD LD (3)	+(LL+I) (4)	-(LL+I) (5)
1	0	.00	-2.052	.000	.000	.000	.000
1	1	7.50	-2.097	.453	.063	.278	.000
			-2.097	.453	.063	.278	.000
1	2	15.00	-2.135	.805	.113	.484	.000
			-2.135	.805	.113	.484	.000
1	3	22.50	-2.160	1.057	.148	.619	.000
			-2.160	1.057	.148	.619	.000
1	4	30.00	-2.176	1.208	.169	.697	.000
			-2.176	1.208	.169	.697	.000
1	5	37.50	-2.181	1.258	.176	.710	.000
			-2.181	1.258	.176	.710	.000
1	6	45.00	-2.176	1.208	.169	.697	.000
			-2.176	1.208	.169	.697	.000
1	7	52.50	-2.160	1.057	.148	.619	.000
			-2.160	1.057	.148	.619	.000
1	8	60.00	-2.135	.805	.113	.484	.000
			-2.135	.805	.113	.484	.000
1	9	67.50	-2.097	.453	.063	.278	.000
			-2.097	.453	.063	.278	.000
1	10	75.00	-2.052	.000	.000	.000	.000

1 CONTINUOUS PRESTRESS BEAM ANALYSIS SYSTEM

PBEAM EXAMPLE 1 (PCI EXAMPLE 3)

TABLE 6.8B SUMMARY OF BOTTOM STRESSES AT SERVICE LOAD (KSI)
****

SPAN NO	INCR. NO	DIST. (FEET)	TOTAL WITH +(LL+I) (6)	ALLOW. (7)	CHECK (8)	TOTAL WITH -(LL+I) (9)	ALLOW. (10)	CHECK (11)
1	0	.00	-2.052	-2.000	NOT OK	-2.052	-2.000	NOT OK
1	1	7.50	-1.303	-2.000		-1.581	-2.000	
			-1.303	-2.000		-1.581	-2.000	
1	2	15.00	-.733	-2.000		-1.217	-2.000	
			-.733	-2.000		-1.217	-2.000	
1	3	22.50	-.337	-2.000		-.955	-2.000	
			-.337	-2.000		-.955	-2.000	
1	4	30.00	-.102	-2.000		-.799	-2.000	
			-.102	-2.000		-.799	-2.000	
1	5	37.50	-.036	-2.000		-.746	-2.000	
			-.036	-2.000		-.746	-2.000	
1	6	45.00	-.102	-2.000		-.799	-2.000	
			-.102	-2.000		-.799	-2.000	
1	7	52.50	-.337	-2.000		-.955	-2.000	
			-.337	-2.000		-.955	-2.000	
1	8	60.00	-.733	-2.000		-1.217	-2.000	
			-.733	-2.000		-1.217	-2.000	
1	9	67.50	-1.303	-2.000		-1.581	-2.000	
			-1.303	-2.000		-1.581	-2.000	
1	10	75.00	-2.052	-2.000	NOT OK	-2.052	-2.000	NOT OK

EXAMPLE DESCRIPTION

I. Job Title: <u>1-Span Truss Bridge</u>

II. General Information:

1.	Number of Spans	1
2.	Number of Panel Points	10
3.	Number of Truss Members	34
4.	Span Length	189.72 ft.

III. Loading Conditions:

1. Uniform Loads:

Floor Steel	0.80 K/f
Slab + W. S.	0.53 K/f
Railing and Curb	0.15 K/f

2. Bracing Loads: (see Input Data Card 0702)

3. AASHTO Live Load: HS-15

4. Group Loading (see Input Data Cards)

* Output in the manual contains Group Loading III for Table 17.3.

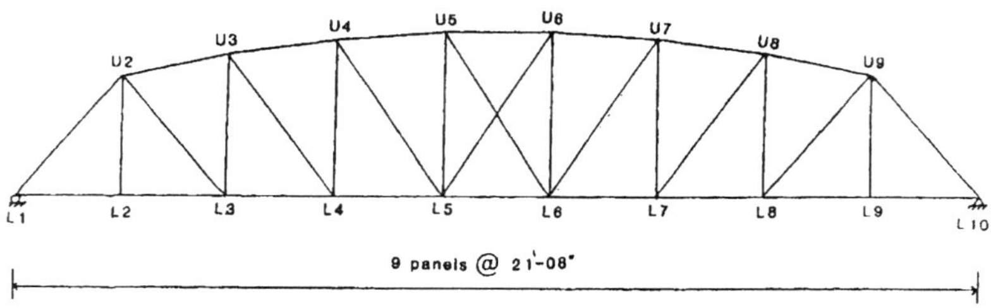

Truss Layout

Figure 3-3 (continued)

INPUT DATA

```
0101 EXAMPLE NO. (1)
0101 STATICAL INDETERMINATE TRUSS
0102 GROUP LOADING ANALYSIS
0103
0103       3        4     29000.          0.075           0
0104
0104 189.72                                           23.0              26.25
1.2
0301
0301HS 15 .55 .0 HS 15 .75 .0      0
0401
0401   1   0.0      0.0     1
0401   2  21.08     0.0           21.08  23.0
0401   3  42.16     0.0           42.16  27.5
0401   4  63.24     0.0           63.24  30.5
0401   5  84.32     0.0           84.32  31.5
0401   6 105.40     0.0          105.40  31.5
0401   7 126.48     0.0          126.48  30.5
0401   8 147.56     0.0          147.56  27.5
0401   9 168.64     0.0          168.64  23.0
0401  10 189.72     0.0   1 1
0501
0501   1L  1L  2      1          .469 12.   19.5    16.57 36.   1  1.0 25.0
0501   2L  2L  3                 .469 12.   19.5    16.57 36.   0  1.0 25.0
0501   3L  3L  4                 .721 12.   30.     25.5  36.   0  1.0 25.0
0501   4L  4L  5                 .829 12.   34.5    29.32 36.   0  1.0 25.0
0501   5L  5L  6                 .685 12.   28.5    24.22 36.   0  1.0 25.0
0501   6L  6L  7                 .829 12.   34.5    29.32 36.   0  1.0 25.0
0501   7L  7L  8                 .721 12.   30.     25.5  36.   0  1.0 25.0
0501   8L  8L  9                 .469 12.   19.5    16.57 36.   0  1.0 25.0
0501   9L  9L 10     10          .469 12.   19.5    16.57 36.   0  1.0 25.0
0501  10U  2U  3                3.277 15.   28.67   24.37 36.   1  1.0 20.0
0501  11U  3U  4                3.1   15.   32.35   27.5  36.   0  1.0 20.0
0501  12U  4U  5                2.97  15.   35.25   30.   36.   0  1.0 20.0
0501  13U  5U  6                2.97  15.   35.25   30.   36.   0  1.0 20.0
0501  14U  6U  7                2.97  15.   35.25   30.   36.   0  1.0 20.0
0501  15U  7U  8                3.1   15.   32.35   27.5  36.   0  1.0 20.0
0501  16U  8U  9                3.277 15.   28.67   24.37 36.   0  1.0 20.0
0501  17L  2U  2      2 23.     1.54  12.   10.3     8.75 36.   1  1.0  0.0
0501  18L  3U  3      3 27.5    2.48  12.   15.6    13.26 36.   0  1.0  0.0
0501  19L  4U  4      4 30.5    1.54  12.   10.3     8.75 36.   0  0.5  0.0
0501  20L  5U  5      5 31.5    1.54  12.   10.3     8.75 36.   0  0.5  0.0
0501  21L  6U  6      6 31.5    1.54  12.   10.3     8.75 36.   0  0.5  0.0
0501  22L  7U  7      7 30.5    1.54  12.   10.3     8.75 36.   0  0.5  0.0
0501  23L  8U  8      8 27.5    2.48  12.   15.6    13.26 36.   0  1.0  0.0
0501  24L  9U  9      9 23.     1.54  12.   10.3     8.75 36.   0  1.0  0.0
0501  25L  1U  2                2.97  15.   35.25   30.   36.   1  1.0  0.0
0501  26L  3U  2                1.96  15.   14.7    12.5  36.   0  1.0  0.0
0501  27L  4U  3                1.54  15.   10.3     8.75 36.   0  1.0  0.0
0501  28L  5U  4                1.54  15.   10.3     8.75 36.   0  0.5  0.0
0501  29L  5U  6                1.54  15.   10.3     8.75 36.   0  0.5  0.0
0501  30L  6U  5                1.54  15.   10.3     8.75 36.   0  0.5  0.0
0501  31L  6U  7                1.54  15.   10.3     8.75 36.   0  0.5  0.0
0501  32L  7U  8                1.54  15.   10.3     8.75 36.   0  1.0  0.0
0501  33L  8U  9                1.96  15.   14.7    12.5  36.   0  1.0  0.0
0501  34L 10U  9                2.97  15.   35.25   30.   36.   0  1.0  0.0
0601
0601  1  .8      0.0  189.72
0602
0602  1  .53     0.0  189.72
0603
0603  1  .15     0.0  189.72
0604
0604  1  .1      0.0  189.72
0702
0702 2  2.3 3  2.3 4  2.3 5  2.3 6  2.3 7  2.3 8  2.3 9  2.3
```

TRUSS RATING AND ANALYSIS PROGRAM (TRAP)
GROUP LOADING ANALYSIS

TABLE 1.5 PANEL POINT DATA

PP NO	LOWER PANEL POINT X-COOR (FT)	LOWER PANEL POINT Y-COOR (FT)	LOWER PANEL POINT RE XY	UPPER PANEL POINT X-COOR (FT)	UPPER PANEL POINT Y-COOR (FT)	UPPER PANEL POINT RE XY	ADDITIONAL POINT A X-COOR (FT)	ADDITIONAL POINT A Y-COOR (FT)	ADDITIONAL POINT A RE XY	ADDITIONAL POINT B X-COOR (FT)	ADDITIONAL POINT B Y-COOR (FT)	ADDITIONAL POINT B RE XY
1	.00	.000	01	.00	.000	00	.00	.000	00	.00	.000	00
2	21.08	.000	00	21.08	23.000	00	.00	.000	00	.00	.000	00
3	42.16	.000	00	42.16	27.500	00	.00	.000	00	.00	.000	00
4	63.24	.000	00	63.24	30.500	00	.00	.000	00	.00	.000	00
5	84.32	.000	00	84.32	31.500	00	.00	.000	00	.00	.000	00
6	105.40	.000	00	105.40	31.500	00	.00	.000	00	.00	.000	00
7	126.48	.000	00	126.48	30.500	00	.00	.000	00	.00	.000	00
8	147.56	.000	00	147.56	27.500	00	.00	.000	00	.00	.000	00
9	168.64	.000	00	168.64	23.000	00	.00	.000	00	.00	.000	00
10	189.72	.000	11	.00	.000	00	.00	.000	00	.00	.000	00

TRUSS RATING AND ANALYSIS PROGRAM (TRAP)
GROUP LOADING ANALYSIS

TABLE 1.6 MEMBER DATA

MEM NO	LCTN FRM	LCTN TO	MEM REL	LOADED MEM SEQ NO	LOADED MEM DEK LCN (FT)	MIN R GYRATON (IN)	MEM DEPTH (IN)	MEMBER AREA GROSS (SQIN)	MEMBER AREA NET (SQIN)	YIELD STRSS (KSI)	INF LIN OPT	EFF LNG FAC	TEMP CHNG
1	L 1	L 2	0	1	.00	.47	12.00	19.5	16.6	36.0	1	1.00	25.0
2	L 2	L 3	0	0	.00	.47	12.00	19.5	16.6	36.0	0	1.00	25.0
3	L 3	L 4	0	0	.00	.72	12.00	30.0	25.5	36.0	0	1.00	25.0
4	L 4	L 5	0	0	.00	.83	12.00	34.5	29.3	36.0	0	1.00	25.0
5	L 5	L 6	0	0	.00	.69	12.00	28.5	24.2	36.0	0	1.00	25.0
6	L 6	L 7	0	0	.00	.83	12.00	34.5	29.3	36.0	0	1.00	25.0
7	L 7	L 8	0	0	.00	.72	12.00	30.0	25.5	36.0	0	1.00	25.0
8	L 8	L 9	0	0	.00	.47	12.00	19.5	16.6	36.0	0	1.00	25.0
9	L 9	L10	0	10	.00	.47	12.00	19.5	16.6	36.0	0	1.00	25.0
10	U 2	U 3	0	0	.00	3.28	15.00	28.7	24.4	36.0	1	1.00	20.0
11	U 3	U 4	0	0	.00	3.10	15.00	32.3	27.5	36.0	0	1.00	20.0
12	U 4	U 5	0	0	.00	2.97	15.00	35.3	30.0	36.0	0	1.00	20.0
13	U 5	U 6	0	0	.00	2.97	15.00	35.3	30.0	36.0	0	1.00	20.0
14	U 6	U 7	0	0	.00	2.97	15.00	35.3	30.0	36.0	0	1.00	20.0
15	U 7	U 8	0	0	.00	3.10	15.00	32.3	27.5	36.0	0	1.00	20.0
16	U 8	U 9	0	0	.00	3.28	15.00	28.7	24.4	36.0	0	1.00	20.0
17	L 2	U 2	0	2	23.00	1.54	12.00	10.3	8.8	36.0	1	1.00	.0
18	L 3	U 3	0	3	27.50	2.48	12.00	15.6	13.3	36.0	0	1.00	.0
19	L 4	U 4	0	4	30.50	1.54	12.00	10.3	8.8	36.0	0	.50	.0
20	L 5	U 5	0	5	31.50	1.54	12.00	10.3	8.8	36.0	0	.50	.0
21	L 6	U 6	0	6	31.50	1.54	12.00	10.3	8.8	36.0	0	.50	.0
22	L 7	U 7	0	7	30.50	1.54	12.00	10.3	8.8	36.0	0	.50	.0
23	L 8	U 8	0	8	27.50	2.48	12.00	15.6	13.3	36.0	0	1.00	.0
24	L 9	U 9	0	9	23.00	1.54	12.00	10.3	8.8	36.0	0	1.00	.0
25	L 1	U 2	0	0	.00	2.97	15.00	35.3	30.0	36.0	1	1.00	.0
26	L 3	U 2	0	0	.00	1.96	15.00	14.7	12.5	36.0	0	1.00	.0
27	L 4	U 3	0	0	.00	1.54	15.00	10.3	8.8	36.0	0	1.00	.0
28	L 5	U 4	0	0	.00	1.54	15.00	10.3	8.8	36.0	0	.50	.0
29	L 5	U 6	0	0	.00	1.54	15.00	10.3	8.8	36.0	0	.50	.0
30	L 6	U 5	0	0	.00	1.54	15.00	10.3	8.8	36.0	0	.50	.0
31	L 6	U 7	0	0	.00	1.54	15.00	10.3	8.8	36.0	0	.50	.0
32	L 7	U 8	0	0	.00	1.54	15.00	10.3	8.8	36.0	0	1.00	.0
33	L 8	U 9	0	0	.00	1.96	15.00	14.7	12.5	36.0	0	1.00	.0
34	L10	U 9	0	0	.00	2.97	15.00	35.3	30.0	36.0	0	1.00	.0

Figure 3-3 (continued)

```
           TRUSS RATING AND ANALYSIS PROGRAM (TRAP)
 GROUP LOADING ANALYSIS

TABLE 1.7  UNIFORM DEAD LOAD DATA
           ----------------------

UNIFORM LOADS DUE TO FLOOR STEEL
--------------------------------

NO LOAD     LOCATION       NO LOAD     LOCATION       NO LOAD     LOCATION
          --------------             --------------             --------------
           FROM     TO                FROM     TO                FROM     TO
   (KLF)   (FT)    (FT)       (KLF)   (FT)    (FT)       (KLF)   (FT)    (FT)

1  .8000      .00 189.72   0  .0000      .00     .00   0  .0000      .00     .00

UNIFORM LOADS DUE TO SLAB + W.S.
--------------------------------

NO LOAD     LOCATION       NO LOAD     LOCATION       NO LOAD     LOCATION
          --------------             --------------             --------------
           FROM     TO                FROM     TO                FROM     TO
   (KLF)   (FT)    (FT)       (KLF)   (FT)    (FT)       (KLF)   (FT)    (FT)

1  .5300      .00 189.72   0  .0000      .00     .00   0  .0000      .00     .00

UNIFORM LOADS DUE TO RAILING + CURB
-----------------------------------

NO LOAD     LOCATION       NO LOAD     LOCATION       NO LOAD     LOCATION
          --------------             --------------             --------------
           FROM     TO                FROM     TO                FROM     TO
   (KLF)   (FT)    (FT)       (KLF)   (FT)    (FT)       (KLF)   (FT)    (FT)

1  .1500      .00 189.72   0  .0000      .00     .00   0  .0000      .00     .00

UNIFORM LOADS DUE TO UTILITY + ACCESS
-------------------------------------

NO LOAD     LOCATION       NO LOAD     LOCATION       NO LOAD     LOCATION
          --------------             --------------             --------------
           FROM     TO                FROM     TO                FROM     TO
   (KLF)   (FT)    (FT)       (KLF)   (FT)    (FT)       (KLF)   (FT)    (FT)

1  .1000      .00 189.72   0  .0000      .00     .00   0  .0000      .00     .00
1
           TRUSS RATING AND ANALYSIS PROGRAM (TRAP)                 PAGE   6
 GROUP LOADING ANALYSIS

TABLE 1.8  BRACING DEAD LOAD DATA
           ---------------------

UPPER PANEL POINT BRACING DEAD LOAD
-----------------------------------

PP  LD   PP  LD   PP  LD   PP  LD   PP  LD   PP  LD   PP  LD   PP  LD   PP  LD   PP  LD
NO (K)   NO (K)   NO (K)   NO (K)   NO (K)   NO (K)   NO (K)   NO (K)   NO (K)   NO (K)

 2  2.3   3  2.3   4  2.3   5  2.3   6  2.3   7  2.3   8  2.3   9  2.3   0   .0   0   .0
```

1

TRUSS RATING AND ANALYSIS PROGRAM (TRAP)
GROUP LOADING ANALYSIS

TABLE 10.1 DL+LL+I FORCE SUMMARY FOR LOWER CHORD MEMBERS

AASHTO LOADING RESULTS

MEMBER	TOTAL DEAD LOAD (K)	MAXIMUM TENSION LL+I (K)	TYPE	MAXIMUM COMPRESSION LL+I (K)	TYPE	MAXIMUM MEMBER FORCE DL+LL+I (K)	TYPE	MINIMUM MEMBER FORCE DL+LL+I (K)	TYPE
L 1 L 2	161.00	58.94	L	.00		219.94	L	161.00	
L 2 L 3	161.00	58.94	L	.00		219.94	L	161.00	
L 3 L 4	236.59	86.26	L	.00		322.85	L	236.59	
L 4 L 5	274.93	100.00	L	.00		374.93	L	274.93	
L 5 L 6	293.56	103.56	L	.00		397.13	L	293.56	
L 6 L 7	274.93	100.00	L	.00		374.93	L	274.93	
L 7 L 8	236.59	86.26	L	.00		322.85	L	236.59	
L 8 L 9	161.00	58.94	L	.00		219.94	L	161.00	
L 9 L10	161.00	58.94	L	.00		219.94	L	161.00	

1

TRUSS RATING AND ANALYSIS PROGRAM (TRAP) 1
GROUP LOADING ANALYSIS

TABLE 10.2 DL+LL+I FORCE SUMMARY FOR UPPER CHORD MEMBERS

AASHTO LOADING RESULTS

MEMBER	TOTAL DEAD LOAD (K)	MAXIMUM TENSION LL+I (K)	TYPE	MAXIMUM COMPRESSION LL+I (K)	TYPE	MAXIMUM MEMBER FORCE DL+LL+I (K)	TYPE	MINIMUM MEMBER FORCE DL+LL+I (K)	TYPE
U 2 U 3	-241.92	.00		-88.21	L	-241.92		-330.12	L
U 3 U 4	-277.70	.00		-101.01	L	-277.70		-378.71	L
U 4 U 5	-296.69	.00		-107.70	L	-296.69		-404.39	L
U 5 U 6	-299.15	.00		-106.68	L	-299.15		-405.83	L
U 6 U 7	-296.69	.00		-107.70	L	-296.69		-404.39	L
U 7 U 8	-277.70	.00		-101.01	L	-277.70		-378.71	L
U 8 U 9	-241.92	.00		-88.21	L	-241.92		-330.12	L

Figure 3-3 (continued)

TRUSS RATING AND ANALYSIS PROGRAM (TRAP)
GROUP LOADING ANALYSIS

TABLE 10.4 DL+LL+I FORCE SUMMARY FOR VERTICAL MEMBERS

AASHTO LOADING RESULTS

MEMBER	LCT	TOTAL DEAD LOAD (K)	MAXIMUM TENSION LL+I (K)	TYPE	MAXIMUM COMPRESSION LL+I (K)	TYPE	MAXIMUM MEMBER FORCE DL+LL+I (K)	TYPE	MINIMUM MEMBER FORCE DL+LL+I (K)	TYPE
L 2 U 2	T	35.47	46.81	HS	.00		82.28	HS	35.47	
L 2 U 2	B	18.82	.00		.00		18.82		18.82	
L 3 U 3	T	-45.22	27.14	HS	-28.26	HS	-18.07	HS	-73.47	HS
L 3 U 3	B	-78.53	10.50	HS	-34.03	L	-68.02	HS	-112.55	L
L 4 U 4	T	-12.57	32.24	HS	-24.64	HS	19.67	HS	-37.21	HS
L 4 U 4	B	-45.88	20.17	HS	-29.57	HS	-25.71	HS	-75.45	HS
L 5 U 5	T	3.09	16.38	HS	-12.20	HS	19.47	HS	-9.11	HS
L 5 U 5	B	-30.22	12.10	HS	-35.36	HS	-18.12	HS	-65.58	HS
L 6 U 6	T	3.09	16.38	HS	-12.20	HS	19.47	HS	-9.11	HS
L 6 U 6	B	-30.22	12.10	HS	-35.36	HS	-18.11	HS	-65.58	HS
L 7 U 7	T	-12.57	32.24	HS	-24.64	HS	19.67	HS	-37.21	HS
L 7 U 7	B	-45.88	20.17	HS	-29.57	HS	-25.71	HS	-75.45	HS
L 8 U 8	T	-45.22	27.14	HS	-28.26	HS	-18.07	HS	-73.47	HS
L 8 U 8	B	-78.53	10.50	HS	-34.03	L	-68.02	HS	-112.55	L
L 9 U 9	T	35.47	46.81	HS	.00		82.28	HS	35.47	
L 9 U 9	B	18.82	.00		.00		18.82		18.82	

1

TRUSS RATING AND ANALYSIS PROGRAM (TRAP)
GROUP LOADING ANALYSIS

TABLE 10.5 DL+LL+I FORCE SUMMARY FOR DIAGONAL MEMBERS

AASHTO LOADING RESULTS

MEMBER	TOTAL DEAD LOAD (K)	MAXIMUM TENSION LL+I (K)	TYPE	MAXIMUM COMPRESSION LL+I (K)	TYPE	MAXIMUM MEMBER FORCE DL+LL+I (K)	TYPE	MINIMUM MEMBER FORCE DL+LL+I (K)	TYPE
L 1 U 2	-238.29	.00		-96.09	L	-238.29		-334.38	L
U 2 L 3	111.87	51.44	L	-14.25	HS	163.31	L	97.62	HS
U 3 L 4	63.03	39.29	L	-25.41	HS	102.32	L	37.62	HS
U 4 L 5	37.68	37.01	HS	-28.65	HS	74.69	HS	9.03	HS
L 5 U 6	5.02	18.87	HS	-14.46	HS	23.89	HS	-9.44	HS
U 5 L 6	5.02	18.87	HS	-14.46	HS	23.89	HS	-9.44	HS
L 6 U 7	37.68	37.01	HS	-28.65	HS	74.69	HS	9.03	HS
L 7 U 8	63.03	39.29	L	-25.41	HS	102.32	L	37.62	HS
L 8 U 9	111.87	51.44	L	-14.25	HS	163.31	L	97.62	HS
U 9 L10	-238.29	.00		-96.09	L	-238.29		-334.38	L

TRUSS RATING AND ANALYSIS PROGRAM (TRAP)
GROUP LOADING ANALYSIS

TABLE 18.1 CRITICAL LOADING FOR LOWER CHORDS

LCT	Group Load (KIPS)	Govering Case
L 1 L 2	219.94	I
L 2 L 3	219.94	I
L 3 L 4	322.85	I
L 4 L 5	374.93	I
L 5 L 6	397.13	I
L 6 L 7	374.93	I
L 7 L 8	322.85	I
L 8 L 9	219.94	I
L 9 L10	219.94	I

1

TRUSS RATING AND ANALYSIS PROGRAM (TRAP)
GROUP LOADING ANALYSIS

TABLE 18.2 CRITICAL LOADING FOR UPPER CHORDS

LCT	Group Load (KIPS)	Govering Case
U 2 U 3	-330.12	I
U 3 U 4	-378.71	I
U 4 U 5	-404.39	I
U 5 U 6	-405.83	I
U 6 U 7	-404.39	I
U 7 U 8	-378.71	I
U 8 U 9	-330.12	I

1

TRUSS RATING AND ANALYSIS PROGRAM (TRAP)
GROUP LOADING ANALYSIS

TABLE 18.4 CRITICAL LOADING FOR VERTICAL MEMBERS

LCT	Group Load (KIPS)	Govering Case
L 2 U 2 T	86.06	IA
L 2 U 2 B	18.82	I
L 3 U 3 T	-73.47	I
L 3 U 3 B	-112.55	I
L 4 U 4 T	-41.23	IA
L 4 U 4 B	-75.45	I
L 5 U 5 T	23.90	IA
L 5 U 5 B	-67.29	IA
L 6 U 6 T	23.90	IA
L 6 U 6 B	-67.29	IA
L 7 U 7 T	-41.23	IA
L 7 U 7 B	-75.45	I
L 8 U 8 T	-73.47	I
L 8 U 8 B	-112.55	I
L 9 U 9 T	86.06	IA
L 9 U 9 B	18.82	I

Figure 3-3 (continued)

TRUSS RATING AND ANALYSIS PROGRAM (TRAP)
GROUP LOADING ANALYSIS

TABLE 18.5 CRITICAL LOADING FOR DIAGONAL MEMBERS

LCT	Group Load (KIPS)	Govering Case
L 1 U 2	-334.38	I
U 2 L 3	163.31	I
U 3 L 4	102.32	I
U 4 L 5	74.69	I
L 5 U 6	28.50	IA
U 5 L 6	28.51	IA
L 6 U 7	74.69	I
L 7 U 8	102.32	I
L 8 U 9	163.31	I
U 9 L10	-334.38	I

4

BRIDGE AESTHETICS

LITERATURE REVIEW

The aesthetics of bridges has been addressed in recent years by a number of investigators. The topic is recognized to be particularly difficult due to the subjectivity of judgments offered by different observers. Concerning the differentiation of observers, a basic distinction is made between the role of professional bridge designers, who depend on their aesthetic values derived from their experience, and the role of the public in general, constituted of laymen who depend on their own subjective experiences independent of specific design knowledge. Other potential critics, who at times may have even more relevance in their criticism, are experts in landscape design or the fine arts.

The main intent of researchers in this field is to create guidelines that bridge designers may follow, which at the same time would also be guidelines for juries assessing design proposals in bridge competitions. These guidelines are created in a coherent common language understood equally by designers and critics.

Leonhardt (1982) addresses some basic rules of aesthetics applicable to bridge design in a very straight forward manner and a refreshing approach to such a subjective topic. His recommended rules of proportioning various components of bridges of different types can be of major use in an area where there are no guidelines. The bridge types that he discussed included: arch bridges, masonry arch bridges, reinforced concrete arch bridges, steel arch bridges, beam bridges, hunched beam bridges, beam viaducts, cable-stayed bridges, suspension bridges, and truss bridges.

Reich (1989) points out three proposals for incorporating aesthetic criteria into structural competitions. Stiny and Gips (1978), Makkuni (1987), and Minai (1989), address *rationalism* in aesthetics. In contrast, Reich (1989) proposes two aspects of aesthetics, romanticism and the combination of romanticism with rationalism, the latter of which produces a different effect than the sum of the effects of the individual components (i.e., a synergetic effect). He further stated the impossibility of static criteria in aesthetics, since such criteria evolve with the evolution of culture.

The model proposed by Reich takes into account the evolution of aesthetic parameters applying the model to a cable-stayed structure through a system referred to as BRIDGER. Such a system is used in the preliminary design of cable-stayed bridges.

Rationalism with in aesthetics is defined by Reich by means of examples, avoiding the confusion that usually arises when a specific terminology is used without adequate definition. Similarly, he describes *romanticism* and provides the reader with original work cited in the bibliography. The proposed model is an iterative process consisting of romanticist generations followed by rationalist adaptations (a discussion which goes beyond the limits of the present text). The use of knowledge-based systems (KBS) for the aesthetics of bridges have been the objectives of several researchers who have concentrated their efforts on tangible qualities such as form. However, less tangible aspects of design should be incorporated if such systems could have an effective meaning. Miles and Moore (1991a; 1991b) address the study of structural form for conceptual bridge design.

Following these studies, Miles, Moore, and Evans (1993) describe their work which focused on small highway bridges most typically expressive of the countryside, rather than on large-scale prototypes that dominate the public attention. In fact, such small structures, which are dominant in the natural environment, would benefit more from a series of guidelines than the large structures, which tend to express their own identity on an individual basis.

Previous studies aimed at similar objectives include Wengenroth (1971), Murray (1982), and Leonhardt (1982a). While all these works attempt to quantify various parameters affecting the aesthetics of bridges, much is left to be done in this area.

While most of the literature reflects the concepts of individual bridge designers, excluding the general opinion of the laymen, Crouch (1969) took into account public reactions on proportions involving girder bridges. By surveying people's opinions, he found that increased slenderness in bridges was not necessarily associated with an increase in aesthetics. People expressed their judgment on the preferred number of spans, favoring one or three-span solutions over two spans. Zuk (1974c), did a study on bridge colors that took into account the public opinion. Connor, et al. (1980) suggested procedures to be followed for assessing public response. He found that there was not a major difference when using sketches of bridges or using photographs when surveying the opinion of the general public. Furthermore, he noted that age and gender among the people being surveyed did not indicate any significant trend differences.

Reflecting the architectural approach to the design of bridges, Hamlin (1980) discusses the basic criteria of design composition that include proportion, unity, balance, scale and rhythm. Elliot (1983) proposed fundamental principles for contemporary bridges that include: (a) dominance of proportioning bridge components over shape itself, (b) relevance of scale for harmonizing the structure with the surroundings, (c) precedence of the compatibility of bridges with their immediate surroundings over the individuality of the bridge identity, and (d) priority to be given to simplicity over the complexity of schemes.

Miles, et al., (1991) propose a series of rules derived from a number of interviews with bridge design experts that include:

- Preliminary design requires first an overall scheme that includes basic requirements, however, the general scheme should be finalized only after having compared a number of different design proposals.
- Slenderness of forms is associated with positive aesthetic values, and in designing bridge decks there are ways to give the impression of slenderness without really reducing the mass.
- Examples for achieving visual slenderness for decks include cantilevering the deck over the beam which projects a shadow over the beam making it disappear from view.

Specific relationships are follows:

$$c = 2a \text{ to } 4a$$
$$c > d \text{ and } c > b/2$$
$$d = 2a \text{ to } 4a$$

where:

a = thickness of flange
b = thickness of web
c = projection of flange from face of web
d = depth of web.

In addition:

1.The cantilevered portion of the deck can have a uniform depth or the bottom part of the slab may be sloped, increased that the depth of the deck in the direction going toward the beam. It is recommended that the slope be 1 in 4 or less.

2.It is recommended that a drip groove be included on the underside of the cantilevered slab which will prevent water from following the underside of the slab, and the beam leaving stains over the beam face. Such provision will also prevent the draining water mixed with salt to reach the underside of the slab, the beam, and the piers avoiding erosion of these parts.

3.Setting back the bridge supports from the fascia would contribute to increase of the visual effects of slenderness.

4.Round columns, when used, are recommended to have a diameter smaller than the thickness of the carriage way, and their number should be limited to the minimum required.

5.Avoid making the space between columns to be equal to the diameter of the columns themselves by increasing or decreasing the spacing.

6.Reduce visually the mass of the supporting piers by indenting a chamfer along the edges.

7.Finish the surface of piers with vertical grooves that would convey water down without producing stains.

8.Rural bridges should harmonize with the surrounding landscape by minimizing the mass of the abutments and seeding the slopes with grass so that the bridge flows naturally in the landscape.

9.Abutments mostly in urban areas are recommended to be textured using for instance vertical grooves that would minimize the boldness of the smooth surface and at the same time minimize the maintenance and the effect of water stains.

10. Minimize the surface of the ground below bridges by using toned-down colors such as gray over pavements or encourage the growth of suitable vegetation wherever possible.

11. Scale models of bridges are recommended to be assembled in the preliminary design for assessing three-dimensionally the proportion between individual components and for judging the effects of shadows by the use of a heliodome.

Further findings from these surveys confirmed Crouch's findings (1969) concerning the preference of one or three spans over two span solutions. The new survey by Miles, et al., was based on a scale system that varied from 0 to 6, with 0 indicating the minimum value of aesthetics and 6 the maximum value. In this survey people confirmed their preference for the number of spans as previously indicated for arch bridges or beam bridges with haunches. Further opinions expressed by the general public included the dislike for wide bridges projecting large dark shadows below and for bridges displaying too many supports.

Another finding by the same researchers was that a background of poor quality shown in the photograph would have the effect of increasing the aesthetics of the bridge due to the sharper contrasts. Thus such effect would require some considerations concerning the validity of judgments in such cases. Moreover, in using photographs, it was found that major differences of judgments about the aesthetics of a bridge would occur according to the perspective point used in taking the picture.

The authors suggest using several photographs taken from different angles to reach a more objective assessment. In conclusion, it was found that the use of sketches was preferable to that of photographs for different reasons. The background would be eliminated, avoiding an influence in judging the bridge itself. Sketches would not exhibit any state of deterioration that a photograph of the real bridge would. This would guarantee that the effect of poor maintenance would not affect their perception of the quality of the bridge. In addition to sketches and photographs, video tapes taken from a moving car were considered also to be a meaningful tool to be used during public surveys, because they would simulate the viewing conditions under which a bridge is more commonly observed by motorists.

The findings of surveys such as the above can only be used to suggest flexible guidelines. This is because of the impossibility of defining aesthetic values within rigid parameters. Nevertheless, structuring such guidelines is a step forward in facilitating preliminary conceptual design phase. To this effect knowledge-based systems (KBS) can be formulated with particular features, including:

- a graphic library of existing bridge designs with visual representations and commentary,
- three-dimensional simulation of bridges through virtual reality renderings,
- renderings of views of bridges through videotapes taken from a moving vehicle, and
- generating suitable code for CAD software (such as AutoCAD)

Reich (1993) suggested aesthetic criteria that follow the rationalistic and romanticist movement of aesthetics. The model referred to in this work is derived from BRIDGER, a system to be used in the preliminary design of cable-stayed bridges. This was first introduced by Reich in *Building and Improving Design Systems: A Machine Learning Approach,* PhD thesis, Department of Civil Engineering, Carnegie Mellon University, Pittsburgh, PA).

Criteria of the rationalist movement of aesthetics include laws that relate different components of the bridge to each other. Among these, the following ratios are of particular concern: (main span):(crossing length); (tower height):(main span); (deck span):(deck depth); and (side span):(main span). Among the aesthetic criteria, Reich attached particular importance to balance, redundancy, and unity of form, which he discussed at great length.

Neural networks for bridge images

To introduce the new techniques based on neural nets for the aesthetic design of bridges, the concept of artificial intelligence must be grasped first, at least in its generalities.

Artificial intelligence

Artificial intelligence uses a digital computer to perform along a pattern similar to that of the human mind involving the ability of reasoning, identifying the significance of concepts, and learning from previous experiences. Research on this topic started in the 1940's. Although since the early stage it was predicted that the computer could indeed operate like the human mind, there was nothing tangible to this effect prior to the 1980s. Yet some practical applications in artificial intelligence had already been reached. These included *decision making, natural language comprehension,* and *pattern recognition.* Software based on elements of knowledge (i.e., "expert systems") would allow computers to reach decisions of complex problems based on nonnumerical data. Such systems consist of thousands of logic rules in the form of a computer program. These programs are considered *heuristic*, provide aid and direction for the solution of problems, and, in so doing, explore new procedures utilizing self-educating techniques.

Neural nets

Cichocki and Unbehauen [1993, "Artificial neural networks (ANNs), also called artificial neural systems (ANSs), neurocomputers, parallel-distributed processors (PDPs), or connectionist models are an attempt to mimic (simulate) at least partially the structure functions of the brain and of the nervous systems of living creatures," *Neural Networks for Optimization and Signal Processing,* p. 38] reported that artificial neural networks are simplified models of the human mind.

Yet, because the human mind is much more complicated, it is not realistic to compare ANNs to the primitive nervous system of some animal species. The human brain contains nets connecting astronomical numbers of neurons of nerve cells. Neurons are elements that consist of: *soma* (body cells), *axon* (a nerve fiber attached to the soma acting as the output medium of the neuron), *dendrites* (a bunch of irregularly shaped nerve fibers attached to the soma), and *synapses* (contacts at the end of each axon that connect to the next neuron).

Neuron interaction

The working mechanism of all these components can be summarized as follows:

1. The cell body receives signals from other neurons.
2. The signals pass though synaptic connections.
3. The output signals being transmitted consist of nerve impulses.
4. Such a signal travels along a branching axon.
5. Passing through the synapses, the signal reaches the other neuron.
6. The neuron receiving the impulse is excited and produces, in turn, nerve impulses of its own.
7. Such new nerve impulses are also called trains of pulses and reach other neurons following the same mechanism as described.
8. The output pulse rate of a neuron depends on the output signals and of the strength of the synaptic connections.
9. The maximum impulse rate is 1000 per second.
10. The pulse rate of the signal can increase or decrease according to synapses.
11. The communication between neurons is transmitted in the form of nerve impulses that could be quantified as digital signals.

In conclusion, the process of communication among neurons that receive inputs and eventually transmit impulses has been briefly synthesized above, leading to simulation of the process among artificially created neurons.

Artificial neurons

These artificial neurons are also referred to as "neuron cells," which include in themselves "nodes" trying to duplicate the action of biological neurons. Such artificial neurons are represented by a variety of models, such as: McCulloch-Pitts model, Fukushima model, Adaline model, Hopfield model, and Grossberg model etc.

Artificial neural networks

Artificial neurons can be connected by different types of networks varying in the degree of complexity. The connections between artificial neurons are *architecture* and *circuit structures*. The architecture of artificial neural networks include three categories: feedforward (multilayer) networks, feedback (recurrent) networks, and cellular networks.

Neurons receive messages from the exterior or from other neurons but do not generate any feedback. In feedback networks, the neurons include dynamic building blocks that operate in the feedback mode. In cellular networks the neurons are of a type specially built called "cells." They are arranged in regular geometric patterns (such as triangles, rectangles, and hexagons). They communicate only with other neurons within their proximity by means of mutual lateral interconnections. Yet, cells not directly interconnected can also influence one another.

The differentiation among artificial neural networks is based not only on the three different types of architecture mentioned above, but also on the type of neurons, the learning or training procedure, and the form of operation.

The learning possibility of a neural network, using past examples of bridge designs, were reported Yasuda, et al., (1995). The authors observed 140 small-scale arch and girder bridges within the period of 1988 to 1992. These structures were evaluated by a group of observers looking at photographs considering the following factors: (a) impressive, (b) unique, (c) sophisticated, (d) friendly, (e) harmonious, and (f) beautiful.

The scale used to evaluate each bridge in accordance to the factors mentioned above, included the following five values: +2 (mismatching and ugly), +1 (unfriendly), 0 (not impressive and not sophisticated), -1 (n/a), and -2 (ordinary). The observers included designers, bridge designers, and university students, but excluded the general public. This method was based on the use of RHINE (a commercially available neural network software).

Another major contribution on the topic of bridge aesthetics is a formal publication of the American Concrete Institute (whose authority in the design of any major significant structure built of concrete is such that its influence in the design profession affects structural engineers all over the world). This publication (1990) is organized in a series of papers prepared by different authors upon the request of the ACI Committee 124, on concrete aesthetics. The fact that this publication is a compendium of papers (rather than a unified codification of rules on aesthetics) confirms that even such an authoritative source has not been able to treat the subject in an objective manner, instead presenting a series of subjective and uncoordinated opinions.

This lack of unification of format further enforces the practical inability to organize the subject of aesthetics in a structured format. Nevertheless, this is a relevant document describing the state of the art at present.

The papers were presented at two sections of the symposium held in Seattle, Washington, on November 12-13, 1987, and at two sections of the symposium held at Orlando, Florida on March 24, 1988. The publication *Esthetics in Concrete Bridge Design*, starting with the foreword by Stewart C. Watson and with the preamble by Fritz Leonhardt, includes a variety of papers. E. Cohen discusses his opinion on bridge aesthetics and also opinions by Amman, Billington, and Whitney. Among required qualities for good bridge design, reference is made about structural efficiency, unity, and simplicity, qualities applied to both bridges and any other structure of aesthetic qualities. R. Lacroix stresses the need for a joint combination between engineers and architects in the design of bridges subordinating the latter to the former.

Burke (1990) refers to an interesting number of papers, including his bibliography (1988). Reviewing the works included in such a bibliography, Burke pointed out the number of rules and formulas for the aesthetics of bridge design proposed by different authors varying from one basic rule suggested by Whitney up to 27 rules recommended by Steinman.

It would be interesting to review such an abundant amount of material on this topic, even if a most everybody agrees that art cannot be learned by rules and regulations. Nevertheless, it is recommended that the reader review these suggested rules using Burke's bibliography.

Because of the limited space in this section it is impossible to mention the various other invited papers included in this publication. Yet it would be relevant for the reader to be informed about the authors the ACI selected to investigate present concepts on the aesthetics of bridges. These include: Kruckmeyer, Ritner, Liebenberg, Grant, Finsterwalder, Burke, Stevens, Kopetz, Lee, Gloyd, Elliott, Schlaich, Muller, McCallister, Gottemoeller, Seim, Lin, Mays, Loveall, Buckland, Smith, Strasky, Kulicki, Snyder, Mertz, and Roberts.

The aesthetics of California bridges

Of major significance in reviewing the literature on this topic is the realization of how far the subject of bridge aesthetics has progressed. This is clear when consulting official publications such as California Department of Transportation (1993) which dedicates an entire section to this topic, entitled "Bridge Design Aesthetics." The guidelines described in this publication by the California Department of Transportation (CALTRANS) concerning bridge aesthetics are useful for designers outside of California.

Starting with the general definition of aesthetics and a definition for what constitutes a "beautiful bridge," the document points out the three fundamental considerations to be given in the design of bridges: cost, construction, and maintenance. Then, introducing the process followed by CALTRANS, the document describes the formation of the project development team, which consisted of by professionals from various relevant disciplines, and engineers, architects, environmentalists, landscape architects, and legal experts. The selection of structural type is eventually controlled by a group of structural engineers.

Then the basic design process consists of modifying and refining the most basic structural type first proposed in the process. More specifically, the document indicates the basic parameters that are part of such modifications, i.e.; grooves, offsets, texture, and cross-sectional changes. Part of the process is also another group referred to as Aesthetics and Models (A&M), which interacts with bridge engineers to identify and resolve critical areas from the aesthetic point of view using visual aids.

The selection of structural types is reviewed at meetings attended by representatives from Specifications, Maintenance, Construction, Estimating, Design, and A&M. Professional consultants are also participants of meetings. In such meetings the engineer is in charge, while A&M acts as a consultant to the engineer. The engineer presents advanced planning studies to A&M which reviews them. The objects of the reviews are the selection of column types, girder edge treatment, and surface treatment. Then architectural sketches of such modifications are prepared, including elevations, sections, and perspectives. Following this last phase the team produces cost estimates and evaluation of structural systems and architectural qualities.

Public meetings or district meetings are organized. At these hearings proposals concerning the project are presented and ideas from the public are also discussed. At these meetings A&M uses visual aids including various types varying from sketches to photographic montages. Time allotted for such visual aids spans from one week to two months. More sophisticated models can also be made available by A&M varying from Styrofoam models to complete detailed models, but in this case the preparation time could vary from a few months up to a year.

Basic concepts

Basic concepts derived from an overall scanning of the documents are briefly listed as follows:

1. Bridges have an influence on their surroundings, including: size, shape, and color.
2. Bridges should be compatible with their environment and should blend in rather than making strong statements.
3. Bridges should be considered from two points of view: (a) considering an observer that looks at the bridge while traveling over it (a good bridge should not distract its user), and (b) considering an observer looking at the bridge from different points of view as part

of the landscape (the bridge is judged in terms of its elevation or profile).

4. Definition of sound design includes a consideration of function and aesthetic treatment.
5. Order of priorities in the design process include: (a) geology of the area for deciding the site of piers, (b) determination of the consideration of spans according to the location of the piers, (c) consideration of structural types according to spans.
6. Keeping to a minimum of difficulties the methods and procedures of design.
7. Appropriateness of structural systems is considered a major factor in aesthetics of bridge design.
8. Bridges within a given route should be compatible in appearance with one another to create a sense of order.
9. Bridges overcrossing a given route should remain homogeneous with one another.
10. Bridges undercrossing a given route do not need to be homogeneous.
11. Bridges crossing rivers and viaducts are not required to be homogeneous because they are not observed in profile by the user.
12. When possible, all bridges interacting with a given route should be compatible with one another to create a sense of order.
13. Bridge parts usually considered for aesthetic values include: railings, decks, girders, soffits, columns, abutments, and wingwalls.
14. Evaluation of the above-mentioned components reflect the views of A&M gathered during public hearings and by accommodations of the districts.
15. Special considerations on aesthetic characteristics are discussed in very specific terms, including illustrations for railings, girders, decks, and piers.
16. Retrofitting for improving the seismic resistance of bridge components is discussed in the document.

Another recent work in this area is that of Yasuda, et al., (1995). In this paper, the authors refer to a neural network trained with over 100 past examples of bridge design. This eventually produced a data base that included images from *Bridge Annals*. From this, a system was eventually developed that had practical applications. The authors refer to constructing the system on a personal computer accessible to anybody from almost anywhere. The database was constructed using *Microsoft Access 1.1*, which can handle image data. In this database, girder and arch bridges were part of the input gathered from *Bridge Annals* using color photographs.

Specifically, the database includes bridge type, spacing, road width, color and background, etc., including also elevations attained from the neural network. Typically, when spatial conditions are entered, the program brings out all bridge examples that would fall within the same conditions being specified. Part of the display also includes the evaluation results, furnishing a complete picture of the various similar cases that the software had recalled. For more specific data on this topic, the reader is referred to the bibliographical source.

AESTHETICS IN BRIDGE RETROFITTING

Maintenance of steel structures requires at times often additional reinforcing components. Such a process of retrofitting bridges that have been weakened by different causes often applies to bridges that suffer damages due to earthquakes. Repairing bridges damage seismic loads is definitely a major task for the Department of Transportation in California. Thus, the experience derived in this state after numerous earthquakes constitutes a significant source of information for professionals in other states, specifically within the realm of structural aesthetics. The CALTRANS document previously mentioned has significant points regarding this topic:

1. Basic philosophy expressed in this document asserts that any retrofitting to bridges should not alter the character of the structure and should therefore be kept to a minimum even when additional members are required. Aspects to be kept in consideration include shape, texture and color.
2. The use of retrofitting cables and fittings are usually found aesthetically unpleasant, except for cables above the girder bottom flange which are less objectionable than others. Also objectionable are cables applied around columns and other structural members, worse yet are cables shown in silhouette.
3. The retrofitting of columns to repair them and to increase the structural capability can be obtained by enclosing the retrofitted column with one of the following three types of steel casing: circular, elliptical, and formfitting.

4. Another method of encasing retrofitting columns can be achieved by using fabric wrap casing.
5. Use of plastic-coated steel cables tightly wound about columns is another method of retrofitting existing columns that can be aesthetically acceptable. In fact, such a seismic reinforcement can only be viewed from a given distance; the coiled cables are practically invisible.
6. Welding steel plates and tie rods for protecting weak columns or for altering dynamic frequency by changing the stiffness should be accomplished without changing the appearance of the structure.

Regionalism influence on aesthetics

The regionalism influence on the architecture of structures, which has traditionally characterized all regions of the world, was gradually disappearing worldwide with the spreading of an international style applied to buildings as well as to engineering structures such as bridges. Especially in the United States, regionalism in the aesthetics of bridges in recent decades was practically unnoticeable. Under more scrutiny, regional differences are eventually becoming apparent.

Watson and Hurd, (1990) point out that after observing predominant bridge design and construction types throughout the United States certain types of structures are favored in various regions. For instance, steel bridges, seem to be most prevalent in the middle west and the northeastern regions. Similarly in the southeast the predominant bridges are of the pre-cast, concrete box-girders and girder types. Furthermore, in the north west and on the west coast the predominant types of bridges consist of cast-in-place, post-tensioned concrete box-girder structures erected on falsework. This last type had been reported by the same authors to have constituted 80 percent of all bridges built in one year in the 1980s in the California.

Bridges and buildings

The *de facto* separation between buildings and bridges which has developed side by side in parallel tracks occasionally shows some points of confluence which are of particular interest to those exploring architectural concepts shared by both types of structures. Historic bridges dedicated only to pedestrian traffic in Florence (Ponte Vecchio) and Venice (Ponte Rialto) crowded with little shops busy with vivacious crowds.

More recently, in Italy, several bridges crossing the main highway *autostrada del sole* consist of one-story buildings housing restaurants and coffee shops offering a dynamic view of the highway traffic to the patrons of these establishments. All these examples, showing bridges housing various activities within enclosed spaces, have also an equivalent counterpart in buildings that have used bridge architecture as the generatrix of form.

In this latter case, reference is made specifically to an imaginative building by Pier Luigi Nervi. The edifice was dedicated to house a paper mill (Cartiera Burgo) in Mantova, in 1961. The structure whose shape was elongated rectangle housing a series of machinery operating in sequence, was 249 m x 30 m, supported above ground by cables, and carried by four pylons of reinforced concrete. Although clearly a structure inspired by suspension bridges, this building dramatizes in strong, delineated shapes made of concrete, the image of the pylons with a rendering absolutely new. The pylons, with anthropomorphic features standing in a dynamic pose, seem to be engaged in a tug of war, pulling the suspension cables. This little-known building opened the way to a new approach in structural art that could have inspired a new generation of bridges searching for a new identity of expression. From the practical point of view, this structure should be of extreme interest to designers in search of new sources for the identification of forms applicable to bridges (Figures 4-1 through 4-3).

Continuing on the same theme of interconnecting the design of bridges and buildings, one could quickly recognize the effect of scale in such a correlation. In fact, for particularly long spans over hazardous sites, structural factors would definitely control. For short spans that could use most any kind of structural system, their morphology would explore structural aesthetics with total adherence to architectural principles. In other words, for small bridges, the architectural input should probably dominate the conceptual design of the structure, regardless of whether the principal designer is a structural engineer or an architect.

Pedestrian bridges, for instance, lend themselves more than anything else to be structural artifacts, where geometry and inventive use of materials could find the most valid stage that will communicate daily to a flux of public much larger than that flowing through any conventional building. It is inconceivable that any new pedestrian bridge would not include an architectural delineation of forms that would speak in the same language as other urban structures.

The aesthetics of box girders

Among the various structural members that have gradually advanced in acceptance in the structural morphology of modern bridges is by all means the box girder, fabricated in steel or in concrete, including two or more cells acting individually or in composite action with the deck. Its appearance has a definite positive contribution to the aesthetics of the overall structure due mostly to the simplicity of the form and the smoothness of its shape. Thus, considering the geometrical relationships of the box-girders with the other structural components, one could consider right from the start some general observations taken from the *Standard Specification for Highway Bridges* (1996, pp. 269 and 287), "the distance

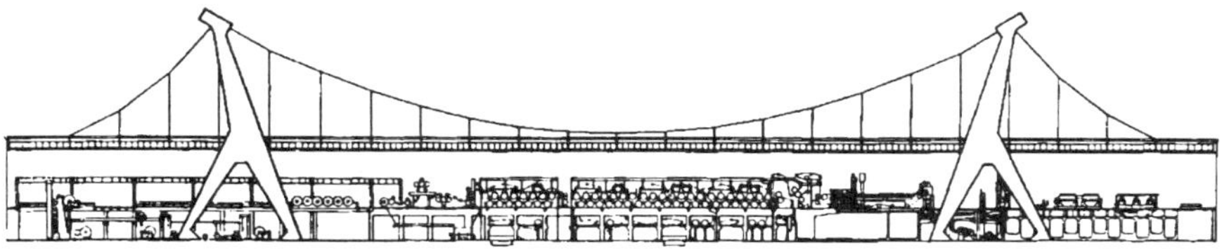

Figure 4-1 "Cartiera Burgo" (Burgo Paper Mill; 1962, Mantua, Pier Luigi Nervi), roof structure with paper manufacturing machinery. An industial building enclosing a paper mill designed to interpret the theme of a suspension bridge (reproduced by permission of the copyright holder, Editalia).

Figure 4-2 "Cartiera Burgo," Mantova, detail of supporting portal (photograph reproduced by permission of the copyright holder, Editalia).

Figure 4-3 "Cartiera Burgo," exterior view, under construction (reproduced by permission of the copyright holder, Editalia).

center-to-center of flanges of each box should be the same and the average distance center-to-center of flanges of adjacent boxes shall be not greater than 1.2 times and not less than 0.8 times the distance center-to-center of flanges of each box. In addition to the above, when nonparallel girders are used, the distance center-to-center of adjacent flanges at supports shall be not greater than 1.35 times and not less than 0.65 times the distance center-to-center of flanges of each box. The cantilever overhang of the deck slab, including curbs and parapets, shall be limited to 60 percent of the average distance center-to-center of flanges of adjacent boxes, but shall in no case exceed 6 feet."

Considering the practical aspects of these enclosed structures, it is interesting to note the problem of condensation and inside inspection. Thus to this effect one notices that indeed the specification by the AASHTO requires that: "consistent with climate, location, and materials, consideration shall be given to the providing of manholes or other openings, either in the deck slab or in the steel box for form removal, inspection, maintenance, drainage, etc." The inside surface of box girders cannot be left unprotected against the corrosion action of the weather; periodical painting is required, necessitating appropriate access. To reduce the maintenance costs, the best coatings should be used. For example, the Saint Nazaire Bridge in France requires a frequency of only every ten years for the painting of the inside surface of the box girder, using zinc-based paint which was applied first on the steel surface after it was sand blasted.

DESIGN PROCESS: THE ARCHITECTURAL PERSPECTIVE

The design of bridges has never excluded the participation of architects. This has been true throughout the history of bridge design and construction whether the engineer and the architect existed separately or blended into one. Bridges, in fact, were part of the architectural scope being structures at the service of mankind. In fact, limiting architecture just to the design of buildings would be a myopic view of the discipline.

Architecture has been traditionally involved with buildings *per se*, with their exterior volumetric design, with interior, with the design of the exterior spaces around buildings, with the design of the landscape as well as the design of monuments and elements of urban spreads. In the middle of the 20th century, architecture emerged as a problem solving discipline, oriented toward the design of the human habitat in general. This definitely included bridges, even if these have become highly specialized, structural problems.

The process of design as it emerged in the1950s had taken a special rationalized format not only for engineering projects but also for architectural processes which have expanded considerably. Architectural designs have become processes that did not depend any longer on experience and intuition alone. Yet, the architectural approach still remained characterized by several individual formats that distinguished them from the general approaches used by other disciplines. Since this work is addressing mostly civil engineers desiring to enter in the area of bridge design with a basic understanding of the architectural components, it is essential to present to them the architectural perspective itself.

It has been found useful to review basic documents defining elements of architecture through the abundant literature on this topic. One of the first references of major interest is A.I.A. (1987) from which several observations should be noted. For instance, it is quite true to emphasize that any architectural design is not based only on moments of inspiration but most of all on hours, months, and years of hard work.

To this effect it is extremely interesting to take note of other statements in this writing such as: "Someone has estimated that a typical large building requires two million 'yes/no' decisions by the decisions team." Therefore, such a design required a process that could either be formulated in a standard format or by each individual designer as a expression.

At the beginning of any building design process, the first essential document is the program that may vary extensively in size according to the complexity of the project. In order of magnitude the program may vary from one page to thousands of pages. Programs for hospitals, for instance, may very well be as long as in the latter case. The extreme variations of design factors that differentiate one project from another cannot be overemphasized. Yet most design processes in architecture include the following activities: *analysis, synthesis, refinement,* and *implementation*.

Analysis may be subdivided into the following: (a) program analysis, (b) site analysis, (c) zoning, (d) codes, (e) documentation of existing conditions, (f) scheduling, (g) cost, (h) construction industry practice, and (i) architectural precedents.

Synthesis is defined as the phase in which analysis generates the project concept. The analysis includes the sections listed above, which are all examined by the designing architects prior to synthesizing them and producing the project concept. Refinements is the phase in which the project concept generates a detailed design solution. Implementation is the phase in which the design documents are produced, including drawings and specifications.

Design process: a new development

As previously mentioned, the new approach to design emerged in the mid-1950s and is presented in more detail by Jones (1970). In this work, the author presents new design methods which have been derived from other disciplines such as operational research, ergonomics, and work study, indicating the beginning of such new evolution in design, starting from three basic events.

These include three conferences on design methods that were held in London 1962, in Birmingham in 1965, and in Portsmouth in 1967. Besides these conferences in Britain, others have been held in the United States and Czechoslovakia.

In describing the new methodologies applicable to designing and planning professions, Jones points out that other disciplines also share equal interest on new methodologies including "industrial activities such as management, production engineering, accounting and marketing, as well as in non-industrial activities like acting, painting, musical composition, literary composition, philosophy, science, librarianship, social work, teaching and military planning."

Also interesting is the list of the new methods that have been derived from this new approach into various disciplines. In turn, these include: operational research, work study, discounted cash flow, market research, "method" acting, drip painting, serial music, stream-of-consciousness novel writing, linguistic philosophy, the science of science, coordinate indexing, group dynamics, programmed learning, and war gaming. Interestingly, Jones describes the traditional methods of design by pointing out two different approaches: "design by drawing" and "design by craftsmanship." On the other hand, he indicates in his work the genesis of 35 new methods which he illustrates with examples in Part II of his book. There, such methods, which he also calls techniques, are individually introduced, specifying that they have all been selected, by applying five different criteria such as: effectiveness, relevance, convenience, familiarity, and criticism.

Included in the list are the following:

1. systematic search (decision theory approach)
2. value analysis
3. systems engineering
4. man-machine system designing
5. boundary searching
6. Page's cumulative strategy
7. CASA (collaborative strategy for adaptable architecture)
8. strategy switching
9. Matchett's fundamental design method (FDM)
10. stating objectives
11. literature searching
12. search for visual inconsistencies
13. interviewing users
14. questionnaires
15. investigating user behavior
16. systemic testing
17. selecting scales of measurement
18. data logging and reduction
19. brainstorming
20. synectics
21. removing mental blocks
22. morphological charts
23. interaction matrix
24. interaction net
25. AIDA (analysis of interconnected decision areas)
26. system transformation
27. innovation by boundary shifting
28. functional innovation
29. Alexander's method of determining components
30. classification of design information
31. checklists
32. selecting criteria
33. ranking and weighting
34. specification writing
35. Quirk's reliability index

Each one of these methods or techniques is in turn assembled into different groups. Including: (a) prefabricated strategies (1-7), (b) strategy control (8-9), (c) methods of exploring design situations (10-18), (d) methods of searching for ideas (19-22), (e) methods of exploring problem structure (23-30), and (f) methods of evaluation (31-35). These includes examples illustrate some of the various methods, the reader is referred to the original bibliographical source for more details. Thus, following the examples by Jones, a process specifically adaptable to the preliminary design of bridges can eventually be structured individual basis.

Probably the most significant and synthetic expression of Jones's methodology can be found in one of his several diagrams illustrating various processes, "Input-Output Chart for Selection Design Methods," (Table 4-1).

The diagram includes six input entries on the vertical axis and five on the horizontal axis. This matrix produces a series of interaction statements which catalogs the activities that can take place in the design process. See the reference for more precise information.

A lesson in bridge aesthetics

Zuk (1976) describes practical aspects in the aesthetics of bridge design with promises of new inventive approaches of special interests in this context. He begins by stating that one of the goals of Ludwig Mies van der Rohe was indeed creating "an architecture that anybody could do" which in Zuk's point-of-view has basically been accomplished as verified by thousands of buildings that have been patterned on Mies's simple concepts. Transposing this concept to the design of bridge structures, he points out that these could also be designed following the same pattern. This is the basic concept from which the author starts developing a design process applicable also to bridges.

Addressing the engineering approach to design he notices the following time typical concerns of major priority: safety, maintenance, construction, cost, and strength, form, proportion, and details are often neglected, which indeed constitutes the basic criticism found in the literature.

The basic content of the book consists of reporting the previously mentioned Crouch paper (1974). In so doing, the book reinforces the major contribution that this researcher has brought to the field of structural aesthetics with special regard to bridges.

Zuk's personal contribution to the topic in question is a direct consequence of Crouch's original work of 1974. Zuk (1975) has written practical tool on aesthetics for structural engineers. It mainly addresses girder type bridges in the short to medium span range without including truss, arch stayed-girder, or suspension systems, which could also be included when needed. As the author explains, the bridge structure is analyzed in its basic design components that include: girders, parapets, curbs, sidewalks, railings, piers, pier caps, abutments, bridge terminals, and slope

OUTPUTS ► INPUTS ▼	2 Design Situation Explored	3 Problem Structure Perceived or Transformed	4 Boundaries Located Sub-solutions Described & Conflicts Identified	5 Sub-solutions Combined into Alternative Designs	6 Alternative Designs Evaluated and Final Design Selected
1 Brief Issued	3.1 Stating Objectives 3.2 Literature Searching 3.3 Visual Inconsistency Search 3.4 Interviewing Users 4.1 Brainstorming	3.2 Literature Searching 3.3 Visual Inconsistency Search 3.4 Interviewing Users 4.1 Brainstorming 4.2 Synetics	3.3 Visual Inconsistency Search 4.1 Brainstorming 4.4 Morphological Charts	3.3 Visual Inconsistency Search 4.1 Brainstorming 4.2 Synetics	2.1 Strategy Switching 2.2 Matchett's FDM
2 Design Situation Explored		3.1 Stating Objectives 3.9 Data Reduction 5.1 Interaction Matrix 5.2 Interaction Net 5.8 Classification 6.4 Specification Writing		5.4 System Transformation 5.6 Functional Innovation 5.7 Alexander's Method	
3 Problem Structure Perceived or Transformed	3.2 Literature Searching 3.5 Questionnaires 3.6 Investigate User Behavior 3.7 Systemic Testing 3.8 Select Measurement Scales 3.9 Data Logging		1.5 Boundary Searching 3.7 Systemic Testing 4.1 Brainstorming 4.4 Morphological Charts 6.2 Selecting Criteria 6.3 Ranking and Weighting 6.4 Specification Writing	4.1 Brainstorming 4.2 Synetics 5.4 System Transformation 5.5 Boundary Shifting	1.1 Systemic Search 1.2 Value Analysis 1.3 Systems Engineering 1.4 Man-Machine System Designing 1.5 Boundary Searching 1.6 Page's Strategy 1.7 CASA
4 Boundaries Located Sub-solutions Described & Conflicts Identified		4.2 Synetics 4.3 Remove Mental Blocks 5.3 AIDA 5.4 System Transformation 5.5 Boundary Shifting 5.6 Functional Innovation 5.7 Alexander's Method		4.1 Brainstorming 4.2 Synetics 5.4 System Transformation 5.5 Boundary Shifting	5.3 AIDA
5 Sub-solutions Combined into Alternative Designs					1.2 Value Analysis 3.5 Questionnaires 3.6 Investigate User Behavior 3.7 Systemic Testing 3.8 Select Measurement Scale 3.9 Data Logging & Reduction 6.1 Checklists 6.2 Selecting Criteria 6.3 Ranking and Weighing 6.4 Specification Writing 6.5 Quirk's Reliability Index
6 Alternative Designs Evaluated and Final Design Selected					

Table 4-1 Input-Output Chart for Selecting Design Methods (reproduced from Jones, J.C., 1970, Human Futures, by permission of John Wiley and Sons).

protection, as well as details such as: bracing, stiffeners, drains and light poles, with a final section on color and texture.

The various components listed above are presented in a series of proposed solutions containing notes that have been evaluated by various groups. Each solution contains intern notes explaining the rationale followed during the evaluation which can furnish a better understanding of the guidelines. See original reference for more detailed information.

A second lesson on the aesthetics of structures

Addressing structural engineers on the aesthetics of the structure that they design is a topic that usually emerges in sporadic papers that have not been officially codified in textbooks. As we have seen, aesthetics has never been an official subject in civil engineering in any curriculum so that the fragments in which the subject is discussed herein represent what the official literature contains.

The previously mentioned paper by Zuk (1976) was followed by Reese (1976) which is another lesson directed to designers with a series of practical recommendations. The author presents a graphic representation of the typical situation of the designer facing a blank sheet of paper going through his or her mind in search of a point of beginning when the hand starts moving, delineating graphically a visual representation of an idea.

This is indeed a crucial point for anyone and the suggestion being given is that of overcoming any hesitation and to start drawing just to keep the ball rolling. In a sense, it is like jumping in the water without knowing how to swim in order to begin the process of learning. Whatever the sketch is, its value is definitely not in the sketch *per se* but in the fact that it provides intangible entities that could be used for subsequent modifications. In fact, as the first sketch is continuously revised in a series of modified versions its practical value at the end is down to nothing and it could probably be trashed. All this is to emphasize how important it is to start drawing something without any hesitation.

Particularly colorful in this paper is the adherence to reality that the author projects in his writings. For instance, he suggests rejecting the use of the hard graphite pencils typically used for drafted drawings, recommending the use of a very soft lead pencil that moves with minimum resistance in a fluid flow of spontaneous tracings reflecting the creative process in the mind. Yet, the practicality of the suggestion, which anybody working in design must have experienced at one time or another, is very much a true reflection of the state of mind of a designer at the beginning of any design process.

The first phase of drawing sketches, are recommended to be done to an approximate scale to keep the results in a form easily understandable. A suggestion is made to use graph paper for the original sketches so that it would be easy to draw in scale, but this is really not necessary when the designer is familiar with the use of engineering or architectural scales in his or her daily practice. Alterations of the sketches with other sketches are easily executed by overlapping tracing paper and proceeding in this step-by-step fashion. The sketches are evaluated not only by the designers themselves but by others who are equally involved in the project, so that their constructive criticism can be utilized in improving the product.

The use of colored pencils is also a good suggestion, especially to emphasize special points. Sketches can present plan and elevation views, but occasionally they could also be made in perspective or in axonometrics when more clarity is necessary. Reese makes the following statement in his paper, which may sound naive and obvious, but is exceedingly useful to keep in mind, "Good designs do not just happen, but are the result of thought, study, and the discarding of unpleasing solutions."

Reflecting general pedagogical notions often heard and typically disregarded, Reese points out one of the important processes aiming at acquiring a critical taste. This involves basically the creation of a data bank in the mind, derived from observations of all kinds of buildings and other structures, analyzed in terms of their positive as well as negative aspects that make them pleasant or unpleasant. Only by using a large data bank and by having acquired a sense of criticism can the designer be efficiently equipped for analyzing and rating his or her design solutions.

In concluding his lesson on the aesthetic components of design, Reese presents a list of basic characteristics which are part of any meaningful design. Within that list are the following: (a) scale, (b) proportion, (c) simplicity, (d) unity, (e) symmetry, (f) truth, (g) relation to surroundings, (h) color, and (i) collection of ideas. Obviously by now the reader has become well accustomed to the terminology used in design so that an individual explanation of these various terms is no longer needed. Yet the reader is recommended to go to the original source listed in the bibliography and go over the brief discussions that Reese includes for each term, including the examples that he uses to illustrate his ideas.

A third lesson on the aesthetics of structures

Another interesting lesson on structural aesthetics has been gained Law (1976). Aesthetics in structures applicable to the practicality of any design, including the conception of bridges, can also be subjectively derived by reasoning prior to obtaining any specific step-by-step process. The first step in understanding aesthetics would be that of collecting meaningful definitions of it that have been presented by many scholars within the field of philosophy and within the areas of various art forms. To this effect, this paper selects three different definitions of the term: "Art is the capacity to make, involving a true course of reasoning" (Aristotle), "Beauty is the splendour of truth" (St. Augustine), and "Beauty is truth, truth is beauty, that is all ye know on earth and all ye need to know!" (Keats). The statements suggest that the degree to which a structure is beautiful depends on the degree to which it is rational or truthful.

Two major qualities seem to be at the base of aesthetically pleasing structures: rationality and truth. Rational structures are defined as being functionally, technically, and economically truthful, which are statements upon which one wants to reflect and are therefore specifically noted in this context. Furthermore, expanding the concept of rationality within structures, Law clearly states that implied within this term is the possibility of reaching aesthetically pleasing structures, reinforcing this concept from quotations attributed to Pier Luigi Nervi.

Other statements are definitely worth being remembered, including references pointing to three basic design principles that constitute pleasantness and unpleasantness in art, such as balance, continuity or rhythm, and the principle of emphasis.

A fourth lesson on the aesthetics of structures

When humans emerged from caves to build their own dwellings, it may have been the expression of that animal instinct to construct a nest, or (more likely) it may have been just the result of mental evolution. In the anthropological evolution, when humans became builders of structures capable of housing them, they started also to create a critical experience of their products which through the long ages of history had developed a taste in the aesthetics of the structures they built. In fact, when aesthetics combined with the utilitarian meaning of a building became a systematic way to assess structures built by man, one could say that this is the point where architecture started.

Architectural artifacts that eventually reached a level of aesthetics qualifying them as works of art in the newborn discipline that we refer to as architecture, were built by men who had not been trained at all but acted spontaneously without being conscious of their intrinsic knowledge of aesthetics. In reality, then, we are talking of an architecture that emerged spontaneously without the participation of anyone that could call himself an architect.

We are talking, in effect, about an architecture without architects, at least when we are talking about the beginning of this art form. If we accept this point, architectural artifacts do not necessarily come from the skill of architecturally trained professionals, but indeed they could come from the refined skills of craftsmen and builders who have developed spontaneously those basic rules that are implicit in architecture.

We have already seen through the course of this text the various qualities that are necessary for an artifact to be a work of architecture. These have been expressed by terms such as scale, balance, order, and unity which have been defined and described to some extent. Those who have inherited such principles instinctively or those who have acquired them spontaneously through personal experiences constitute a special group of specific interest in this context. At this point, we can talk about the builders of architectural structures who arrived at architecture itself without any special training that made them skillful in the art of building.

These men and their work became the most meaningful source of guidance for engineers aspiring to design bridges that are more than functional structures and achieve the status of works of art. Who are these men? The names of the most prominent that come to mind who have contributed with their work in the architecture of the twentieth century include: Robert Maillart, Eugene Freyssinet, Eduardo Torroja, Buckminster Fuller, Pier Luigi Nervi, Riccardo Morandi, Frei Otto, and Heinz Isler. Again, none of these men had any training in architecture but were educated in engineering only, and also should not be confused with others that started from architecture and eventually reached engineering through experience, such as Filippo Brunelleschi, Domenico Fontana, Antonio Gaudi, Eero Saarinen, Felix Candela, and Santiago Calatrava.

The students of structures, whether in the educational realm or in the professional arena, who are aspiring to reach aesthetic values in their work should concentrate their attention on the legacy that those men have left. Their works, dissected in their principal components, constitute the most efficient guidance from which to learn the pattern of thinking.

Many of them have left documents specifically concerned with bridges, while others have dealt with buildings. Yet even if the topic in question focuses on bridges themselves, any other structural artifact can also be referred to bridges through a vocabulary that is common to all structures built at the service of mankind.

Thus, in practical terms, the bridge designer facing a blank sheet of paper at the point of beginning should have in front of him or her a compendium of drawings and notations that expresses the key points derived from the study of the works of these masters. Today, in fact, with the large availability of photocopy machines, it is easy to scan the architectural literature and compile an inventory of images.

These resources are extremely valuable, not for copying individual details, but because they could be absorbed into the memory by looking at them and analyzing them through the course of time. In fact, as in a kaleidoscope, where the images are a reflection of the chips that have been inserted, the human mind definitely needs an input of ideas that eventually will germinate other ideas. Thus, the larger the input in the memory bank, the larger and more complex will be the output, so that a major emphasis is given to this visual resource to which we have referred.

Also, another useful tool is to have something that indicates the meter that is used by official juries of so-called experts, when they grade on a yearly basis those bridges that have been built in that period. In fact, several organizations have been grading every other year the most prominent bridges, including for instance the American Institute of Steel Construction, the Federal Highway Administration, and recently the Portland Cement Association. A full documentation of these projects, combining drawings and reviews, would definitely be a practical tool to be developed by individual engineers and engineering firms as part of their library.

A fifth lesson on the aesthetics of small bridges

Bridges covering long spans within the various structural categories obviously attract the most attention from the engineering point of view based on structural analysis and construction techniques that make the structure possible. Bridges that reach spans close to the limits attract the interest and admiration by members of the profession. However, aesthetic factors have entered into the overall process of judging bridges with an increasingly dominant value. Therefore, although engineering aspects will always be hierarchically high, they are no longer sufficient to win a competition unless they include aesthetic values. Therefore, small bridges have an equal chance to compete for aesthetics.

In this context, the architectural aesthetics of small bridges is discussed using examples to validate the point. Let us consider the case study of a pedestrian bridge recently designed and built in Japan. The designer is Professor Mamoru Kawaguchi from the Department of Architecture of Hosei University, Koganei, Tokyo, Japan. The granite pedestrian bridge located in Beppu, Japan, on the northeast coast of the Kyushu Island, is the structure in question described in the IABSE's magazine *Structural Engineering International* (March,1996). This city, famous for its hot springs and scenic beauty, had commissioned the building of a pedestrian bridge for providing access across a river to a park (Figures 4-4 through 4-9).

Figure 4-4 Inachus Bridge, Beppu, Kyushu, Japan (photograph by M. Kawaguchi).

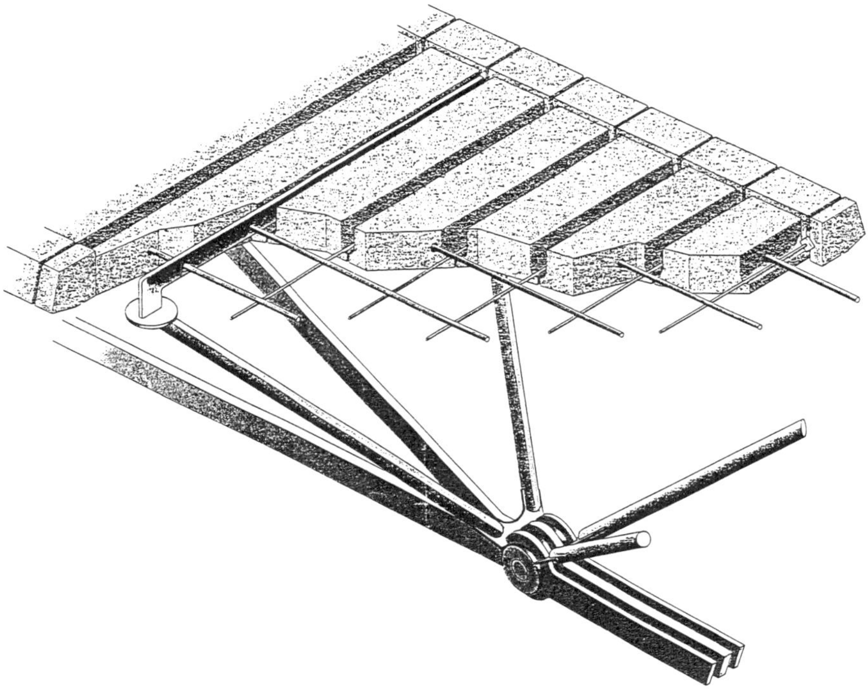

Figure 4-5 Inachus Bridge, cutaway drawing showing the structural elements (drawing by M. Kawaguchi).

The dominant factor in the program emphasizes the aesthetic qualities for a structure that would harmonize with a scenic landscape that included a view of the Beppu Bay and the mountains. The response to such parameters was to design a bridge of particular simplicity articulated with an explicit structural system. The bridge, 34 m long, has a lenticular structure with the top chord consisting of granite blocks and a bottom chord of steel. The granite for the top chord had a symbolic value because it came from a sister city in China named Yantai which had already furnished granite for the pavements of the city of Beppu.

The blocks of the top chord, perforated to receive the cables, were installed prestressing the granite, forming a monolithic chord interposing concrete joints between the blocks. The bottom chord instead was made of steel plates and has a configuration of the funicular shape symmetrical to the top chord. The web members connecting top and bottom chords consist of steel tubing arranged in a pyramidal configuration.

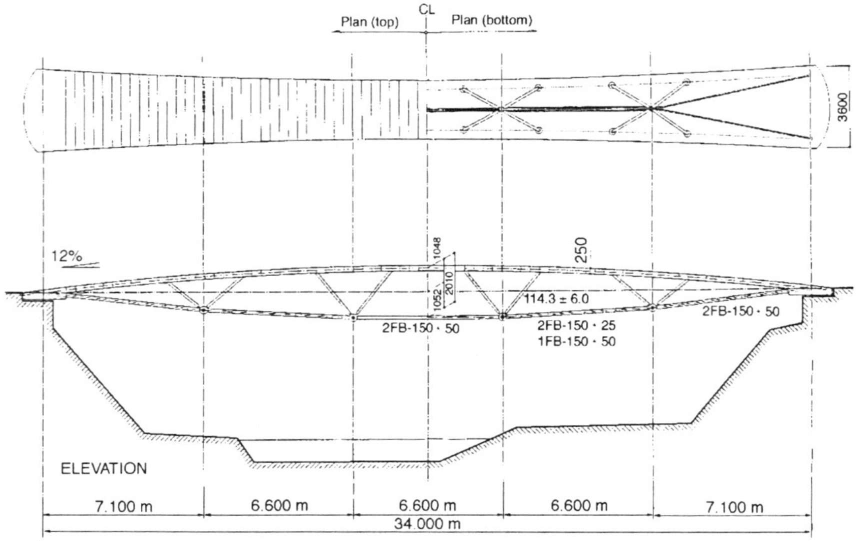

Figure 4-6 Elevation and plan of the Inachus Bridge (drawing by M. Kawaguchi).

Kawaguchi's work is an outstanding example of pedestrian bridges within the background of a landscape of particular aesthetic values. Pedestrian bridges of this kind could probably form a category of their own. This depends not only on the fact that they are dedicated just to pedestrian traffic, but also on the fact that they are harmoniously inserted within landscapes of great significance such as are usually found in parks, gardens and other landscaped forms. Examples of the typical pedestrian bridges of this type can be very commonly found, such as that in Marshall Park, Charlotte, NC (Figures 4-10 through 4-12).

Lesson 6: bridge awards

Of major importance to any designer entering a bridge competition is without doubt a panoramic view of those projects that have been evaluated in the past and have succeeded in receiving an award. The awards to which reference is made in the present context are the biennial American Institute of Steel Construction/National Steel Bridge Association (AISC/NSBA) Prize Bridge Competition that was initiated in 1928 and the Portland Cement Association bridge awards which started in 1988.

Any documentation concerning the awards will be the most important guidelines that any designer should consult in preparation of any design proposal. The basic criteria that the jury adheres to in the evaluation process include aesthetics, economics and innovation in structural design. It would be of major importance to

Figure 4-7 Inachus Bridge, detail (photograph by M. Kawaguchi).

analyze in detail the official comments made by the jury for each individual project on the basis that such comments may give a further notion on how the jury thinks. Furthermore, each individual project speaks on its own, furnishing ideas that could eventually be applied to other cases or extended forward to new applications. The steel bridges that have received the previously mentioned award is listed in detail in Appendix 9, which includes: (a) the category in which the bridge has been classified, (b) bridge name, (c) jury comments, (d) designer, (e) consultant, (f) general contractor, (g) fabricators, (h) erectors, and (i) owner.

These data are extremely valuable for anyone who wants to go deeper into the analysis of each one of these bridges. For instance, an individual could document on their own any bridge in question by taking photographs of the structures from different angles. By personal inspection of each bridge, the individual may want to expand his or her notion about the structural system used in general or in detail by annotating any visual observation. Knowing the name of the fabricator and erectors, more investigations could be extended to acquire first-hand details

Figure 4-8 Inachus Bridge (photograph by M. Kawaguchi).

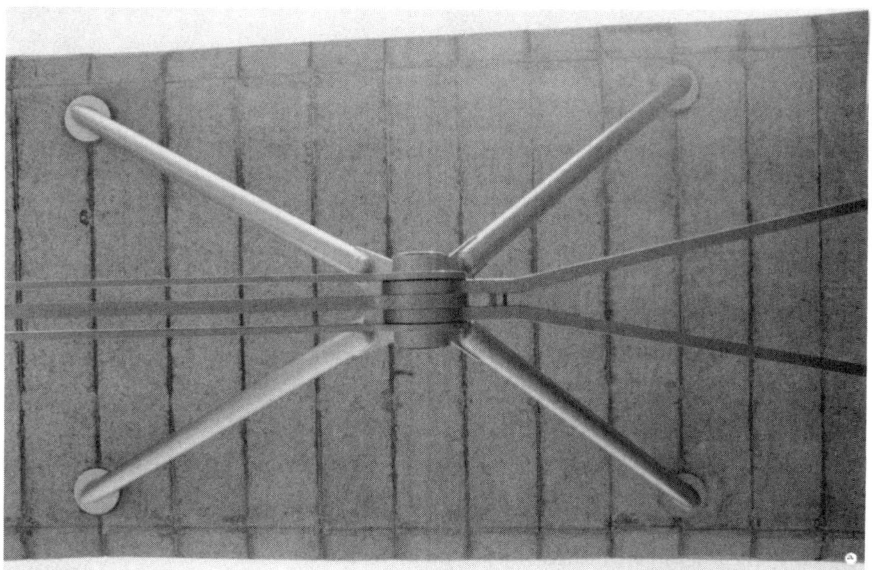

Figure 4-9 Inachus Bridge, truss hinge and granite deck from below (photograph by M. Kawaguchi).

Figure 4-10 Marshall Park (Charlotte, NC), with pedestrian bridge in background, fountain in foreground, and seating area over the water between the two (photograph by D. Melaragno).

Figure 4-11 Marshall Park, fountain adjacent to bridge and seating area (photograph by D. Melaragno).

Figure 4-12 Marshall Park, bridge with seating area in background (photograph by D. Melaragno).

about fabrication and erection that could be used in predicting similar problems in relation to the structure that one intends to design. In other words, knowledge of these data constitutes an important lesson for anyone interested in the design of a new bridge.

Similar data are also included for concrete bridges in Appendix 9 for the 1996 bridge awards, which includes nine bridges, and for the 1993 awards, which includes eight bridges. Illustrations of the concrete bridge winners for 1996 and 1993 have been procured from the Portland Cement Association and are herein included by their permission (Figures 4-13 through 4-29).

Aesthetics vs. ornamentation

To an audience of architects, any discussion addressing the separation between ornamentation and intrinsic aesthetics in a architectural artifact would definitely be a redundant topic equivalent to the beat of a dead horse. No one would bear to hear concepts that have been repeated over and over in the educational realm. However, to an audience of engineers the subject may not be as clear, therefore it seems appropriate to present a few remarks on this topic.

Any element of aesthetics in a work produced by a craftsman for the service of man is assumed to emerge as part of the product as a result of the craftsman's ability to express the real nature of the artifact. One could probably say that aes-

Figure 4-13 Big Qualicum Bridge (Vancouver Island, BC, Canada; 1996; photograph courtesy of Portland Cement Association).

Figure 4-14 Depot Road Bridge (Coventry, CT; 1996; photograph courtesy of Portland Cement Association).

Figure 4-15 Kellogg CBD Viaduct (Wichita, KS; 1996; photograph courtesy of Portland Cement Association).

Figure 4-16 Log Farm Access Bridge (Ottawa, Ontario, Canada; 1996; photograph courtesy of Portland Cement Association).

Figure 4-17 North Santiam River Bridge (Gates, OR; 1996; photograph courtesy of Portland Cement Association).

Figure 4-18 Pasadena Blue Line, Los Angeles River Bridge (Los Angeles, CA; 1996; photograph courtesy of Portland Cement Association).

Figure 4-19 Mendota Bridge, reconstruction (Mendota, MN; 1996; photograph courtesy of Portland Cement Association).

Figure 4-20 Portage Avenue Culvert at Omand's Creek, rehabilitation (Winnipeg, Manitoba, Canada,1996; photograph courtesy of Portland Cement Association).

Figure 4-21 Rockcut Bridge (Stevens and Ferry Counties, WA; 1996; photograph courtesy of Portland Cement Association).

Figure 4-22 Acosta Bridge over St. Johns River (Jacksonville, FL; 1994; photograph courtesy of Portland Cement Association).

Figure 4-23 Windward Viaduct (Haiku Valley, Oahu, HI; 1994; photograph courtesy of Portland Cement Association).

thetics within an object comes from within rather than from the outside. Following other similitudes, for instance, the tradition of creating a typical Christmas tree starting from an evergreen and completing it by adding a variety of decorations is the most visual example of what would be absolutely forbidden in the field of design. Yet, such a absolute basic rule may not have been a part of engineering education.

Applying this concept to bridges, we occasionally notice several transgressions to such a basic rule of design when we examine many traditional bridges built throughout various historical periods. Probably one of the most vivid examples that comes to mind is a bridge crossing the Tiber River in Rome right in

Figure 4-24 Natchez Trace Parkeway Bridge (Franklin, TN; 1994; photograph courtesy of Portland Cement Association).

Figure 4-25 Oak Canyon Bridge (San Diego, CA; 1994; photograph courtesy of Portland Cement Association).

Figure 4-26 Cross-over, interstate routes 5 and 14, (Los Angeles, CA; 1994; photograph courtesy of Portland Cement Association).

Figure 4-27 Scripps Crossing (San Diego, CA; 1994; photograph courtesy of Portland Cement Association).

Figure 4-28 Semiahmoo Trail Pedestrian Bridge (Surrey, BC, Canada; 1994; photograph courtesy of Portland Cement Association).

Figure 4-29 Upper Middle Road, 16-Mile Creek Bridge (Oakville, Ontario, Canada; 1994; photograph courtesy of Portland Cement Association).

the task for which they have been built. In so doing, their basic forms reflect the logic of pristine prototypes to be analyzed with care in order to capture their basic identity. Most of these structures have been designed following basic standards without the participation of individual designers. They are simple and easy to grasp in their simplicity.

The bridges we are referring to in this context are those that we commonly see crossing over the main interstate highways of the United States, as well as similar counterparts in other countries. To adhere to a sense of order, similar designs are typically used for the bridges crossing over any major interstate highway. Their commonality of forms does not diminish their basic sense of aesthetics, since their repetitiveness is supported by their interrelationship among themselves through the highway system to which they are connected. To this effect these bridges, which are part of the landscape that many motorists experience visually on a daily basis, teach us a lesson on the aesthetics of simple and unpretentious morphological stereotypes. A series of photographs included herein capture the details and the overall configurations of such typical bridges crossing over main highways in and around a typical city, Charlotte, NC. Abutments, piers, girders, parapet walls etc. merge to form individual compositions of bridges that make us read them as the basic building blocks that, put together, create structures easily comprehensible (Figures 4-33 through 4-50).

Lesson 9: the *Bridge of Sighs* and its modern counterparts

The "Bridge of Sighs" is reintroduced in the present context to illustrate a recent modern version that has become popular in the United States. Within the typical repertoire of visual illustrations of the cityscape of Venice, the "Bridge of Sighs" is always present. In the guided tours of the city and in the guidebooks of "La Serenissima," this bridge is never omitted. Banister Fletcher, in his book *A History of Architecture,* refers to it as "a romantic feature, with its elliptical arch, rusticated pilaster and heraldic devices." Built in 1595, this structure connects the Doge's Palace with the prison building called "Piombi." The latter term, meaning leads, referred to the lead roof which made the prison cells hot and particularly uncomfortable for the inmates. The "Bridge of Sighs" carries the name given to the structure because of the imaginary sighs of the prisoners crossing the bridge after receiving their sentences.

This structure spans from building to building high above the intermediate canal, and it seems to be the first prototype of a bridge connecting two separate buildings high above the ground level. Can this be the prototype first to appear in western architecture? Whether this structure is the first or not does not really matter, but the fact is that this structure is the first to come to mind when one looks at the present-day prototypes spanning from building to building. This relatively new urban feature, has become particularly popular for the task that it accomplishes.

front of Castel S. Angelo. This structure, consisting of one main central masonry arch and two lateral ones, is particularly short and narrow, yet it was retrofitted at one time with twelve oversized statues. Lining the bridge, six on each side, these were carved of marble. Two of them were the work of Bernini himself and the remaining ten were the work of some of the master's pupils. Representing the figures of angels with prominent wings, each of the statues is particularly overwhelming to the pedestrian crossing the bridge. At the same time they also dwarf the entire structure, creating a crowded effect that suggests confusion and lack of scale. In fact, judging from these elements, it is obvious that the statuary were not part of the original design but were posthumous to the completion of the bridge, proving the point in question.

This example illustrates the need of removing redundancies in appropriate cases. In simple terms, a drastic but effective solution would be the removal of such overwhelming decorative elements, bringing the bridge back to its pristine original silhouette of a simple masonry arch bridge. Acquiring again its original silhouette in the proper scale, the bridge would definitely re-acquire its intended character based on simplicity. Furthermore, it appears that such a solution would improve the relationship between the bridge itself and the fortress that faces it. Castel S. Angelo, in fact, in its sobriety of a medieval, severe, and austere delineation of forms, would indeed harmonize much better with the simplicity of the new bridge rehabilitation.

The removal of such statuaries would generate a two-fold advantage, because it would enhance the bridge itself and it could provide twelve aesthetic features for the urban landscape. The individual statues could be relocated throughout the existing parks in the city of Rome or in other singular landmarks, following the same tradition that relocated the several Egyptian obelisks within the topography of the cityscape.

In other words, the essential elements expressed during the conceptual phase of design should be brought out at that stage rather then attaching them retrospectively. Specifically, some of the basic elements that come readily to mind are: (a) the railing in its variations of forms such as balusters, and parapet walls expressed with different delineations, (b) luminaries in terms of their location along the railing that follows the sidewalks or mounted overhead on suspension cables, including any geometrical rhythm, (c) sidewalks subdivided into different categories such as urban bridges, highway bridges, rural bridges, and pedestrian bridges; development of the walkway areas delimited between exterior railing and curb line; consideration of pavement in this area; consideration of the curb itself; separation of pedestrian area from the roadway itself; relationship between curb line and guardrail; consideration of observation points in terms of size, pattern, etc., and (d) special features at the entrance of the structure announcing from a certain distance the bridge entry to the motorist. In other words, delineating in architectural terms the approach to any bridge that will constitute a different experience for the motorists who will experience a sudden change in the surrounding background as well as often a change in the surface of the pavements.

These previously mentioned items are examples of particular features which offer to the designer an opportunity to develop themes that characterize individual projects without incurring any criticism of ornamentation.

Such elements are essential parts of the bridge itself that require special attention to the details often constituting special focal points that, although contributing to the overall perception of the bridge, cannot be disassociated from the main structure as posthumous elements of decoration.

Lesson 7: graphics

A typical approach to instructional design is the philosophy that the content of any project should be the prevailing factor guiding judges in assessing an engineering artifact; yet when elements of aesthetics are involved, juries are often affected by factors other than pure rational ones. In judging aesthetically a work of engineering, such as a bridge which in reality may be considered also an architectural artifact, it is easy to be induced to emphasizing the aesthetic qualities over the rational ones.

When this happens, consciously or not, the factor that dominates is most of all the rendering of the object through graphics or modeling. This is a very well known fact that any student in an architectural curriculum knows about the jury critics. It is seldom, in fact, that well-executed drawings do not capture the attention and the appreciation of jury members. Therefore, it is important for structural engineers presenting ideas for the preliminary designing of bridges to learn from their architect colleagues the basic rules of good renderings.

Some basic rules can easily be borrowed from any text that teaches renderings. However, a few comments should be mentioned: (a) ink drawings are definitely sharper than pencil drawings especially when dealing with elevations and perspectives, (b) use dark shadows, especially for elevations and perspectives, (c) use more than one thickness of lines when producing drawings; probably three different values would be sufficient so that different emphasis could be placed in a single drawing, (d) sometimes accentuate drawings with color, (e) keep drawings extremely simple without crowding them together on the same sheet, and (f) be clear in the notations and identify individual components.

Another factor that is typically ignored, even by experienced architectural draftsmen, is the effect that structural drawings can sometimes achieve, transforming a construction drawing into something that acquires aesthetic values. Structural drawings are in a category all their own and are hardly part of preliminary drawings. Not much emphasis is placed on them since they are not drawn to capture the attention of the viewer. Yet one finds exceptional structural drawings that command attention when these drawings are generated by special individuals who succeeded in creating structural artifacts.

We speak, of course, of those masters such as Maillart, Nervi (Figure 4-30), Torroja, Morandi, and Isler. Figure 4-31 shows the effectiveness of structural drawings in an example by Pier Luigi Nervi showing the formwork for the viaduct at Corso Francian in Rome. This figure was published in a book dedicated to his work and is hereby reproduced with permission from the copyright holders.

Similarly, another exceptional rendition of structural drawings competing in elegance with the best of architectural graphics, is the work of Riccardo Morandi. The illustration in the book *The Concrete Architecture of Riccardo Morandi*, by Giorgi Boaga and Benito Boni (Alec Tiranti, London, 1965, pp. 76-77), illustrates the longitudinal section showing the reinforcement of the inclined piers, of the ties and of the beams, of the viaduct on the Via Olimpica in Rome (Figure 4-32).

Figure 4-30 Pier Luigi Nervi (photograph courtesy of Editalia).

Lesson 8: the *Aurea Mediocritas*

From the highly acclaimed bridges, winners of special awards, to the small bridge culverts, the large majority of typical bridges stand silently in between, mostly unnoticed and taken for granted. Referring to them as "the silent majority" according to a terminology popular in the last few decades, or even better, referring to them as the "aurea mediocritas" from the pages of Latin manuscripts, this large family of structures represents the essence of what constitutes the essential identity of bridges. On this basis, the members of such a large family respond most clearly to the basic demands that bridge structures have to satisfy to accomplish

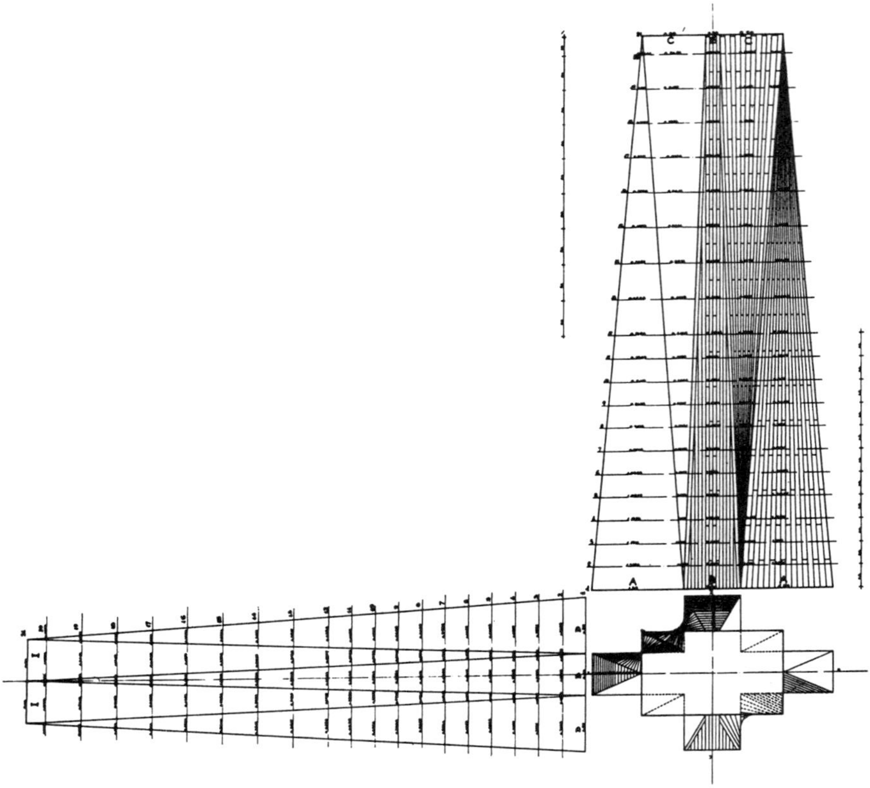

Figure 4-31 Viaduct at Corso Francia, drawing of formwork (1960, Rome, Pier Luigi Nervi; reproduced with permission of copyright holders).

Such tasks, in fact, create a secondary pedestrian circulation above the street level. Pedestrians within a building can comfortably cross over busy streets to other buildings. This comfort includes the thermal control of the bridge space, but even more important, protects the public from the unsafe environment of the city streets.

The morphology of such bridges is not restricted by any statical limits since the relatively short span can be materialized by means of any available structural system. Their own identity, in fact, can be expressed in architectural terms through an unlimited vocabulary depending on any choice of materials and architectural delineations. Such bridges, in fact, being part of the original buildings or retrofitted many years later, can manifest an identity of their own independent of that of the buildings that they connect, providing that they do not contrast with the surroundings. This, of course, can be achieved also with structures that fit not just by conformity but also by contrast that can still harmonize with the surrounding structures.

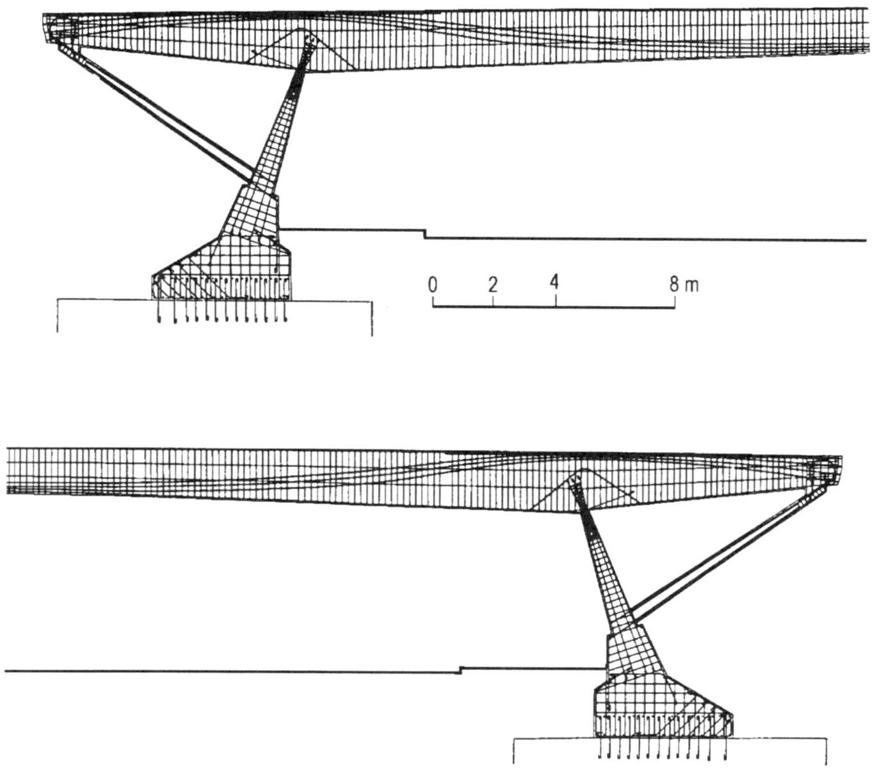

Figure 4-32 Drawing of the steel reinforcement for a viaduct along the Via Olimpica (Rome, Riccardo Morandi; reproduced with permission of copyright holders).

Some examples of the many urban structures of this type that have become popular in the United States' cityscapes have been included in this context to identify in practical terms the visual characteristics of these pedestrian bridges. These examples were taken from downtown areas in the cityscape of Charlotte, NC (Figures 4-51 through 4-53).

THE SILENT MAJORITY

Borrowing the term from a recent political aphorism, the "silent majority" is hereby used to identify the large number of anonymous bridges that constitute a family of structures that the typical motorist driving on a highway looks at daily without noticing. Reference is made to overpasses that cross typical highways and maintain with their uniformity a sense of order along the highway itself. These bridges

Figure 4-33 Approaching view of two consecutive overpasses over Interstate 485 (Charlotte, NC). As part of the same system, they reflect the required homogeneity of all overpasses on the same route (photograph by D. Melaragno).

Figure 4-34 Interstate 485 overpass (Charlotte, NC) showing center support consisting of three closely spaced columns emerging from the ground without pedestals (photograph by D. Melaragno).

Figure 4-35 Interstate 485 overpass detail showing columns with single unifying capital (photograph by D. Melaragno).

Figure 4-36 Interstate 485 overpass (Charlotte, NC) abutment with transitional concrete slope (photograph by D. Melaragno).

Figure 4-37 Short bridge on Interstate 485 (Charlotte, NC) continuing the flow of the highway in a smooth transition free of any particular characterization (photograph by D. Melaragno).

Figure 4-38 Rail overpass on Interstate 485 (Charlotte, NC) differs from the other vehicular bridges on the same route. Notice the monolithic center support instead of triple columns (photograph by D. Melaragno).

Figure 4-39 Rail overpass, closeup detail (photograph by D. Melaragno).

Figure 4-40 Overpasses of four highways showing a composition of engineering rationality. Like an expression of "constructivism," bridges, overpasses, and viaducts can be seen as significant examples of structural art (photograph by D. Melaragno).

Figure 4-41 View of the first level of the above system of overpasses (photograph by D. Melaragno).

Figure 4-42 View of the above system of overpasses showing slender T-shaped supports of the upper-most circular overpass (photograph by D. Melaragno).

Figure 4-43 Detail view of above slender T-shaped supports (photograph by D. Melaragno).

Figure 4-44 Overpass of Interstate 77 over Westinghous Boulevard (Charlotte, NC). Note the double structure of the two opposite lanes of traffic and details of the four-column piers (photograph by D. Melaragno).

Figure 4-45 Overpass over Billy Graham Parkway (Charlotte, NC) showing stiff supporting pier near one abutment consisting of five closely spaced cylindrical columns. This configuration is basically similar to the uniform standards in the area, but differs from the archetypes used along nearby Interstate 85 (photograph by D. Melaragno).

Figure 4-46 Detail of another overpass over Billy Graham Parkway showing the longitudinal steel girders bearing over the top of the concrete beam supported by the cylindrical columns (photograph by D. Melaragno).

Figure 4-47 Overpass over Billy Graham Parkway, detail showing the columns forming a side support near the abutment emerging along the slope (photograph by D. Melaragno).

Figure 4-48 Overpass over Billy Graham Parkway, detail showing the overall configuration from below. Notice the overall simplicity of the integration of the concrete and steel components in the overal assembly (photograph by D. Melaragno).

Figure 4-49 Detail of an overpass showing the steel girders, cross bracing, and corrugated metal deck which supports the concrete pavement above (photograph by D. Melaragno).

Figure 4-50 View from below of another overpass showing the steel superstructure (photograph by D. Melaragno).

Figure 4-51 Urban pedestrian bridge (crossing College Street, Charlotte, NC) connecting building at second floor (photograph by D. Melaragno).

Figure 4-52 Superimposed pedestrian bridges (Charlotte, NC) over parking lot connecting two buildngs. The upper deck is enclosed and the lower deck open (photograph by D. Melaragno).

Figure 4-53 Pedestrian bridge (Charlotte, NC; photograph by D. Melaragno).

execute their functions without the glamour associated with the major bridges around the world. The latter, in fact, capitalizing on ingenious structural systems, can span long distances and capture public admiration. On the other hand, the former short overpasses have nothing to show in terms of engineering originality, and stand on their own as pure expressions of simplicity.

As previously mentioned, on the topic of aesthetics in general, artifacts on a smaller scale can compete with a counterpart on a larger scale without suffering any penalty. Thus, theoretically the design of small bridges should by all means have equal chances of producing a work of art just like the design of megastructures. Therefore, it has been useful to analyze typical small overpasses conventionally visible along highway systems. In so doing, the basic structural components are scrutinized in their essential role and are pointed out as the basic components that, when put together, give a certain identity to the structure as a whole, bringing it out from its anonymity.

Compatibility with other bridges

Among the various requirements that have been discussed concerning the aesthetics of bridges, much has been said about the compatibility of any bridge with its surroundings. More specifically, consider the case of a bridge to be designed in the proximity of an existing one. The new structure becomes visually related to the existing one through the variety of criteria that need to be addressed. To this effect, a recent paper by Matsumoto, Yamanaka, Mori, Suzuki, and Taguchi discusses this specific subject through a case study that illustrates basic design criteria. The case in question is to the Shin-Onomichi Bridge to be constructed at a distance of only 55 meters from the existing Onomichi Bridge over the Onomichi Channel between Honshu and Mukaijima Island in Japan.

The original bridge (1968) consists of a three-span cable-stayed structure with a center span of 215 m that, at the time of construction, represented a milestone in this bridge type in Japan. Because this existing structure constitutes a major landmark in the area, the new bridge had to be designed to be compatible with the older one. How can one bridge be compatible with another? This paper tries to describe following the design process adopted for their new bridge.

Basic design criteria adopted included a series of requirements as follows: (a) harmonizing with the existing bridge without detracting from its appearance, (b) avoiding any negative visual effects, and (c) acting as a gateway symbol for the Nishi-Seto Expressway. The selection of the type of bridge followed a process that included: establishment of basic themes, features of the landscape, survey of existing parallel bridges, and evaluation of bridge types.

In determining the theme of the new bridge, the local seascape and the perception of the general public concerning the existing bridge were considered. On this basis, several designs were prepared and analyzed in terms of harmony and creativity. The concept of harmony was further emphasized with respect not only to the existing bridge but also with the surroundings. Yet, in concomitance with harmony, the new bridge should further emphasize the special characteristics of the previous structure that acted as a focal point in the cityscape.

Furthermore, after examining similar situations where two bridges must relate to one another for reasons of proximity, it was found that two basic bridge design types were suitable: cable-stayed, and girder. In the latter case, the following considerations were made: (a) the new bridge would have a low-key image which would not diminish the character of the existing bridge, (b) the upper structures of the two bridges would not interfere with each other, and (c) the two structures in harmony with one another would reinforce their visual appearance.

Further evaluation for the design of the new bridge also included: the impact of the bridge on the landscape, and the structural aspects. With regard to the structural aspects, the following points were taken into consideration: (a) feasibility of design and fabrication, (b) ease of construction, (c) aerodynamic stability, and (d) economic feasibility.

Signs: functional and aesthetic components of bridges

Graphics, within the spectrum of topical issues in the complexity of present-day structures, occupies a particular place in the design of modern bridges. Its role is dual in that, while its purpose is of a functional nature first, there is also the secondary aspect involving aesthetics. With regard to the functional role, it is important to note that bridge users of today are motorists moving at high speeds who depend on visual instructions to make instantaneous decisions about which lane of traffic to use.

This must be accomplished without making wrong choices that could result in complex detours in order to correct the wrong selection. Lack of clarity in signs can indeed be more drastic than just annoying. They could be the cause of fatal accidents when the motorist becomes confused and eventually makes corrections at the last instant, causing potential collisions with the flowing traffic. Graphics on bridges are probably more important than they are in many other applications. In situations involving pedestrians, the time for decision making is practically not a factor, while in situations involving motorists, the instantaneous choice previously described makes it of much greater importance.

The structures that support signs should be three-dimensionally located for the best possible view. Such structures constitute an integral part of the overall design of bridges. In fact, the AASHTO Code dedicates a section to the design of sign-supporting structures.

In practical terms, the design of signs and the framework supporting them involves first the configuration of the sign itself, including shapes, colors, and sizes and then the supporting structural components. In any design process, trial and error is the conventional methodology used in order to achieve the desired aesthetic qualities and can be extremely time consuming. To simplify the design, some computer software addresses this topic, facilitating therefore the design process. Among these, The Bridge Engineering Software (BEST) Center of the Department of Civil Engineering at the University of Maryland has a program entitled Sign Bridge Analysis and Evaluation System (SABRE).

According to the BEST Center, "SABRE was developed to shorten and simplify the design/analysis process for sign support structures. The program uses AASHTO Code for load application and allowable stresses, and uses the direct stiffness method for three-dimensional space frames for analysis. With the advances made in personal computers and improvements in the SABRE program, it is possible to design and analyze a fairly complicated sign bridge structure in minutes, resulting in a significant increase in production." The overall design of supporting structures of bridge signs can be better understood by dividing the typical whole structure into several components: (a) sign bridge configuration, (b) tower types, (c) beam types, (d) sign bridge types, (e) member types, (f) section types, (g) sign types, (h) walkway types, and (i) vertical attachment member (VAM) types.

Each of these nine components have several options. (a) There are five types of sign bridge configurations. (b) There are two tower types: single- or double-post. The single-post can be prismatic with a uniform cross-section or it could have a tapered variable shape. The double-post instead can be of the trussed type which will make the two posts structurally interacting. (c) There are eight different beam types. (d) There are six sign types (attained by various combinations of beams and towers). (e) There are four member types, (f) eight section types, four tubular shapes, and four general shapes. (g) With regard to sign types, the geometry being selected includes only rectangular shapes. (h) Walkways are rectangular and mounted parallel to the bridge beams, and (i) vertical attachment members (VAM) types are general prismatic shapes only.

A step-by-step illustration on the special features of this program include the following: (a) pull-down menu system, (b) data entry and editing, (c) automatic joint and member renumbering, (d) shape files, (e) on-screen help, (f) hot keys, (g) automatic mesh generation, (h) on-screen graphics, and (i) structural analysis capabilities.

The following complimentary remarks will help in the understanding of the entries of the previous list: (a) the pull-down menu system, helps with navigating about the program, (b) with regard to data entry and editing, the user may move to any cell by using the arrow, tab, and page keys and may move to any screen by using the pull-down menu system, (c) joint and member renumbering, that avoids the manual changes of joints and members, (d) the shape files, feature allows the user to call the file onto the screen and select a suitable section, (e) with respect to on-screen help, there is a key (F1) that displays information on the part of the program in use, (f) hot keys, are single character commands which perform a specified operation, (g) the mesh generator can create all data required to define the joints and members of a bridge sign, (h) with regard to on-screen graphics, the graphic image consists of a wire frame representation of the structure and can be of the current structure being edited or a structure previously analyzed, and (i) with respect to structural analysis capabilities, this program can analyze support structures for a wide variety of configurations, boundary conditions, member types, and different loading conditions.

This program is one of several currently available. They are part of the practical tools that are available for any designers whether structural engineers or architects, who try to achieve meaningful and aesthetic solutions in the competitive market of bridge design.

The aesthetics of box girders

Box girders of steel, concrete, or composite steel and concrete have been discussed previously in Chapter 2. They are reconsidered here with regard to their aesthetic qualities. These structures allow for the first time not only rectilinear longitudinal axis but also a curved longitudinal axis. This is a characteristic unique to box girders. No other structural type can be built with a horizontally curvilinear axis with such simplicity. An axis curved in the horizontal plane creates torsional moments of extremely high magnitude that would be unbearable by other conventional structural members. The box girder, with its tubular cross section, attains torsional stiffness with respect to the shear center that is unattainable by I- or W-shaped sections. The torsional resistance of a box girder can be as high as 100 to 1,000 times that of an equivalent open-section girder.

Because of the complexity of fabricating closed sections, the box girder has been widely adopted for bridges only when welded construction is permitted by governing codes. As the steel box girder emerged, its application to bridges became astonishingly dominant with respect to other systems. Its appearance has captured the imagination of designers for its simplicity and purity of form, especially when compared with more conventional steel construction shapes requiring exposed bracing members. In the box girder, everything is smooth and flows naturally, displaying clean surfaces that reflect architectural qualities.

When designers utilize box girders, (steel or concrete), they often encounter significant criticism about the visual compatibility of the box girder with the surrounding environment. This is true whether the surroundings consist of a natural landscape in the open country or in an urban context, due to the fact that the simplicity of the forms does not make any strong statement that could create contrasts. Elegant in its simplicity, box-girder bridges silently perform their task in structural terms without demanding any attention. Silent in the architectural language of the morphology of structures, the surfaces of box girders allow for the display of other design statements to be made without cluttering the delineation of the bridge.

Furthermore, when applied to curved bridge configurations, the box girder enters in an incomparable architectural realm that transcends any other rectilinear composition in the horizontal plane. The purity of the form previously mentioned, with regard to the surfaces of the box girder, is now displayed in combination with the elegance of the curves that are soaring in the three-dimensional space from support to support. There is extensive literature on box girders in terms of structural design, fabrication, architectural values, and case studies which expand each topic in more specific detail. Suggested readings include: Ichimaru, Ishii, and Shinohara (1996), Richardson, Gordon and Associates (1976), Green and Strevel (1977), Wright (1968 and1969), Mattock, et al. (1967), Lally (1973), Tanner (1970), Strobel (1969), Heins (1978), Tung (1967), and Tung and Fountain (1970).

The aesthetics of piers

As previously noted, the intermediate supports of bridges can include several types. These respond to different situations and express logical solutions in terms of practicality of construction, as well as efficiency of design. They express the direct reaction of the engineering mind without any pretenses of exploring aesthetics. However, when one proceeds to a higher level of design, structural aesthetics manifests itself when the designer reflects a special sensitivity to form.

In the case of piers or columns, exceptional rendition of particular significance is presented in the accompanying figure. This column is part of Pier Luigi Nervi's repertoire of plastic forms that acquire a personality of their own, independent of the buildings of which they are part. In this case, the column is notable for the rationality of its form.

The shape of this column clearly responds to the two different conditions at the top and bottom. On the top, the cruciform shape indicates a moment resistant connection in two directions. At the bottom the square cross-section indicates another condition which could be a hinged connection or a moment connection with smaller moments. The top and the bottom are connected by the shaft which gradually transforms its cross section along transitional lines. The result is a gradual mutation of forms without any feeling of forceful manipulation. There is no distress or sense of distortion, but instead a sense of a fluid flow of forces from one end to the other of the column (Figure 4-54).

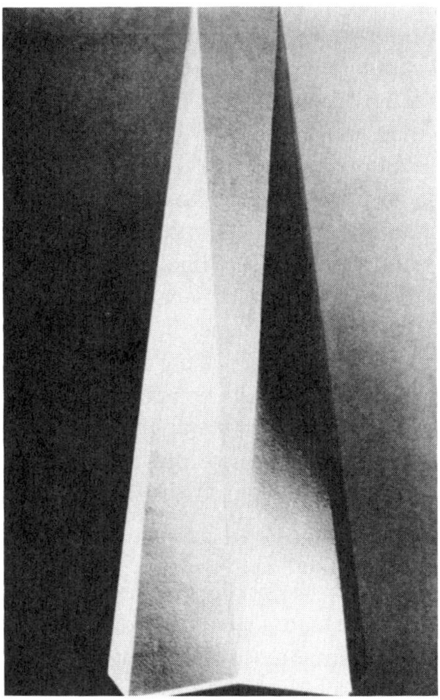

Figure 4-54 Port Authority Bus Terminal (1962, New York, NY; Pier Luigi Nervi) detail of concrete column (photograph reproduced by permission of Editalia).

Nervi discussed these specific columns in an open debate at the end of a lecture he gave at the auditorium at Massachusetts Institute of Technology (MIT) during a visit to the United States in the early 1960s. Asked whether the intermediate sections of the column were equally efficient as the sections on the top and on the bottom, Nervi confirmed that at any section the column reflected comparable efficiency. This column bears close study and reflection about other potential forms that this particular example can inspire.

There are many other examples of columns incorporated in buildings that can inspire designers who are searching for forms applicable to bridges. Yet another example of an articulated form of piers applied to bridges are the columns designed for the proposed bridge over the Nile, southwest of the town of Sherbeen, and near the Mediterranean Sea. The piers in the approaches of each side of the river, have been shaped in the form of trees to reflect the agricultural character of the surrounding area. The piers appear are Y-shaped in both directions due to the four branches stemming out from the top of the shaft (Hosny, Hosny, and El-Metwally, 1996).

A lesson on sketching

The representation of the physical realities through sketching and drawing is one of the first means of communication among humans dating prior to speech and writing. The sketches left by prehistoric man verify the typical proverb that a figure is indeed worth a thousand words.

Prior to the advent of photography, developed with the invention of the *camera obscura* and light-sensitive chemical coatings, sketching and drawing were the only means of expression for the visual arts. In the hands of some artists, sketching is a spontaneous skill that develops without any training; in others, sketching and drawing are learned through educational and practice.

Traditionally, the training of gentlemen in the 19th century, was not complete without a tour of Europe. With a diary and a sketch pad, the young gentlemen traveled, taking notes and sketching the works of both man and nature that impressed him in his sojourns through the major centers of Europe. Sketching was a mandatory skill that had to be acquired before completing a final education curriculum and, in the case of young architects, the skill of sketching was naturally of even more importance.

In the preliminary design of bridges, it is easy to recognize the warmth and the fascination that hand sketches project to the observer. Skillful architects, versatile in sketching, can capture the emotional feelings of the observer with a seduction that hard-line drawings and photography do not possess. Thus, the importance of sketching in the preliminary design of bridges cannot be over-emphasized.

Structural engineers are usually not particularly trained in sketching and are seldom spontaneously inclined to this type of skill. Furthermore, most engineering curricula may include training in drafting but seldom have courses in freehand drawing. Because of this, some suggestions on learning sketching techniques may be of benefit to engineers and other consultants.

It is recommended that a slide photograph be taken of the subject to be sketched. The subject can be any object, structure, other manmade features, or products of nature including plants and other vegetation, as well as the scenes of natural landscapes. After taking a number of photographs, these are projected on a wall or any other convenient surface, enlarging the projection to any scale one desires. A sheet of paper is applied over the surface, and by hand, anyone would be able to trace the relevant parts of the pictures being projected. The result is a vibrant freehand drawing that selectively emphasizes some parts and excludes (or de-emphasizes) others which may compete for attention, reducing the clarity of the focal points. This ability to control emphasis is a major advantage sketching has over photography.

At this point, one may have sketched the image of a bridge or may have sketched the landscape in which a new bridge would be inserted. With such sketches, at the proper scale it is easy to superimpose structures over the landscape choosing different observation points. The structure itself can also be modified in a sketch producing a variety of drawings to be used for comparative studies.

With such a simple technique, many designers, like structural engineers, environmentalists, and others can achieve immediate results without any long-term training, resulting in a sense of self-gratification.

Structural integrity

Typical curricula in civil engineering do not cover courses in design leaving a large vacuum in the cultural back-ground of students. Such a vacuum is reflected specifically in the absence of basic courses in design and also in history of architecture and architectural criticism. Upon graduation, the students entering into professional practice find themselves unable to sustain a dialogue with architects.

The frustration can be easily understood. A common language which is the basis for mutual understanding is non-existent, not only the terminology unknown on the part of the engineering graduates, but often confusion arises when terms used in architecture do not conform exactly to the definitions given in dictionaries. Especially when abstract concepts are addressed, communications become even more difficult.

In absence of special preparation, the engineering graduate in his or her design practice of projects in civil engineering would probably research examples of similar projects that could be found in magazines dealing with architecture or other sources. Typical situations that often occur following this process are honest misunderstandings conducive to solutions that are totally erroneous. A typical approach in the design of civil engineering structures can also include total disregard for aesthetics and architectural values, since the concept of engineering design is still valid within the scheme of functionality and deficiency. In this case, the aesthetic values are totally ignored.

When attempts are made to attain reach the approval of an audience through aesthetic values, it is extremely important that the designing engineer realizes his or her limitations. Architectural values can definitely apply to civil engineering projects, even structures such as buildings, factories, bridges, and even highways. The latter case, the functionality of the highways, does not exclude the inclusion of aesthetic factors as it may appear at first glance, rather the relationship of the roadway with the surrounding landscape is an essential part of highway design that takes into consideration the natural land forms that the highway crosses and the views that the motorist will experience while driving.

Aesthetic values relative to the views that the motorist experiences as he or she drives along the roadway can be effectively designed in such a way that the view could change from time to time, creating an interest through a variety of views in the landscape. This can usually be achieved by selecting a highway delineation that gradually changes directions using very gradual curves with large radius of curvature in the order of magnitude of railroad tracks. Typical applications for this approach to the aesthetics of highways are, for instance, the famous highway programs of Germany that pioneered such a design features starting in the 1930s.

Focusing on the consequences that derive from the cultural void in basic design and architectural criticism in engineering curricula, a special example is presented in this context. In the August 1997 issue of the Journal of the International Association For Shell And Spatial Structures (IASS) published periodically by the IASS, an article was published by the title of "Postscriptum to the IASS Symposium in Stuttgart, 1996." The IASS is a very prestigious professional society founded by Eduardo Torroja in the 1950s.

The association, includes a majority of structural engineers and a smaller number of architects from around the world, representing in practical terms all the continents. The writer of the article is a renown member of the association, specializing in structural engineering with a large engineering staff in Stuttgart, Germany. Among various structural typologies his firm specializes in bridge design, as indicated by eighteen bridge structures that are included in part of the curriculum of the firm.

Described in the article, is a recent case in which a group of structural engineers working for the German railroads had just designed a railroad bridge of which they were particularly proud. The writer explains further that such a group of engineers had invited him to express his observations on the structure in question. Surprisingly, the comments were absolutely negative. The bridge was constituted by two parabolic arches, one on each side of the roadway. These carried the roadway itself, only at two points in each arch.

Therefore, the bridge, supported at the abutment and at two intermediate supports, by the arch, did not justify at all the use of arches. In fact, it was pointed out that the portion of the arch extending way above the roadway was an unnecessary structure that was not needed to support the bridge. Furthermore, the parabolic shape of the arch, by all means, did not reflect the catenary curve of the loads. Only uniformly distributed loads generate funicular polygons with a parabolic delineation. The structure was therefore basically a negation of the principles to be satisfied in architectural design.

The arches in this case could not be justified from any point of view, but the engineers responsible for the design brought out a new aspect of their design parameters. These consisted in the intent of creating a glamorous structure which would impress the public at large. If the arches did not work in logical terms their existence was more than justified by the visual impact on the observer glamorizing the image of the railroads. The engineers commented, this was not the first case in which a structure fulfilled only symbolic purposes without any other reasons and other uses. Special reference was made to the parabolic arch in St. Louis, Missouri adjacent to the Mississippi River, designed by the architect Eero Saarinen.

The structure had been erected as a monument to the conquest of the west, symbolizing the gateway from which the pioneers started their long march toward the western shores of the United States. In a sense, one could understand as this professional group of engineers could be misguided by a misinterpretation of the comparison between the two structures. The arch in St. Louis is just a monument and should be judged as such. The bridge with two faked arches was instead a utilitarian structure which failed to express itself in logical terms and purposely tried to mystify the public presenting a configuration that did not respond to structural requirements.

Trying to attain architectural and aesthetic values in such a manner is a complete transgression of the basic rules of design. More specifically the criticism of the writer explained that in bridge design the form derives from the behavior of the structures under the applied loads. Furthermore, the criticism suggested that the symbolism of the arches would probably be more in tune with that used by a famous fast-food enterprise rather than create a positive image for the railroads. Furthermore, the critic explained that "... structural art is an integral part of culture and has something to do with responsibility, veracity, appropriateness, modesty, longevity as against kitsch, showing-off and fast depletion were in vain." The criticism proved the point and the project was later scrapped.

BRIDGES AND THE NATURAL ENVIRONMENT

The highway systems typical in most industrialized countries, have impacted the national environment over particularly large zones. The right of ways of present highways that develop several lines of traffic occupy a much larger surface area than those of the roadways of the past. These corridors dedicated to high-velocity traffic, contribute to create zones of environmental degradation.

Tunnels, bridges, and culverts that constitute an integral part of highway systems manifest themselves with the same relationship to the natural environment as in the case of highways interfering with thousands of miles of occupied land on the national scale. When considering a bridge within the reference of present day criteria, it has become of particular importance to relate the bridge structure to the surrounding landscape, harmonizing the two with one another following basic concepts of preservation of the natural habitat. So in the design of highways system, bridges included, it is necessary to consider the aesthetic values of the landscape surrounding the bridge within the parameters of present-day ecological requirements.

Fundamental points addressing the conceptual relationship between a bridge structure and the natural landscape, include topical issues related to conservation of flora and fauna with special attention given to wildlife preservation. Establishing guidelines to achieve such goals may leave the designer of a bridge in an uncomfortable situation because of the imprecise interpretation of harmonizing the bridge with an undisturbed natural environment. To solve such a problem it should be necessary that the various agencies in charge of regulating highways and bridge structures are the federal, state and local governmental levels. They should be very explicit in giving and exact definition of the terminology used and clear statements of the procedure to be followed. Basic conservation criteria derived from design, construction, operation and maintenance of highway and bridges can work in an integrated manner that support each other.

Procedures to be followed during the preliminary design, construction, and maintenance phase of a bridge structure should conform to guidelines organized into a logical order. In this regard, a useful reference is the1976 publication, *Design Guide For Wildlife Protection and Conservation for Transportation Facilities* published by the American Association of State Highways and Transportation Officials. Basic topics addressed include: a variety of pertinent subjects exploring basic data.

There are no doubts that structures like bridges erected along the paths of highways passing through an existing natural environment, can disrupt the original equilibrium not only generated by the bridge itself but also by the approaching highways. Further damage to the preexisting natural equilibrium can be caused during the construction phase when clearing the site, and grubbing the vegetation may have major effects disturbing the site to a much wider area than that strictly occupied by the structure itself. The first goal in the project is to specify construction procedures that minimize the disruption of the existing landscape involving the flora and fauna. One of the major causes responsible for the damage along the construction site of the bridge and of its approaches is the action of fuel and other toxic materials that come in contact with the vegetation as well as the wildlife.

When disruption cannot be avoided, it would be logical to specify remedial actions for restoring the original site conditions. Further precautions are suggested to be included in the specifications with regard to the protection against the erosion of the disturbed soil. Temporary procedures for controlling the erosion include the covering of slopes and ditches with plastic as well as the construction of drainage ditches that divert the surface run-off. The establishment of permanent vegetation over a disturbed site may be considerably improved by reducing the inclination of slopes, and by serrating the contour of the slope as well as rounding the top edge of the incline.

Typical sites involving the construction of bridges in rural areas include a variety of conditions such as the crossing of ravines with steep slopes, the crossing of gentle valleys covered with a permanent vegetation of different botanical species as well as water surfaces of rivers, streams, lakes, and wetlands that comprise swamps and marshes. Such natural habitat demand special attention for protecting the site altered by the construction of bridges and their approaches using appropriate methodologies that may vary according to the different types of sites just mentioned. To this effect, basic notions of ecological concerns are an essential necessity for the design team of highways and bridges.

With regard to the preservation of the existing wildlife in a natural habitat, one should recognize the basic definitions of specific terminology. The terming habitat related to wildlife it is intended to indicate a natural territorial space characterized by specific types of vegetation and other conditions that promote the survival of organisms. Basic characteristics of the habitat also include a food source, cover, water, and favorable climatic conditions. Furthermore, basic characteristics of a habitat extend to the size of the site, the type of soil, the contour of the land, the vegetation type, and the source of drinking water. In restoring a disturbed habitat that had been affected by the construction of highways and bridges, it is important to notice that it may not be necessary to reconstitute the existing habitat characteristics.

This is particularly important due to the fact that many species of wildlife adapt themselves to a variety of conditions, therefore leaving the option of choosing among different types. On the practical application, it is logical to require specific information of the existing ecological conditions, which should be part of the preliminary phase in the design of bridge structures and its approaches. Specific factors addressing the profile of a natural habitat, includes: territorial ranges of the existing wildlife; typical habits; populations census of the wildlife species; rate of mortality; distribution of ages within the wildlife population; change of conditions of seasonal climate affecting the wildlife.

As bridges cross waterways, their impact on the natural existing aquatic habitat is one of the most relevant effects to be considered among basic conservation strategies. Increases and basic changes of the aquatic habitat have a direct effect on the animal population, especially when changing the type of soil cover, the source of food, and areas for breeding. Similar concepts apply not only to dry land sites, but also to aquatic habitat that when they change, also produce a change in the animal population.

In general, basic knowledge of different ecological systems is a prerequisite for understanding the protective measures for reducing the impact of bridge structures on the existing habitat. Such knowledge is comprehensive of the trophic structure, which consists of the food energy level, and the interaction among the factors characterizing the ecosystem itself. Such factors deal with the nutrient types and quantities, food production sources, the animals consuming forage, the decomposition of organic matter and the quantification of solar energy absorbed. Among such factors one of the most typical correlation is that which relates the quantity of forage generated on the site, and the requirement of the wildlife which should be in a condition of equilibrium.

Images of wildlife affected within aquatic sites includes examples presented in the reference source *A Design Guide For Wildlife Protection and Conservation For Transportation Facilities*. The presence of alligators in marshes and swamps in the southern regions of the United States, specifically in the Everglades, plays an interesting role for the preservation of certain animal species. When the water level in the wetland is reduced to a minimum, by instinct, alligators dig a small pool of water that is referred to as "gator holes." The drought that will endanger the survival of several aquatic organisms, including fish and invertebrates that are part of the food chain supporting higher-level wildlife such as birds, reptiles, and small mammals. These pools have proven to sustain the existence of fish and other organisms which in turn are part of the food chain that support other wildlife species, including birds, reptiles, and small mammals. Under these conditions, alligators creating these pools may be the only source of water left to satisfy the demands of the existing animal population.

Considering the basic nutrients, that an ecosystem depend on for its animal conservation, typical components extend to carbon dioxide, oxygen, nitrogen, minerals, and salts. The major role of the producing factors concerning the nutrients involves the synthesis of plants and bacteria that change inorganic substances into organic products that are necessary to initiate any animal food chain. The decomposition of organic matter depends upon bacteria and fungi that in turn release nutrients into the ecosystem. Within the aquatic environment draining marshes would have long-term negative repercussions, while the creation of artificial ponds can benefit the environmental conditions also on a long-term basis.

SPECIAL BIBLIOGRAPHY: AESTHETICS OF BRIDGES

Considering the high specialized area of interest related to the aesthetics of bridges, it is useful to compile a special bibliography concerned just with this topic. To this effect it should be pointed out that a special bibliography on this subject had already been prepared by Martin P. Burke, Jr. whose work is also cited in the present bibliography. His work however was limited to 1986, while in this book the topic is further investigated up to the present date of publication. Also notice that although the two bibliographies overlap one another, to a certain extent they compliment one another, since they have been based often on different sources.

A.I.A., "The Project," *The Architects' Handbook of Professional Practice*. Washington, DC: American Institute of Architects, 1987.

Allan, B. J., "Aesthetics in bridge engineering." *Journal*, Institute of Highway Engineers, London, United Kingdom, May 1971, pp. 7-16.

______, "Aesthetics in highway engineering." *Journal*, Institute of Highway Engineers, London, United Kingdom, Nov. 1968, pp. 17-24.

______, "Bridgeworks in the new town of Cumbernauld," Proceedings of Institute of the Civil Engineers., November 29, 1964, pp. 497-534.

American Institute of Steel Construction, *Prize Bridges/1980* and *Prize Bridges/1983*. AISC, 1980 and 1983.

______, "Steel bridges 1984." *Engineering News-Record*, Special Advertising Supplement, McGraw-Hill, Inc., New York, New York, Nov. 22, 1984, pp. 518-531.

______, "Esthetics in concrete bridge design" A.C.I. MP-1, 1990, pp. 334.

Anheim, R., *A Review of Proportion*. 1955, referred to by Pandovan, R., *A Necessary Instrument? The Architect*. London, April, 1986, p. 29.

Bache, A., et al., "The aesthetic treatment of bridge structures" (correspondence). *Proceedings* (ICE). Vol. 145, 1901, pp. 216-244.

Bakht, B., and Jaeger, L. G. "Bridge aesthetics." *Canadian Journal of Civil Engineering*, National Research Council, Ottawa, Canada, Vol. 10, 1983, pp. 408-413.

Ballo, G., *The Critical Eye*. Heinemann, London, 1969.

Bard, A. S., "Aesthetics and the police power." *The American Journal of Economics and Sociology*, Robert Schalkenbach Foundation, Lancaster, Pennsylvania, Vol. 15, no. 3, 1956, pp. 265-275.

Berthelsen, G., "Beauty and the bridge." *Civil Engineering* (ASCE), June, 1968, pp. 68-70.

______ , "Beauty and the Bridge." *Going Places*, California Department of Transportation, Sacramento, CA, November–December, 1985, pp. 15-19.

Billington, D. P., "Bridge design and regional esthetics." *Journal, Structural Division, ASCE*, Vol. 107, no. ST3, 1981, pp. 473-486.

______ , "Bridges and the new art of structural engineering." *American Scientist*, Vol. 72, no. 1, 1984, pp. 22-31.

______ , "History and esthetics in concrete arch bridges," *Journal of the Structural Division, ASCE,* Vol. 103, November, 1977, pp. 2129-2143.

______ , "Bridge design and regional esthetics," *Journal of the Structural Division, ASCE,* Vol. 107, November. 3, Mar., 1981, pp. 473-486.

______ , *Bridge Aesthetics: 1925 - 1933*. Final Report 11th Congress of International Association of Bridge Structural Engineers, Vienna, 1980, pp. 115-120.

______ , "The new art of structural engineering," *Civil Engineering*, August, 1984, pp. 54-57.

Braithwaite, R. G., and Davies, D. J., "Aesthetics: Welded Highway Bridges." *Journal* (ICE), Vol. 34, 1950, pp. 151-153.

Buckland, P. G., "The inherent beauty of cable-stayed bridges," ACI MP-1, 1990, pp. 233-246.

Burke, M. P., "Aesthetics - rules, formulas and principles- the negative view," ACI MP-1, 1990, pp. 71-79.

______ , "Bridge Aesthetics Bibliography," *Bridge Aesthetics Around the World*. Transportation Research Board, Washington, DC, 1988.

______ , "Bridge design and the bridge aesthetics bibliography." *Journal, Structural Engineering* (ASCE), Vol. 115, no. 4, 1989, pp. 883-889.

Cohen, E., "In the eye of the beholder," ACI MP-1, 1990, pp. 1-11.

Connor, et. al.,"Bridge Appearance," ASCE, Vol. 106, no. 1, 1980, pp. 1678-1687

Crouch, A. G. D., "The aesthetics of bridges," Thesis (B. E.), University of Queensland, 1969.

______ , "Bridge aesthetics: a sociological approach." *Transactions*, Institution of Engineers - Australia, Barton, Australia, Bol. CE16, no. 2, 1974, pp. 138-142.

Day, H. B., *Aesthetics of Concrete Bridges*. Eleventh Biennial Conference -Sydney, Concrete Institute of Australia, Sydney, Australia, 1983, 12pp.

Elliot, A. L., *Aesthetics of Highway Bridges*. Civil Engineering ASCE, June, 1968, pp. 64-69.

______ , "Aesthetics in structures." In Special Report 138: *Environmental Considerations in Planning, Design, and Construction*, HRB, 1973, pp. 101-105.

______ , "Esthetic development of california's bridges." *Journal of Structural Engineering* (ASCE), Vol. 109, no. 9, 1983, pp. 2159-2174.

______ , "Embellishment returns to american bridges." *Roads and Bridges*. Scranton Gillette Communications, Inc., Des Plaines, Illinois, March, 1985, pp. 30-31.

______ , "California's way to aesthetic bridges," ACI MP-1, 1990, pp. 119-131.

Faber, O., *The Aesthetic Aspect of Civil Engineering Design (First Lecture)*.

Finsterwalder, U., "Esthetic principles in bridge design - a memoir," ACI MP-1, 1990, pp. 67 - 69.

Ghaswala, S. K., "Bridge aesthetics," (discussion and response). *Canadian Journal of Civil Engineering*, National Research Council, Ottawa, Canada, Vol. 11, 1984, pp. 368-370.

Gloyd, C. S., "Some thought on bridge esthetics," ACI MP-1, 1990, pp. 109-117.

Goodkind, M., "Architectural considerations in bridge design." *Journal* (ACI), Vol. 32, no. 1, 1935, pp. 29-38.

Gottemoeler, F., "Esthetics and engineers," ACI MP-1, 1990, pp. 167-187.

Grant, A., "Beauty and bridges," ACI MP-1, 1990, pp. 55-65.

Hamlin, *Forms and Functions of Twentieth Century Architecture*. 1982.

Harley-Haddow, T., "Structures in architecture," *Structural Engineering*, Vol. 59A, no. 6. June 1981. pp. 193-201.

Hawkins, H. E., "Have you beautified a bridge lately?" *Traffic Engineering*, Institute of Traffic Engineering, Washington, DC, January, 1970, pp. 42-47.

Hosny, A. H., et. al., "The design of a bridge over the Nile - a challenge for esthetics, quantities, and durability," *Proceedings Conceptual Design Of Structures International Symposium*, University of Stuttgart, October 7-11, 1996, Vol. II, pp. 609-615.

IABSE, "Nine engineers discuss aspects of bridge aesthetics," *IABSE 11th Congress Final Report*, 1980.

Ichimaru, Y., Ishii, N., and Shinohara, O., "Structural design of S-curve bridges, *Proceedings*. Conceptual Design of Structures, International Symposium, University of Stuttgart, October 7-11, 1996, Stuttgart, Germany, pp. 616-623.

Inglis, C. E., et al., "Aesthetics of engineering structures" (discussion). *Journal* (ICE), Vol. 16, 1941, pp. 182-197.

______ , *Institution of Civil Engineers*. London, 1945, pp. 1-23.

Irwin, D.,"The visual arts, taste and criticism," Blackie and Son, Glasglow, 1969.

Jones, C., Design Methods. New York: John Wiley and Sons, 1970.

Joyce, J. R., and Pryor, A. J., "Development of the roadside and aesthetics in road design," *Journal of Industrial Engineering,* Australia, Vol. 40, no. 7-8, July-August, 1968, pp. 139-153.

Kavanaugh, T. C., "Some esthetic considerations in steel design." *Journal, Structural Division* (ASCE), Vol. 101, no. ST11, 1975, pp. 2257-2275; and closure, Vol. 103, no. ST4, 1977, p. 920.

Kirjassoft, G. L., "Esthetics for the highway designer." *Civil Engineering* (ASCE), January, 1966, pp. 69-71.

Kruckemeyer, K. E., and Bacow, A. F., eds., *Bridge Design: Aesthetics and Developing Technologies*. Conference and Workshop Proceedings, Massachusetts Council on the Arts and Humanities, Boston, Massachusetts, 1986, PP. 171.

Kruckemeyer, K. E., "Bridge design process—achieving the goal of esthetic quality" ACI MP-1, 1990, pp. 13-28.

Kulicki, J. M., Snyder, H. H., and Mertz, D. R., "Bridge aesthetics as practiced by Modjeski and Masters," ACI MP-1, 1990, pp. 275-292.

Kuphal, H. E., "Simplicity marks esthetic California bridge," *Civil Engineering* (ASCE), November. 1945, p. 536.

Lacroix, R., "Fundamental principles of bridge esthetics," ACI MP-1, 1990, pp. 29-31.

Law, F. M., *Aesthetics Consideration in Civil Engineering: A University Course*. ASCE Preprint 82-008. ASCE, 1982, 10pp.

______ , "Aesthetics: a part of civil engineering." *Engineering Issues* (ASCE), Vol. 98, no. PP2, 1972, pp. 275-285.

______ , *Aesthetics Considerations: A First Workshop for Civil Engineers*. ASCE Preprint 84-014, ASCE, 1984, 12pp.

______ , "Thinking aesthetically is a part of structural engineering," *Methods of Structural Analysis,* ASCE, Vol. 1, *Proceedings*. National Structural Engineering Conference, Madison, WS, 1976, pp. 1-18.

Le Corbusier, (Jeanneret Gris). *Modulor 2*. 1955, Faber Faber Limited, London (English edition 1958). p. 335.

Lee, D., "Bridging the artistic gulf," ACI MP-1, 1990, pp. 101-108.

Leonhardt, F., "Aesthetics of bridge design," *P.C.I. Journal,* Februrary, 1968, pp. 14-31.

______ , "Looking back on 45 years as a structural engineer." *The Structural Engineer*, Institution of Structural Engineers, London, UK, Vol. 54, no. 3, March, 1976, pp. 87-89.

______ , *Bridges: Aesthetics and Design*. MIT Press, Cambridge, MA, 1982a, 308□pp.

______ , *Brucken Bridges*. Deutsche Verlags-Anstalt mbH. Stuttgart, Germany, 1982b.

Lewis, H. M. "Esthetic responsibilities of the engineer." *Civil Engineering*, March, 1941, pp. 154-156.

Liebenberg, A.C., "The aesthetic evaluation of bridges." *Handbook of Structural Concrete*. Pitman, London, 1983, pp. 36-46.

______ , "The controversy about rules for the aesthetic design of bridges," ACI MP-1, 1990, pp. 47-53.

Loveall, C. L.,"Tennessee bridges the color line," ACI MP-1, 1990, pp. 229-231.

Ludlow, W. S., Jr., and Elliott, A. L., "Aesthetics in Bridge Design," *Manual of Bridge Design Practice,* California Department of Transportation, Sacramento, CA, Section 16, 1970, pp. 3-26.

Mainstone, R. J., "The aesthetics of civil engineering structures." *Proceedings of the Institute of Civil Engineers*, Vol. 22, 1962, pp. N32-N33.

Makkuni, R., "A gesture representation of the process of composing Chinese temples," *Proceedings*. IEEE Computer Graphics and Applications. 7(12), 1987, pp. 45-61.

Mays, R. R., "Aesthetic rules should not be set in concrete - a bridge architect's view on bridge design," ACI MP-1, 1990, pp. 203-228.

______ , "Beautiful bridges." *Civil Engineering* (ASCE), Aug. 1989, pp. 72-74.

McMahan, J. E., "Toward improved aesthetics in bridge design." *Proceedings,* Wichita, AASHTO, 1966, pp. 63-65.

Miles, J. C., and Moore C. J., "The development and verification of a user-orientated KBS for the conceptual design of bridges," *Proceedings*. Knowledge-Based Systems for Civil and Structural Engineering Conference in Edinburgh, UK, 1991a.

______ , "An expert conceptual design of bridges," *Proceedings*. Knowledge-Based Systems for Civil and Structural Engineering Conference in Edinburgh, UK, 1991b.

______ , "Deriving rules for bridge aesthetics," *Proceedings*. Knowledge-Based Systems for Civil and Structural Engineering Conference in Edinburgh, UK, August 17-19, 1991b.

——, and Evans, "Deriving rules for bridge aesthetics," *Proceedings*. Knowledge-Based Systems for Civil and Structural Engineering Conference in Edinburgh, UK, August 17-19, 1991.

Minai, A. T., Design as Aesthetic Communication. Peter Lang, New York 1989.

Ministry of Transport, *The appearance of bridges*. Her Majesty's Stationary Office, London, 1964.

Mock, E. B., *The Architecture of Bridges*. Museum of Modern Art, New York, New York, 1949, 127 pp.

Moore, H., *Philosophy of Modern Art*. Faber and Faber, London, 1969, p. 209.

Muller, J. M., and McCallister, L. F., "Aesthetics and concrete segmental bridges in the United States." *Concrete International (ACI)*, Vol. 10, no. 5, May 1988, pp. 25-33.

——, "Esthetics for Concrete Segmental Bridges," ACI MP-1, 1990, pp. 150-166.

Murray, J., "Visual aspects of motorway bridges," (discussion), *Proceedings of the Institute of Civil Engineers*. Vol. 70, 1981, pp. 755-778.

——, "Visual aspects of motorway bridges," *Proceedings of the Institute of Civil Engineers*. Vol. 72, part 1, 1982, pp. 675-690.

National Research Council, "Bridge aesthetics around the world," Transportation Research Board, 1991, p. 276.

Neuman, E. S., and Lyles, R. W., "Why include esthetics in civil engineering education?" *Engineering Issues* (ASCE), Vol. 104, no. E11, 1978, pp. 1-8.

North, B. H., "Appearance matters," *Concrete*, Cement and Concrete Association, London, UK, Vol. 7, no. 5, May, 1973, pp. 18-23.

O'Coner, C., Burgess, M. D., and Egan, M. J., "Bridge appearance," *Journal of the Structural Division, ACSE*, Vol. 106, no. 1, January., 1980, pp.331-349.

Ohta, T., "Studies on aesthetic design method of double deck bridges with different types constructed contiguously and its application," *Proccedings of the Japan Society of Civil Engineers*, Vol. 344, no. I-1, April., pp. 205-213.

Ohtsuka, H., et al., "Drift capacity and failure mechanism of toll reinforced concrete piers," *Proceedings*. 25th Japan Congress on Materials Research, Kyoto, Japan, 1982, pp.261-265.

Pandovan, "Measuring and counting," *The Architect*. London, May, 1968. p. 25.

Parr, J. G., "The social impact of engineers," *Engineering Journal*, December, 1968.

Pearson, B. J., and Fisher, J., *The Aesthetics of Bridges*. Department of Main Roads, New South Wales, Sydney, Australia, 1987, 111pp.

Pierce, L. F., *Esthetics and Standardization in Bridges*. ASCE Preprint 3788. ASCE, New York, New York, 1979, 22pp.

Potyondy, J. G., *Aesthetic Problems in Contemporary Concrete Bridge Design*. ACI Publication SP-23. ACI, 1967, pp. 7-18.

Read, H., *The Meaning of Art*. Penguin/Faber and Faber, London, 1949. p. 29.

Reekie, F. R., *Design in the Built Environment*. London, Edward Arnold, 1972.

Reese, R. C., "Aesthetics is a do-it-yourself project," *Methods of Structural Analysis*, ASCE, Vol. 1, *Proceedings*. National Structural Engineering Conference, Madison, WS, 1976, pp. 33-38.

Reich, Y., "A model of aesthetic judgement in design," *Computational Mechanics Publications*. 1989, Vol. 2. no 2. , pp. 141-153.

______ , "A Model of Aesthetic Judgement," *Artificial Intelligence in Engineering*, Vol. 8, 1993, pp. 141-153.

______ , *Building and Improving Design Systems: A Machine Learning Approach*, dissertation, Department of Civil Engineering, Carnegie Mellon University, Pittsburgh, PA.

Ritner, J. C., *Bridges produced by an architectural engineering team*. Transportation Research Record 1044, TRB, 1985, pp. 26-34.

______ , "Creating a beautiful concrete bridge," ACI MP-1, 1990, pp. 33-45.

Roberts, J. E., "Aesthetics and economy in complete concrete bridge design," ACI MP-1, 1990, pp. 293 - 329.

Rothman, H., et al., "History and esthetics in suspension bridges" (discussion), *Journal, Structural Division* (ASCE). Vol. 104, no. ST1 (pp. 246-249), ST2 (pp. 378-380), ST3 (p. 619), ST4 (pp. 732-733), ST6 (pp. 1027-1035), and ST7 (pp. 1174-1176), 1978.

Russell, P., et al., "Some notes on significance of form in bridge engineering," (discussion). *Proccedings of the Institute of Civil Engineers*. Vol. 60, 1976, pp. 547-562.

Schlaich, J. M., "Of the aesthetics of pedestrian bridges," ACI MP-1, 1990, pp. 133-148.

Schmid, W. A., "Proportionen in der Matur un im Menschemwerk - wir messen gehen und hören," *Final Report 11th Congress*. International Association of Bridge Structural Engineers, Vienna, August-September, 1980.

Seim, C., and Lin, T. Y., "Aesthetics in bridge design accent on piers," ACI MP-1, 1990, pp. 189-201.

Shimokawa, H., et al., "Hitsuishijima and Iwakurojima bridges," presented at the 12th Congress of IABSE, Vancouver, Canada, September, 1984.

Slater, R. E., "Bridge aesthetics," *Final Report 11th Congress*. International Association of Bridge Structural Engineers, Vienna. August-September, 1980, pp. 115-120.

Smith, P., "An audience participation research project on bridge esthetics," ACI MP-1, 1990, pp. 247-255.

Snowden, W. H., *Formulas for Beauty*. Institute of Transportation and Traffic Engineering, University of California, Berkeley, California, 1966, 16 pp.

Sparkes, S. R., "Engineering aesthetics: an informal discussion." *Proceedings of the Institute of Civil Engineers*. Vol. 42, 1968, pp. 435-437.

State of California, Department of Public Works, *Manual of Bridge Design Practice*. 1971.

State of California, Department of Transportation, *Manual of Bridge Design Practice*. Febuary, 1993, pp. 7-i-7-26.

Stevens, D., and Kopetz , J., "Integration of function, form, and materials in today's bridges," ACI MP-1, 1990, pp. 81-99.

Stiny, G., and Gips, J., *Algorithmic Aesthetics: Computer Models for Criticism and Design in the Arts*. University of California Press, Berkeley, CA, 1987.

———, *Pictorial and Formal Aspects of Shape and Shape Grammars*. Birkhauser, Basel, Switzerland, 1970.

Strasky, J., "Precast stress ribbon pedestrian bridges," ACI MP-1, 1990, pp. 257-273.

Tarnos, T. P., "Relativity and Optimisation of Aesthetic rules for Structures," *Final Report of the 11th Congress*. International Association of Bridge Structural Engineers. Vienna, August-September, 1980, p. 59.

Task Force on Design Art and Architecture in Transportation, *Design Art and Architecture in Transportation*. Report to the Secretary of Transportation, U. S. Department of Transportation, 1980.

Tedesko, A., "Some esthetic considerations in design" (discussion). *Journal, Structural Division (ASCE)*. Vol. 102, no. ST11, 1976, pp. 2238-2246.

Tiller, R. M., and Ward, F. W., *Concrete Finishes for Highway Structures*. Cement and Concrete Association, Wexham Springs, Slough, United Kingdom, 1982, 24 pp.

van der Laan, H., *The Plastic Number.* 1960, as cited by Watson, S. C., and Hurd, M. K., eds., *Esthetics in Concrete Bridge Design*. ACI, 1990, 334 pp.

Watson, S., and Hurd, M., *Esthetics in Concrete Bridge Design*. ACI MP-1, 1980.

Wengenroth, R. H., "Bridge engineer looks at esthetics of structures," *Journal of the Structural Division, ASCE,* Vol. 97, no. 4, Apr., 1971, pp. 1227-1237.

Yasuda, K., Furuta, H., and Kobayashi, T., "Aesthetic design system of structures using neural network and image database," *Proceedings*. International Symposium on Uncertainty Modelling and Analysis and Annual Conference of North American Fuzzy Information Processing Society, September 17-20, 1995, College Park, MD, pp. 115-120.

Yasuji, T., and Yoshio, "On the manual for aesthetic design of bridges," *Final Report*. 11th Congress International Association of Bridges Structural Engineers, Vienna, August-September,1980, pp. 101-108.

Zuk, W., "Esthetic considerations for steel bridges," *Proceedings*. Specialty Conference on Bridges, St. Louis, MO, ASCE, 1974a, pp. 29-41.

______, *Esthetic Enhancement of Concrete Structures Using Ferro-Cement Panels*. Virginia Highway and Transportation Research Council, Charlottesville, VA, 1974b, 29 pp.

______, "Public response to bridge colors," *Transportation Research Record* 507, TRB, 1974c, pp. 1-5.

______, *Bridge Esthetic Guidelines*. Virginia Highway and Transportation Research Council, Charlottesville, Virginia, 1975, 28 pp.

______, "How almost anyone can design a good looking bridge in one easy lesson," *Methods of Structural Analysis*, ASCE, Vol. 1, *Proceedings*. National Structural Engineering Conference, Madison, WS, 1976, pp. 19-32.

______, *A Rating System for the Aesthetics of Bridges*. Virginia Highway and Transportation Research Council, Charlottesville, VA, 1980, 25 pp.

______, "The architecture of bridges," *Progressive Architecture*, March, 1983, pp.□96-99.

______, *An Expert System for the Esthetic Rating of Bridges*. Virginia Transportation Research Council, Charlottesville, VA, 1990, 22 pp.

______, "Methodology for evaluating the aesthetic appeal of bridge designs," Highway Research Report no. 428.

5

REHABILITATION AND IMPROVEMENTS OF BRIDGES

GENERALITIES

Thousands of bridges presently in service throughout countries worldwide are in need of rehabilitation through major works of repairs and additions. In the United States alone, over one-half million bridges are currently in existence and out of these, 100,000 are in need of rehabilitation. Modern bridges of this century may probably be considered more vulnerable to aging by reason of the vehicular traffic to which their older counterparts were not subjected. Such traffic constantly increases in volume and speed, justifying the fast rate in aging, and many modern bridges require modifications and enlargement due primarily to the intensification of traffic volume.

Another cause that demanded the rehabilitation of many modern bridges of this century was the large number of bridge structures damaged or destroyed during war periods.

In the past centuries of bridge construction, survivals of two millennia of existence indicate that certain building techniques and construction materials indeed have the capacity of such a prolonged life, which definitely cannot be compared to that of any bridge structures built since the Industrial Revolution. An example of that is "Ponte Milvio" over the Tiber in Rome, which since its construction in 110 AD has been in service without interruptions up to the present. In this case, there are two factors accounting for such durability. One is the longevity of the bricks in general, and the other is the mortar that bound them. This latter material, made of lime and pozzolana, constituted a very rudimentary type of concrete that cannot be compared to the present equivalents, yet their amazing resistance to aging poses a major point of reflection.

In the future, the rehabilitation of existing bridges will constitute an exceptionally large field of operation that will extend for many years. It will involve thousands of projects that will keep a large number of engineers and consultants busy as never before. To this effect, this chapter aims to pointing out the basic components that constitute the bulk of design proposals that will be commissioned by the proper bridge authorities.

While historically masonry bridges still in existence constitute a group that requires special attention, the majority of bridges in question today are concrete and steel structures. Focusing first on concrete bridges, their presence numbers in the thousands. In fact, Cope (1987) indicates that the first reinforced concrete bridge in the United Kingdom dated back to 1902 and was then followed by 50,000 other bridges either in plain reinforced concrete or prestressed concrete. Naturally, when the reinforced concrete bridges in the United States and the rest of the world are added to the 50,000 just mentioned, one could imagine the extraordinary number of bridges built of concrete that could become the object of rehabilitation.

On the basis of the concrete mortar used by Roman building technology, the durability of present-day concrete was originally predicted to extend into centuries, yet although present-day concrete mixtures have been gradually improved to strengths up to 10,000 lb/in^2, it seems that concrete bridges already built do not respond to such an anticipated duration. Among the agents that negatively affect the durability of concrete is the porosity and consequential moisture absorption that takes place when the concrete mixture is not sufficiently dense as indicated by a slump test.

However, many other factors, including the following, affect the durability of concrete:

- the design and composition of the aggregates constituting the mixture,
- the amount of cement in bags per unit of volume, including the type of cement,
- the volume of the water/cement ratio,
- degree of compaction using vibrators,

- the quality of the aggregates and their proportions in the batching,
- curing of the concrete,
- effects of admixtures, and
- control of cracking

Focusing on porosity and permeability of concrete mixtures, one first notices that the penetration of water in the pores requires the presence of interconnecting canals that reach the exterior surface of the material. Obviously therefore, the presence of pores without such interconnecting links does not allow the penetration of either water or air. Significant also is the size of the pores. Small pores, in fact, would require a higher pressure for the air or water to penetrate their cavities.

To this effect, the size of pores is classified in three different categories: air pores, capillary pores, and gel pores. Small air pores are usually artificially entrained by chemical mixtures and in so doing the mixture is more easily workable and the concrete will have higher resistance against frost. When concrete is not properly poured and compacted, large irregular pores are produced. Then considering capillary and gel pores, it should be noticed that these form spontaneously when water and cement react during the process of hydration.

Obviously, these latter cracks cannot be avoided completely since hydration is the fundamental part of the process. The cracks could be influenced by the water/cement ratio. By lowering such value, that is reducing the amount of water, the capillary and gel pores can be effectively reduced.

DESIGN TEAM EFFECTIVENESS

In the case of rehabilitation and upgrading of old bridges the composition of the design team may still be the same as it was for the design of new bridges. Such a team included structural engineers, architects, environmentalists, and at times, additional consultants such as landscape architects and transportation experts. However, in the case of rehabilitation and upgrading projects the team participants may not play a similar role. At first glance the structural engineer may seem to play the most important part because of the very nature of these types of projects. This, of course, depends on the fact that the structure to be repaired is a tangible reality and the major part of the project consists in evaluating this existing structure and in deciding about proper procedures for the restructuring of the bridge. This consists of basic operations that do not require any particular creativity. On this basis, therefore, the role of the other participants in the team may also appear to be redundant.

On the other hand, this may not be the case in all projects. In fact in most cases, the project does not stop at the reconstruction of the bridge by replacing the damaged parts bringing back the structure to its original state. Many projects may require not only the rehabilitation, but also the modernization of the bridge, adding additional features that were not included in the original design. This may often be the case because the original bridge may have been designed with criteria that are no longer considered sufficient. So after the first phase of the project aims at reconstructing the original bridge the way it was conceived, a second phase may include the retrofitting of additional features and these may include minor or major changes. In this case, besides the role of the structural engineers, the other participants that include architects and environmentalists, as well as additional consultants, still have a role to play as in the case of new bridges.

For instance, the most typical situation may include the concept of aesthetics that in the original structures may have been totally ignored. As we have seen, in fact, the new generation of bridges in the last few decades have required aesthetic factors to be of major importance together with environmental concerns aimed at making the bridge harmonize with the surrounding landscape in the countryside and with the surrounding buildings in urban sites.

In the projects of rehabilitation saying that the programs aim only at the restructuring of the bridge to bring it back to the same condition it was originally designed is definitely a myopic approach that will seldom be applied. The authorities in charge of the bridges and the general public would have no difficulties in recognizing how unwise such an approach could be. So, it is assumed that in a majority of the cases any project of rehabilitation of existing bridges will go further than the restructuring itself and will definitely aim at creating new structures that are in line with the criteria of the present and of the immediate future. On this basis therefore, while the restructuring phase may only require the intervention of structural engineers, the modernization and upgrading phase will definitely need the participation of the other members of the team.

From the architectural point-of-view, the retrofitting of an existing structure aiming at creating a new personality that was not expressed before, is by all means a challenging task that is much more complex than in the case of a new structure conceived from the very beginning. Further challenging tasks confronting the architect would be the use of proper finishes such as gunite for concrete and paint for steel bridges, but even more challenging would be the addition of new structural members reinforcing the existing ones or the restructuring of some components. Furthermore, in phase of upgrading the existing bridge major changes may include the enlargement of sidewalks or the substitution of parapet walls or railings. Capitalizing on the design of these additions, the architect should try to unify them and integrate them with the original structure obtaining a design that uses a universal language.

From the environmentalist point of view, the newly rehabilitated bridge is scrutinized with regard to its relationship to surrounding landscape as well as the adjacent building structures in the city. The role of the environmentalist in these cases of rehabilitation does not vary that much from its original role in new bridges. Yet the scope of the work in these latter cases is more restricted, since the structure and the site are already there, therefore restricting the sphere of action. However, if the original structure had obvious faults in harmonizing with the surroundings, this is the opportunity to correct them.

A well-organized design team that aims to produce preliminary designs for new bridges can be equally effective in the case of rehabilitation of these structures, even if at first glance the design for the rehabilitation seems to be much different from that of new bridges. As we have seen, the role of the structural engineer in these cases may have appeared to be the dominant one; but considering in more detail the processes of rehabilitation, modernization, and upgrading, the role of the other participants of the team is definitely not a secondary one, and the team therefore still retains its effectiveness in competitions for the commission of design works.

Consulting engineering work in bridge rehabilitation

The present situation concerning the rehabilitation and upgrading of existing bridges offers a large number of opportunities to consulting firms of engineers, architects, and environmentalists both in the present and near future. This is not an unexpected condition, since the very nature of bridge structures requires constant maintenance which cannot keep up with the rate of degradation that both concrete and steel bridges experience, until major operations of restoration are implemented to prolong their lives before eventually being demolished and rebuilt. Since aging is implicit in any structure, all existing bridges will eventually reach this critical stage. Consequently, the process of rehabilitation applies to hundreds of thousands of existing bridges.

Examples of programs implementing plans for the rehabilitation of bridges can be found in all 50 states of the United States. In Ohio, for example, 1,500 bridges have been either rehabilitated or replaced; about 500 were rehabilitated. To perform such a large number of projects, the state of Ohio employed about 100 consulting firms to prepare the necessary plans. For each project, the state determines the scope of the project and then selects the consulting firms to which the work is assigned. The Ohio Department of Transportation also hired a consulting firm to evaluate preliminary and final design plans for all of the projects.

The State of Ohio organized for this work a special plan, "Bridge Replacement/Rehabilitation Program," to be implemented by the Department of Transportation. Rehabilitation of most of the structures that were involved at least the replacement of the deck, yet the spectrum is very broad and may involve many components. Key elements for the successful rehabilitation of bridges include good engineering judgment, knowledge of construction practices, and familiarity of the behavior of the existing structure.

One of the first steps in the preparation of plans for the rehabilitation of bridges is to survey and measure in the field all details of the existing structure. The field survey is essential even if the original construction plans are available, because of changes that may have occurred. These include deviations during the original construction caused by unforeseen conditions, changes during the life of the structure such as settlements and distortions, and previous rehabilitations which have not been recorded. The preparation of plans follow general procedures.

One relevant observation is that the consulting firm is not required to identify which components of the bridge should be replaced. Such a decision is left with the State Department of Transportation which will make the decision on the basis of the records made by the department itself. Such a procedure relieves the consulting firm of a major responsibility, which should be greatly emphasized to those firms intending to do consulting work in the rehabilitation of bridges in Ohio.

Consulting firms intending to work in other states should be advised to check with the local department of transportation to obtain specific guidelines controlling the rehabilitation of bridges in that state.

Assessing the integrity of bridges

With the extensive number of bridges to be evaluated for their structural integrity, computer programs have been developed for accelerating the assessment of bridge integrities. Also to this effect, special engineering groups involved in the evaluation of bridges have been formed. In particular, examples of consulting firms working in this field could be found through literature searching within the ranks of structural engineering firms.

One of such firm, for instance, displays a presentation of their services such as: "Graphic displays of bridge mode deflection shapes together with intellegent AI look up of frequency tables which enables accurate determination of bridge capacities." The method allows the measurement of bridge deflections without direct measurements on the original prototype. Measurement of such deflections on the real prototype in the field is very hard to obtain especially when the quantities are in the order of 0.01(mm).

The software for the Dynamic Bridge Testing System can generate measurements of EI and frequencies more accurate then the deflection system then could produced. With an accuracy larger then five orders of magnitude.

Typical causes for bridge deteriorations

Bridges to be restored and rehabilitated have been damaged by different causes. These can first be subdivided into two different categories. One of these is the dynamic forces which impact the bridge suddenly, including the effect of seismic forces during earthquakes, the aerodynamic action of wind, the impact of avalanches, and the damages inflicted to the superstructure by vehicles going out of control, impacting with bridge parts such as railings and parapet walls. Finally, in this category are damages incurred by the impact of navigating vessels or heavy debris against the piers of bridges which cross over rivers and other waterways.

The other category includes deteriorating forces that have been acting on a bridge for long lengths of time, and sometimes centuries. In these cases the processes are so slow that they escape the human eye. These damages, in fact, can be detected only after long periods of time. For instance, in the case of bridges in the U.S., inspections are conducted on bridges with a frequency of two-year intervals. This is a good procedure that allows repairs before the damage gets out of control; however, in other countries inspections may not be as frequent and damages could proceed at a faster rate. Yet, even in the U.S., the process of decay is not totally overcome, and as a result thousands of bridges are on the list of projects due for restoration.

The damages considered in the first category produce an impact on the public opinion through the coverage of the press, and consequently demand much more attention than the structures in the other category. Usually, emergency funds can quickly become available from proper agencies of the state and federal governments and restoration projects can be underway in record time.

Retrofitting for seismic damage

Among the several causes of damaged bridges are the effects of seismic forces. In the United States, codes have been updated in recent years to include more stringent criteria for the design of structures and bridges that reflect the progress made by engineering science in this area. Starting in 1971, the effects of the San Fernando earthquake indicated the inadequacies of the existing codes controlling seismic design. Following this, the California Department of Transportation (CALTRANS) issued more stringent criteria in 1973. Furthermore, in 1975 the AASHTO interim specifications included the criteria previously issued by CALTRANS in a slightly modified version. Such additions to the AASHTO Standard Specifications for Highway Bridges were eventually followed on a national scale by all states.

The Federal Highway Administration (FHWA) also reacted to include the design standards by giving a contract to the Applied Technology Council (ATC) for developing new guidelines on seismic design. The criteria used for the formulation of such guidelines included the state-of-the-art data developed in the United States, New Zealand, and Japan. In 1981 the ATC-6 Report was published by the FHWA under the title "Seismic Design Guidelines for Highway Bridges, Final Report." Following this publication, AASHTO adopted the report in 1983 as Guide

Specifications for Seismic Design of Highway Bridges. They were later adopted as seismic provisions within the Seismic Specifications in 1990. After earthquakes in California (1989), Costa Rica (1991), and the Philippines (1991), AASHTO asked the Transportation Research Board to review specifications and prepare revisions for new specifications. Such new guidelines were eventually prepared by the National Center for Earthquake Engineering Research (NCEER) with funds through the National Cooperative Highway Research Program (NCHRP) under the NCHRP Project 20-7/45.

According to AASHTO (Division I-A SEISMIC DESIGN Sec. 1.3 Basic Concepts), the development of these specifications was predicated on the following basic concepts:

- hazard to life to be minimized,
- bridges may suffer damage but have low probability of collapse due to earthquake motions,
- function of essential bridges to be maintained,
- ground motions used in design should have low probability of being,
- exceeded during normal lifetime of bridge,
- provisions to be applicable to all of the United States, and
- ingenuity of design not to be restricted.

In the meantime, in the 1990s, the individual states initiated their programs to evaluate the feasibility and the cost of implementing the retrofitting of existing bridges built prior to the new regulations. For instance, the state of New York started such a program in 1990, and in 1992 the New York State Department of Transportation (NYSDOT) issued an Interim Seismic Policy concerning bridges scheduled for rehabilitation. More specifically, the main objective of the NYSDOT Interim Seismic Policy included the following: provide guidance for the potential seismic failure of those vulnerable in the state which could eventually be retrofitted; assess costs of retrofitting; and incorporate retrofitting methods in the preparation of rehabilitation plans.

The size of such programs may not be grasped at first glance, yet data confirm that projects of rehabilitation are in the order of hundreds of thousands. In fact, data for the state of New York alone indicate that there are 14,000 steel bridges in service at present.

Of these, it is expected that more than half will receive severe damage to the steel bearings due to the effect of earthquakes. Important criterion in the practical application of bridge retrofitting, implies in all of these programs that a priority scale be considered. In other words, since it is impossible to conduct a program of this size all at one time, a scheduling program should be set on the basis of a priority criterion. Bridge importance should be evaluated in accordance to their influence on the population needs.

THE DESIGN PROCESS

In the case of rehabilitation and upgrading of bridges, the design process follows the general outline used for the design of new bridges plus some additional steps that are specific for these particular project types. The following is a series of steps that facilitate the organization of the work:

1. Study the documents gathered by the bridge authorities concerning the project in question. In doing so, the design team should familiarize itself with the general content of these documents, organizing them in a suitable manner and compiling a list of data that is not part of these documents but should eventually be attained by inspectors working for a design team. Particularly important is the gathering of the inspection reports that are usually conducted every two years, describing damage observed during the inspection.
2. Act on the data from step 1. This would include a survey of the existing bridge and of the surrounding structures. Such a survey would include a sketch of the whole bridge plus a series of sketches illustrating each individual component of the structure, including structural members, internal connections, external supports, and nonstructural components. Each sketch should be numbered and briefly identified, followed by photographs of each component.
3. Evaluate the damage of each individual component in descriptive terms, using the sketches and photographs previously gathered.
4. Diagnose all damage, describing the causes and the procedures to be followed for the restoration of each part. In this phase, the repair of a damaged member may also include additional reinforcement that can guarantee an additional strength, which would be necessary since the member in its original configuration failed.
5. Specify the overall finishing of the rehabilitated bridge. This would include gunite coverings and painting of concrete surfaces and the painting of steel bridges, capitalizing on colors, shadings and combinations of colors.
6. Prepare a flow chart similar to the critical paths used in construction. Such a chart should indicate the sequences of work to be per-

formed, considering the fact that the bridge will never be totally closed to traffic. Traffic can be blocked in the areas where work has to be performed, and therefore it is essential that the scheduling of work in various parts of the bridge be carefully organized.

In addition to these six steps that apply to the rehabilitation process, additional steps apply in the case of upgrading the bridge. Such steps include the intervention of the architect and the environmentalist which, would be too specific to be categorized in step-by-step general procedures and would be left to the discretion of the architect and the environmentalist.

Corrosion protection generalities

Corrosion in the case of bridges is one of the dominant factors affecting the longevity of the structure. Protecting bridges against corrosion is therefore a major operation to be performed on a systematic basis as part of the maintenance program. The cost of protection against rust in comparison to the cost of major repairs, in the case when maintenance has been neglected, is more than justified.

Technical literature on corrosion and corrosion protection points to the definition of corrosion as an electrochemical process that includes three basic elements: an anode, a cathode, and an electrolyte. These produce an oxidation, in the case of steel, referred to as rust. This consists of the change of iron into compounds that flake out of the basic metal. Such oxidation of iron through electrolysis and chemical attack takes place in the atmosphere as well as underground and under water. Besides iron, other metals are subject to oxidation, and the process is more intense as the metal ranks higher in the electromotive force series. In such a series, iron is high in values and is therefore particularly vulnerable to oxidation.

In examining a rusted surface, it is possible to distinguish between two types. One type of rust is strongly bonded to the metal below and will not be removed easily. In this case, this layer of rust protects the metal and retards further oxidation. The other type of rust is instead porous and loosely attached. In this case, this layer does not offer much protection since it will allow water to pass through, accelerating the oxidation of the metal underneath. Along a steel surface is the existence of different conditions in certain parts. Some parts may act as cathodes and others as anodes, and water may act as the electrolyte between such parts. Furthermore, increasing the conductivity in the process is the presence of salt ions.

The following are aspects of the corrosion process:

- Corrosion requires the presence of oxygen as well as water.
- Corrosion starts at a very high rate and eventually slows down as layers of rust are generated because each layer offers a resistance to oxidation that results in a lower rate of progression.

- When two different materials are in contact with one another, the one with anodic characteristics is corroded at a fast rate.
- Galvanic action, which constitutes a high system of corrosion, takes place when two metals, one being electronegative to the other, are placed in contact and exposed to an electrolyte.
- When anodic spots in the metal are much smaller than the cathodic areas, the oxidation in the anodic spots proceeds at a higher rate.
- Corrosion in cracks or crevices proceeds at a faster rate than it does on open surfaces.
- The products of corrosion have larger volume than that of the basic metal. This expansion in volume may cause the material to warp, bend, or fracture.

In view of such processes of deterioration of steel bridges, the design team involved with projects of restoration should try to eliminate any spots on the bridge where water could accumulate. Furthermore, in the selection of materials used for repairs and substitution of damaged parts, the choice should be based also in choosing materials that are implicitly resistant to oxidation, for example, stainless steel, galvanized steel, aluminum, as well as epoxy and high-quality paint coatings. Additional care for protecting steel bridges may include cathodic protection and protective overlays.

The following list includes suggestions to reduce corrosion in bridges:

- Consider use of various protective systems that include high-quality paint systems, epoxy coatings, zinc coatings, special overlay systems, and cathodic protection systems.
- Reduce the number of expansion joints. Specify the joints to be watertight and correctly installed. Special attention must be given to protect deck joints.
- Install sufficient number of drains with proper grades. Increase the size of pipes, fittings, inlet openings, and catch basins with respect to the minimum necessary.
- Particular attention should be give to accessibility of all components to allow for inspections as well as painting.
- Minimize the generation of cracks in concrete bridges using various methods, including the use of mixtures with minimum ratios of water to cement and the use of steel reinforcement for temperature stresses.

HISTORICAL BRIDGE RESTORATION

Among the thousands of bridges that require restoration on a worldwide basis, one could distinguish a particular category. Bridges of this type may have acquired fame and notoriety through different avenues, including historical events, artistic qualities, representation of a social period, and special notoriety of the designer. Some of these bridges are not particularly old enough to acquire fame; nevertheless, they can also be included among the number of this particular family of structures.

For instance, recently the public press reported the case of the Bridge of Lions in St. Augustine as being part of a group of eleven structures of fame in the United States. requiring restoration. This bridge in particular is not that old in terms of historical values. It was built not later than 1927 and connects the old part of the city with Anastasia Island that faces the ocean. The bridge, ornated by marble lions, was probably named from marble statuary that decorate it.

In the case of bridges of fame that have acquired an identity of their own, a special kind of attention is due to them by the very essence of their nature that reflects an artistic identity. In these cases, the problem of restoration is no longer just an engineering endeavor to reconstitute the structural integrity of the bridge as well as its original morphological identity. The problem falls mostly within the realm of architectural design that by the same nature of any architectural product implies the satisfaction of many design parameters.

In practical terms, however, it is possible to distinguish two different approaches which have been used extensively in the restoration of monuments in the last few decades, especially by particular architects sensitive to the problem of restoration of historical monuments. The two schools of thought imply in one case that the structure to be restored should reconstruct exactly the original prototype both in the delineation of forms as well as in the type of materials used. The ideal effect in this case would be that of denying the existence of any damage and deterioration as if the structure had never been touched. In this case, the reality is camouflaged and, in a sense, the restoration work is a cover-up that reduces the pristine qualities of the original bridge, denying the truth.

A second approach is to instead document the state of deterioration, taking full account of all the details prior to starting any reconstruction. The damaged surface is clearly exposed and prepared to receive the new material that will restore the function of the structure. The new selected material will not try to match the existing structure at all. As a matter of fact, the new material will be in perfect contrast with the original one, emphasizing the difference with particular care.

An existing masonry wall, for instance, partially demolished, may be restored by extending it with a totally new material such as concrete, emphasizing the line of contact between the two structures and underlining the identity of the older part in contrast with that of the new addition. Often sheet glass, constituting the new addition over the old one, has been brought directly to the old masonry with a clear cut that leaves no doubts about what is old and what is new. An excellent example of mastery in this area is the work of Albini in Italy, an architect who has worked on quite a number of restoration projects of this sort that have opened the way to this new modern approach to the restoring of old structures.

The structural engineer or the architect who are part of the team approaching the restoration of famous bridges should definitely organize their thoughts along the lines discussed above, choosing one of the two alternatives and carrying the design all along with clarity of ideas without compromising or mixing the two opposite approaches. In this respect, a word of caution is probably needed. Many historical societies have particular interest in the preservation and restoration of old bridges, and their attitude is often oriented toward the restoration by the first method. Their conservative nature considers the second method too avant garde and too daring. An eloquent and convincing argument must therefore be carefully constructed to support the choice of the second method if this is the selected choice.

DURABILITY OF CONCRETE BRIDGES

Of fundamental importance to the structural engineers involved with projects related to the rehabilitation of bridges is to attain a general perspective of the major factors affecting the durability of concrete:

- structural deficiency,
- corrosion of reinforcement,
- chemical attack,
- frost damage,
- internal reaction with the concrete,
- restrained movement,
- mechanical damage, and
- cracks.

As the term implies, structural deficiencies of the bridge may occur from improper design or from poor execution during the construction process. Naturally, in the process of rehabilitation due to structural deficiencies, both the design documents and the careful inspections of the structure must be processed, identifying the causes and providing viable solutions accordingly.

The corrosion reaction of the steel reinforcement in any given structure should be explained first to attain a more clear perspective of the phenomenon. Metal, of course, has a tendency to corrode in the presence of water. Iron and steel have a higher tendency than other metals to be affected by oxidation. The steel reinforcement imbedded in concrete mixture, however, is highly protected. In fact, a very thin layer of corrosion may form around the steel surface and will adhere to it, preventing any further oxidation.

The presence of concrete surrounding the steel bars in a reinforced concrete structure has a protective action due to the alkalinity of concrete with a pH of 12.5 to 13.0. On the other hand, in the presence of acid gases or acid solutions the pH is reduced. As pH falls below 9.5, the thin layer of oxidation around the surface of the bars breaks down and further oxidation occurs. There are four main factors affecting the corrosion of the steel reinforcement: carbonation or sulphation, chloride attack, inadequate cover, and crack.

The phenomenon of carbonation is that process which generates oxidation of the steel reinforcement by creating a lowering of the alkalinity in the concrete. Such a process is produced by the generation of substances that occur by the reaction of concrete with water, which in turn reacts with carbon dioxide.

Chloride can increase considerably the oxidation of steel reinforcement when small amounts of chloride come in contact with the steel. The practical conditions in which concrete bridges can be damaged by chloride include the action of deicing salts such as sodium chloride. When blown by wind or when the salts are washed sideways and/or over bridge deck parapets, these can be absorbed by the concrete surface in depths up to several millimeters. Even more severe is the action of chloride to structures that are exposed to a marine environment. The process is drastically accelerated by excessive porosity, cracking, and inadequate concrete cover.

The cover protecting any steel reinforcement is determined on the basis of several factors and eventually established by code. Such factors include the compressive strength of the concrete at 28 days, the type of exposure, water/cement ratio, and minimum cement content in the mixture. The most important of these four factors is the water/cement ratio. Experiments clearly support the strong relationship between the two factors (i.e., water/cement ratio and corrosion resistance).

Cracks generated by bending stresses in reinforced concrete members, usually in the range of 0.2 to 0.5 mm, are not considered to be capable of contributing to the reabsorption of water affecting the corrosion of the steel reinforcement. However, other cracking conditions may produce cracks of larger gaps. Such cracks are large enough to activate the factors previously described which affect the corrosion of the reinforcement.

Chemical attack to concrete can be produced by aggressive liquid chemicals that can soften the surface of concrete.

The spalling effect in concrete is a destructive phenomenon generated by freeze-thaw cycles which take place in the pores and cracks of concrete due to the expansion of the frozen water with a volume increase of 9%. Each cycle generates damage with additive effects which can eventually produce major failures.

Internal reactions within the concrete include two basic types involving the aggregates. Such reactions are referred to as alkali-silica and alkali-carbonate. The former takes place between the hydroxyl ions in the pore water and some reactive silica within the aggregates. The reaction produces a volume expansion which produces tensile stresses, creating internal fractures and exterior cracking. The latter reaction is similar to the other except that the aggregates must have a critical amount of reactive carbonate.

When expansion and contractions of concrete are prevented, stresses naturally occur. When such stresses include tension, cracks can develop due to the inability of concrete to resist the tension. Change in volumes occur due to temperature variations and during the shrinking that occurs as the concrete begins to set and harden. Also, in the drying of the concrete, further shrinkage can occur when water evaporates at a high rate.

Mechanical damage is also to be expected on bridges due to potential collisions of vehicular traffic with bridge structures themselves.

REPAIRING OF CONCRETE BRIDGES

In the reparation of concrete bridges, the concrete itself is the basic material in need of repair as well as the steel reinforcement when the latter indicates oxidation. With regard to concrete bridges, the material to be used in the restoration of it is obviously more concrete. In this respect, the addition of concrete used for the repair should be as close as possible to the original concrete being repaired. The similitude between the original concrete and the new concrete includes the matching of the modules of elicity, the concrete strength, and the coefficient of thermal expansion. In this way the bond between the two concrete masses is assured to be most effective.

There are four fundamental types of basic material employed for repairing the deteriorated parts of bridge structures: cements, resins, polymer-modified cement materials, and cement/pozzolanic-modified materials.

Cements

The basic four types of cement readily available on the market include the following: ordinary Portland, rapid-hardening Portland, and sulfate-resisting Portland cement. Typical use of these types of cement for repairs can be distinguished into two categories: (a) preparation of a cement grout used for filling up cracks in the concrete mass, and (b) preparation of a concrete mixture using cement and aggregates to be used for filling up cavities and recovering damaged concrete surfaces. In the latter case, as it has been said previously for cements, the similitude between the two type of concretes (i.e., the existing and the one used for repairs) should be similar.

This implies that fine and course aggregates to be used in the new mixture should also be similar to that of the original concrete. The curing of the new concrete should be carefully monitored so that shrinkage will not occur and consequently, there will be no slippage along the surface of contact that could destruct the bond between the two portions of concrete. In preparing the mixtures it is important to consider special additives that will improve considerably the performance of the concrete used for repairs. Among such improved characteristics, one should include: an increased workability, an increased density, acceleration of the hardening process, and superplaticising effect allowing repair using thin layers of concrete.

Resins

Resin-based materials include two basic types: polyester and epoxy. Polyester are oil-derived substances that are usually applied as a covering mortar or as an adhesive substance. These substances must be used mostly in dry conditions because they deteriorate in damp and moist conditions losing their rapid curing characteristics and their adhesive capabilities. Their applications require the use of several layers to guarantee the elimination of shrinkage. The polyester, in general, generates heat during the curing period which can produce thermal changes and consequently, internal stresses. This requires several thin layers rather than thick lumps as previously mentioned.

Epoxy resins derive also from oil. They are adhesive substances operating well both in dry and in damp conditions. This substance also generates heat during the curing process but the shrinkage and consequential thermal stresses can be reduced. The basic characteristics of epoxies include the following:

- high compressive and tensile strengths
- high adhesive capacity to concrete and steel
- vapor resistance to water and other chemicals
- modulus of elasticity lower than concrete
- coefficient of thermal expansion higher than that of concrete
- fast curing.

The mechanical characteristics of epoxies include higher coefficient of thermal expansion and lower modulus of elasticity of that of the concrete.

Because of such differences, shear stresses are expected to occur between the epoxy and the basic concrete, a condition that is to be considered when assessing the effectiveness of epoxies in the repairing of concrete bridges. In fact, when the shear stresses exceed the tensile strength of the concrete, the bond between the two materials is compromised. When epoxy is used, its typical applications include the following conditions: low-viscosity grouts with low viscosities, mortars, and protective coverings.

Polymer-modified cement-based materials

These consist of a dispersion of plastic spheres within a cement mixture. Such spheres can vary in hardness and adhesiveness which give different properties to the cement paste. The size of these spheres is in the order of magnitude of 0.1 microns in diameter. When the spheres are introduced in the cement mixture they adhere to one another, creating thermoplastic films. Polymer dispersions increase substantially the performance of the cement pastes in the following manner:

- increase adhesion
- increase flexural strength
- increase abrasion resistance
- reduction of shrinkage
- improve chemical resistance
- reduce permeability

The basic types of polymer dispersion include:

- *polyvinyl acetate* — used as a bonding coat for various building materials in dry conditions only.
- *styrene-butadiene rubber* — used for grout, mortar, and concrete with the following characteristics: high adhesion in dry and wet conditions, high bending and tension strengths, high abrasion resistance, limited shrinkage, improved bonding, good chemical resistance.
- *acrylic co-polymers* — exceptional adhesive capacity in dry conditions, high compressive strength mortars, good abrasion resistance, and good chemical resistance.
- *styrene-acrylic co-polymers* — particularly high adhesive capacity to concrete and steel in dry and wet conditions, high compressive and tensile strength, high abrasion resistance, high impermeability to water, resistance to freeze and thaw cycles, low shrinkage, and high alkalinity.

Cement/Pozzolanic-modified materials

These include the following additives to be mixed with Portland cement: micro-silica, PFA, GGBFC. Micro-silica reduces pores in the concrete. PFA and GGBFC have a similar effect in decreasing the permeability.

Methodologies in bridge repair

Basic criteria for repairing a damaged concrete bridge must include the following: restoration of the operational life of the bridge; restoration of the structural integrity; insuring the monolithic action of the restoring materials with the original structures; and inclusion of aesthetic values within the final appearance of the restored structure.

Furthermore, the various methodologies for repair should include the following objectives: mechanical strength; protection of the reinforcement; adhesion between the repair and original concrete; bond between the repair and the steel reinforcement; shrinkage control; thermal expansion and contraction; durability; ease of application; and cost control.

When considering different types of categories for the repairing of concrete bridges, the following should be carefully considered: preparation of the surface, concrete mortars applied by hand to the damaged surface, repairing the damaged concrete by spraying, filling the cracks within the concrete, covering damage by recasting additional concrete; and use of resurfacing concrete mortar coatings.

Surface preparation

To achieve satisfactory bonding between the basic concrete and the new one being applied for the repairs, it is essential that the surface of contact between the old and new concrete be properly prepared. The surface must be scraped removing any loose material. With regard to the concrete reinforcement, it's a good practice to expose the steel rods entirely along their perimeter. Furthermore, the steel rods should be exposed lengthwise a little more past the point where corrosion ended.

In preparing the surface, it is usually recommended to use power tools. Consideration must also be given to the use of propping supporting members when extensive removal of the basic material takes place to avoid potential collapses. Particular attention should be given to the cleaning of the reinforcements by removing rust and scale from the bars. Also, when corrosion has reached 10 percent of the original diameter of the rods, these must be replaced by additional bars.

Hand-applied mortars

In repairing steel and concrete surfaces it is necessary to cover the steel reinforcement once it has been exposed. This is usually accomplished by using strongly alkaline, dense, cement-based coatings, that include polymer dispersion and corrosion inhibitors. The concrete should then be wetted to saturate its mass, leaving a dry surface. If the concrete substrate is particularly porous, then a "bonding bridge" should be applied to the concrete surface.

Such an expression refers to a material consisting of the same components as the repair mortar. The bonding bridge is absorbed into the concrete surface, leaving the surface capable of receiving repair mortar. The repair mortar is applied over the concrete and the reinforcement before the curing of the steel protector and the bonding bridge. Finally, as the patching of the damaged area is completed, aesthetic requirements suggest that the entire prepared bridge member be completely refinished with a surface coating that would totally cover the patched surface.

Sprayed concrete

This technique has been used extensively in the last decades for different kinds of work in concrete. The process consists of spraying cement, fine aggregate, water, as well as fibers and/or admixtures. There are two basic methods referred to: "gunite" and "shotcrete." Gunite concrete spray aggregates are less than 10 mm, while "shotcrete" aggregates are 10 mm or more.

The spraying of concrete is distinguished into two different processes. One of these is referred to as "wet" and the other as "dry." In the wet process, the concrete mixture also includes the necessary water, while in the dry process the dry components and the water come into contact at the nozzle. The dry process is more complicated than the wet, in the sense that the operator needs to have sufficient experience to achieve a satisfactory final product. From the aesthetic point of view, it is important to notice that the final coloring of the sprayed material does not necessarily conform spontaneously, but quite often different coloring shades are attained compromising the aesthetic values of the finished work.

These processes, such as *Gunite* and *Shotcrete*, are by no means new methods; but they have been used extensively in the last few decades for a variety of applications such as in "thin shells" construction, as well as in the construction of many small swimming pools. Their effectiveness has been proven over and over, and the strength of the sprayed concrete is usually higher than that of the same type of concrete cast in conventional fashion. Moreover, for a better workability of the mixture, it is recommended that polymer admixture be added to the ingredients.

Crack filling

Crack repairs is an essential part of the process of repairs. In general, concrete structures sustain cracks due to different causes, varying from extremely thin gaps to substantial gaps of much larger size. The fine cracks may be as small as 15 micrometers in width, and can be of different origin, including structural as well as other causes. The technique of repairing such gaps includes using epoxy grouts with low viscosity. To inject the grout into the cracks, two methods are usually followed. One includes the use of pressure and the other, vacuum. If doubts may exist about the effectiveness of repairing structural cracks by injection of epoxy, such doubts are definitely removed. This conclusion has certainly been established by a variety of tests performed both in the laboratory and in the field.

There may be more than one technique to inject the filler into the cracks, but one is specifically quite effective as demonstrated by many applications that occurred in the last few years. The method consists of installing several injection/bleed nipples in the cracks and then filling up the cracks, starting from the bottom. As the filling progresses, air is expelled from the injectors, providing a compact filling free of voids.

Larger cracks are instead filled with two types of materials that include polymer-cement mortar and flexible filling compound.

Recasting

Recasting of concrete to repair damages is a method used when large portions of concrete have deteriorated and have been removed for an effective restoration. Related to this process are several factors that may cause problems during and after the application of this method. Among these, bond between the old and new concrete along the surface of contact must be guaranteed.

Quality control of the batching of the new concrete has to be carefully controlled. In applying this method, there is a minimum thickness required for a proper execution of casting concrete. The minimum thickness is required to be not less than 7.5 cm to assure the proper compaction. From the point of view of aesthetics, these methods may generate a final product, in which color and texture may not conform to the rest of the structure, creating unacceptable aesthetic qualities. To this effect, however, it is probably possible to eliminate the problem by resurfacing the structure, or part of it, with concrete coating by gunite.

Sealers and surface recoatings

These techniques are used at the end of the process of repairing and rehabilitating concrete bridges that are visible to the observer. Their quality is particularly critical to obtain results that do not compromise with the aesthetic qualities required. Such treatments include different types that can be summarized as follows:

- *Water repellant surface membranes* — the coating is impregnated with substances with hydrophobic qualities which are dissolved in a fluid that eventually evaporates. After the evaporation, the residual substances create a surface impermeable to water. Such substances include: silicone resins, polysiloxanes, silanes, and stearates. Such hydrophobic surfaces are also protecting the surface from chloride in addition to the water-repelling action.

- *Surface hardeners and pore blockers* — consist of materials that include calcium, sodium, and potassium silicates. Their effect on concrete surfaces is somewhat similar to the previous one, i.e., water repellent and protectors against chloride penetration.
- *Cement-modified polymers* — used in concrete coatings may consist of different polymer dispersions to be selected for the proper conditions. These coatings are applicable only when the concrete is saturated but the surface is dry.
- *Membranes with special elastomeric polymers* — constitute protective coating for concrete surfaces in accordance with the following basic conditions: (a) are permanently elastomeric, (b) have high diffusion resistance to prevent carbonation or sulphation of the concrete, (c) are weatherproof, (d) have a permeability capable to allow water vapor to escape from the concrete, (e) have resistance to ultraviolet degradation, (f) are abrasion resistant, and (g) are attractive to view both in texture and color.

Such coatings have been recommended to be used on all existing concrete bridges for protecting them from deterioration and, prolonging the serviceability of the structure.

Evaluation of steel bridges

As the concrete counterpart, steel bridges numbered into the thousands worldwide are vulnerable to decay and require periodical inspection. As a result of such periodical inspections occurring every few years, maintenance programs are conducted. Such programs may be effective, allowing repairs that extend the serviceability of the bridge, or in the case of more significant deterioration, plans may be prepared for full restoration of the structure with partial demolition and restructuring when necessary.

The problem of restoration of steel bridges, as well as of concrete bridges, includes thousands of structures worldwide. The estimated number of steel bridges alone in need of structural rehabilitation in the United States is more than 100,000 (Ghosn, Mosen, and Gobieski, 1986). This gives an indication of the number of bridges in need of repairs worldwide.

In the United States, the AASHTO rating procedures for the evaluation of existing bridges is used as the basic guideline for evaluating the conditions of bridge structures. Such guidelines, however, have been questioned, as illustrated in the previously mentioned paper. According to their work, published by the Transportation Research Board of the National Research Council, the typical evaluation procedures following the AASHTO guidelines underestimate the real strength of bridges being inspected because of their conservative nature.

It has also been reported by the same authors that while the inspection of bridges is usually conducted independently of the testing of the structure underload, more dependable data are obtained by testing the bridge on the vehicular load. Such a procedure has been referred to as a "weigh-in-motion" system and is considered more realistic since it reflects the response of the structure underload.

Bridge bureaus exerting their authority over bridges worldwide follow a system of rating and evaluating bridge structures, reaching final conclusions about proper actions to be taken for guaranteeing the safety of the structures. In general, the final conclusions of bridge inspections may include the following cases: repairing the damaged components; rehabilitating the whole structure; closing the use of the structure; and demolition and reconstruction. The weigh-in-motion method used by these authors is eventually compared to the AASHTO system indicating the more realistic conclusion of this method over the AASHTO (for more detailed information, see the original paper).

Another valuable resource is "Methods of Strengthening of Highway Bridges," report 293, National Cooperative Highway Research Program (Transportation Research Board, National Research Council, Washington, DC 20418).

Bridge painting

All metal components of bridge structures are eventually covered by coatings of paint, whether the structure is new or is undergoing routine maintenance programs. The main purpose of painting the structure is to protect the metal surfaces from the deterioration caused by the weather. First, the protection is intended for preserving the structural members from being corroded and eventually losing their carrying capacity. The secondary purpose of painting is concerned with the aesthetics of the bridge, which is also a significant factor, especially in the rehabilitation of old structures.

In rehabilitating old bridges, the design team, dealing with upgrading the visual aspect of the restructured bridge, would probably be particularly concerned with the potential benefits of using color in a creative way. The architectural aspects of bridge rehabilitation projects would capitalize from the use of painting with particular emphasis, and therefore, adequate knowledge is required on the part of the architect to specify the various components involved in the process of painting. To this respect, it is particularly relevant for the architect to familiarize himself or herself with standard specifications from the AASHTO standards as included in Section 13.

From a broad point of view, the painting of bridges considers a variety of subjects that include the preparation of surfaces to be painted; the application of the paint; the curing of the paint; protection of the painting during the curing stage; and protection of adjacent structures from being damaged during the painting process.

The painting of metal structures implies the identification of several parameters, including the environmental conditions surrounding the bridge to be painted, such as high pollution and coastal zones; mild climate; and repainting in mild climate. The systems to be considered include the application of the primer, intermediate coating, and top coating. The type of paint to be used with regard to such conditions and coating types are as follows: primers include inorganic zinc, organic zinc, and oil/alkyd; intermediate coatings include epoxy, vinyl wash primer, and oil/alkyd; top coatings include epoxy, vinyl, urethane, and oil/alkyd.

The weather conditions required for satisfactory application of paint are of fundamental importance and should be clearly specified. Air temperature and the temperature of the surface to be painted is not adequate if it reaches 40 F or less. Similarly, on the upper limit, the temperature should not be 100 F or more. Another consideration is the temperature of the surface with respect to the dew point to prevent condensation to damage the coating. To this regard, the temperature of the metal to be painted should not be less than 5° F above the dew point nor should the relative humidity of the air be 85 percent or more.

Furthermore, cautions should be given to other damaging factors such as the effects of rain on paint not yet cured, and the effects of fog and dust. In addition to the limits of temperature, care must be given to the possibility of the blistering of paint, the production of a porous film of the paint, and the separation of the vehicle from the pigment that could occur if the metal surface is too hot.

The preparation of the surface prior to painting is of fundamental importance for a successful job. This includes first the cleaning of the surface by a number of methods: blast cleaning, steam cleaning, solvent cleaning, and hand cleaning. Blast cleaning, should be accomplished using clean dry sand, mineral grit, steel shot, or steel grit. Unwashed beach sand is not adequate and therefore should not be allowed for a satisfactory preparation of the surface. In addition, as the blasted surfaces are being completed they should not be left exposed for a long time prior to receiving a primer coating. Usually such coating should be applied within the same day that blasting is completed. More detailed sand blasting specifications can be found in Steel Structures Painting Council Surface Preparation Specification no. 10, SSPC-SP10, "Near-White Blast Cleaning."

Steam cleaning apparatuses are usually employed to clean existing painted surfaces as well as galvanized surfaces from materials such as dirt, grease, loose chalky paint, or other foreign substances. If other operational cleaning should be employed, the steam cleaning should precede it.

When the surface is covered by oil, grease, and other soluble contaminants, then solvent cleaning must be applied and the operation should precede any sand blasting in the case where this is applied. Finally, hand cleaning must be performed to remove all dirt, loose rust and mill scale, using wire brushes, either hand or powered, hand scraping tools, power grinders and sandpaper.

The procedures described up to this point include the painting of metal surfaces, excluding galvanized ones. When galvanized surfaces are to be painted, they should first be prepared by cleaning with mineral spirit solvent in order to remove oil, grease, or other materials foreign to the galvanized surface.

The first coating should be made by applying vinyl wash primer. This should be applied by spraying to obtain a coating of a thickness varying between 0.3 and 0.5 mils. The final coating then will be applied over the cleaned galvanized parts just as has been done for the adjacent metal parts.

The painting of concrete is also controlled by specifications to guarantee acceptable results. As in the painting of steel, concrete surfaces must also be prepared prior to painting. Sand blasting is usually recommended. Completing the preparation phases, the surface should be free of dust and dry.

The paint used to complete is an acrylic emulsion. More specifically, they should comply with Federal Specification TT-P-19 (latest revision), "Paint, Acrylic Emulsion, Exterior." Coloring can be added to the basic paint and not less than two coatings should be applied. The minimum temperature during painting and the process of curing should not be allowed below 50°F.

Recent developments in bridge rehabilition

Topical issues of the rehabilitation of bridges are the subjects of a large number of research papers due to the special interest this subject has created among structural engineers involved in bridge design. The first symposium on this subject ("Bridge Strengthening and Rehabilitation," Des Moines, IA, April 5-6, 1993, sponsored by the National Science Foundation).

Following the first symposium of this topic, a second was recently held, (Second Symposium on Practical Solutions for Bridge Strengthening and Rehabilitation, Kansas City, March 25-27, 1997). The symposium presented recent views on the most efficient solutions for the rehabilitation of existing bridges in need of restructuring. Contributors included a variety of researchers, coming from different professional sectors including academia, government agencies, and consulting practices.

The numerous papers presented included a variety of topics such as: Rehabilitation/Strengthening, Rehabilitation of Railroad Bridges, Seismic Retrofit and Rehabilitation, Rehabilitation of Hiastoric Bridges, Repair and Strengthening Technologies, Rehabilitation of Concrete Bridges, and Rehabilitation and Strengthening II. Other specific topics included: seismic retrofit, bridge rehabilitation, bridge upgrading, bridge decks and dampers, use of composites in rehablitation, bridge repair, and rehabilitation projects.

Steel prestressing in bridge rehibilitation

Similar to the prestressing of concrete structures, a new application of prestressing to steel girders in bridges is part of the development of new techniques which are gradually taking place in these latest years.

The prestressing of steel is not really that different from that of the prestressing of concrete. In fact, like in the prestressing of concrete, the prestressing of steel counteracts the stresses produced by dead and live loads, increasing the capacity of the structural member. The prestressing, however, does not contribute to the increased stiffness of the member, and therefore the deflection remains the same as it was before prestressing, which could eventually exceed the maximum required. Also, in prestressing steel members, the additional costs of cable tendons, anchorages and tendon supports may be considerable and should therefore be taken into account before comparisons are made.

The installation of tendons is usually made by welding a pipe to the web of the girder having a diameter larger than that of the tendon. After inserting the tendon into the pipe and after prestressing it, grout is applied to protect the tendons from corrosion. That the tendons and anchor plates used for prestressing steel members are the same as those used for the prestressing of concrete, so that they are readily available and do not require any special fabrication that would add to the cost.

A recent case study describing the strengthening of an existing bridge proved that the prestressing of steel girders offered a solution that generated a savings of up to 80 percent with respect to any other proposal. Procedures for the installation of the hardware necessary for the prestressing of steel is typically simple and quick; and equally simple is the prestressing operation itself.

Although the prestressing of steel has been proved possible and economical, its practice is not widespread. Furthermore, the methodology for prestressing steel has not been recognized yet by the official codes. However, research in this area is being conducted, especially at Iowa State University. A practical application of such methodology can be found in a case study of the Berkely Bridge, Norfolk, VA, completed in 1991 (Lawrie and Seim, 1993).

Retrofitting columns using fiber/epoxy jackets

Epoxy jackets used for the retrofitting of damaged columns are of two different types. One that was recently tested consists of jackets applied directly on the surface of concrete columns by means of an adhesive. The other type, in use for a number of years, consists of fiberglass jackets, larger than the column to be repaired, which allow the pumping of concrete around the damaged column.

New techniques for the retrofitting of columns, involving the wrapping of existing damaged structural elements with high-strength fiber/epoxy jackets, have been conducted at the University of California at San Diego. Results of testing have proven the system to be valid and have been described in the technical literature. As a result of the positive testing being done on this new methodology, a few projets have applied this technology since 1992.

Early applications of this methodology include projects such as: the retrofitting of a bridge in Los Angeles, CA, at Highway 5 and Highway 2; the retrofitting of two columns off Highway 101, North of Santa Barbara, CA; the retrofitting of a bridge in Sparks, NV; and the retrofitting of a bridge in Tacoma, WA, carrying North 21st Street, designated as a historic structure and entered into the Natiônal Register of Historic Places and also into the Washington Register of Historic Places.

The theory of this new technology of retrofitting columns consists in recognizing a few basic criteria. One is to recognize the "lap splice flexural area" which occurs at the bottom of the column, just above the footing, where the steel in the columns is spliced with a steel dowell coming from the footing. A second area to be considered in the column is at the top of the column where the steel reinforcements run continuously without any splice.

Such an area is referred to as a "no-lap splice flexural area" and is located where the column enters the slab. The retrofitting of these vulnerable areas in the columns consisted of applying an elastomeric bladder around the column at the base, and then filling it with grout. Then the bladder was wrapped up with high strength fiber material, and subsequently impregnated with high-quality epoxy. After the setting of the epoxy, the bladder was eventually filled with grout.

The pouring of the cement grout is accomplished at a pressure varying between 50 and 250 lb/in^2. The system increases the column strength and ductility to a factor of 8 or more (Fyfe and Kuruvilla, 1993).

A further application of the methodology just described involves the use of prefabricated composite jackets. This technology expands to more practical uses. This methodology of retrofitting bridge columns is especially used for columns subjected to seismic damage. Continuing research on this topic is being conducted at the University of California, San Diego. Two prefabricated E-glass fiber-reinforced composite jacketing systems support experimental and analytical evidences of the validity of this methodolgy.

The application of the E-glass fiber-reinforced composite jacketing systems can prevent brittle failures caused by lap-splice or shear deterioration and increase the column load carrying capacity and its ductility. Further application, besides the repairing of seismic damage, include: the repair and protection from environmental damage and exposure, the protection from freeze/thaw and deicing salt corrosion, the deterioration due to water absorption of porous concrete, damage by fire, and for use in harbor wharf and pilings and power poles.

The prefabricated jackets consist of multilayer prefabricated composite shells. The material includes E-glass fibers reinforced in the direction around the circumference, impregnated with isophthalic polyester resin. The shells are are in the shape of cylinders having the same diameter of the columns to be reinforced. Their height is aproximately 4 ft. A vertical cut along the height of the jacket allows it to be opened and then wrapped around the column. During the installation, the columns are first cleaned and then coated with adhesive, so that when the jacket is installed these will be cemented to the column. During the process of bonding, when the adhesive is curing, metal straps are placed around the jackets and kept tight for about one to two days. Then the straps will be removed, leaving the jackets permanently bonded to the concrete column (Xiao, et al., 1997).

Another system for the restoration of columns and piers is provided by other types of jackets which, instead of being bonded to the concrete column, are used as forms that will allow the pumping of concrete around an existing column forming a protective barrier a few inches thick. Among the many patented jackets of this sort are those patented by Symons Corporation (Figures 5-1 through 5-3). These jackets, consisting of fiberglass, have a diameter a few inches larger than the column, which will allow for the pumping of the concrete around the column. To install them, the jackets have a longitudinal cut that permits them to be stretched open and fitted around the concrete column to be rehabilitated. Then the seam is

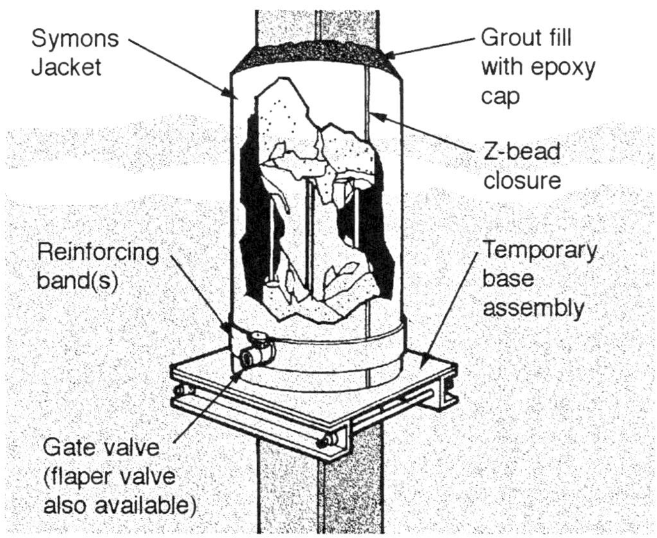

Figure 5-1 The basic Symons system.

Figure 5-2 Fitting the Symons jacket around a column.

closed by means of a patented 2-bead positive locking closure. After centering the shell in place, concrete is pumped from the base of the column with sufficient pressure to make the concrete rise to the top. After the concrete has hardened, the jackets remain, giving a further protection to the exterior of the rehabilitated column. In the practical use of pumping the concrete into the jacket, it is important to know that the fluidity of the grout will allow a satisfactory result. To this effect, the grout is a high cement-sand mixture with a slump between eight to ten inches. For further data consult the Symons Corporation.

Besides the epoxy jackets and prefabricated composite jackets previously discussed, the retrofitting of columns with jackets could also be performed using steel jackets varying aproximately from 2 to 10 millimeters thick. Such jackets have been proven to reduce cumulative seismic damage in at least 50 percent of most cases. However, this is not always the case. The literature on this subject, in fact, indicates that for slender columns, subjected mostly to flexure, having a small diameter, the effect of jackets is zero. Furthermore, it has been reported that in some cases jacketing can actually increase damage by changing the fundamental frequency of the pier (Taylor and Stone, 1993).

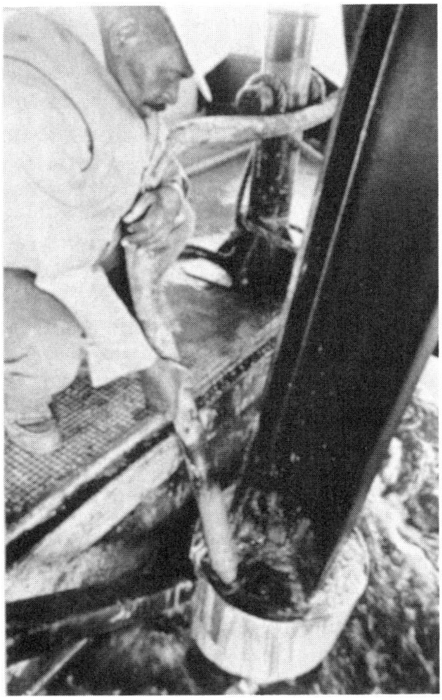

Figure 5-3 Pumping concrete into the Symons jacket.

Steel arch rehabilitation

Many historical steel truss bridges in the United States have been officially restricted to low traffic loads which did not allow present-day loading conditions. On this basis, therefore, they have been demolished and substituted with modern new structures that do not carry any memory of the past. The rehabilitation and upgrading of these old structures to carry HS20 live loads have included a number of methods, yet the latest patent seems to have become the most efficient and economic method. The system is referred to as "steel arch rehabilitation system" and was originally patented by Bucknell University. Basic advantages of this system over the others are briefly summarized as follows:

- It is less labor intensive than other methods.
- The added arches assume most of the load relieving the loads in the trusses.
- It avoids the necessity of substituting truss members by upgrading the old arches with new ones.

- By posttensioning the tied arches, the load capacity of the entire bridge is drastically increased.
- The thermal stresses in using these arches are minimal.
- Arches add redundancies to the whole structure making the bridge less vunerable to potential local failures.
- The cost of the system is lower than other typical restoration processes.
- This system has a higher reliability than comparable other types.
- The process requires shorter construction times in comparison to others.
- The method often allows the work to be done without closing the bridge to traffic.

The typical arch configuration is parabolic. The arches are tied at the base and by posttensioning the ties, the arches are prestressed and therefore provide an uplifting to the trusses connected to them. An application of this system is found in case study number 9 describing the rehabilitation and upgrading of an historic bridge in Stuyvesant Falls, NY (Kim, et al., 1993).

Field welding

The welding process in the repair and rehabilitation of steel bridges is a subject in need of attention due to limited amount of data concerning this process within the context of codes. Specifically, *Structural Welding Code* (AWS D1.1-92) contains a minimum coverage of the strengthening and repair of existing structures. The guidelines governing the retrofitting of field welding, in fact, are so limited that many engineering consultants have a tendency to avoid field welds with respect to other connections. This is definitely a negative point, since in a majority of the cases, welding is the best method for the repairing of steel bridges. When comparing field welding with shop welding, the difficulty of controlling the quality of the work in the field increases substantially due to additional factors that do not occur in the shop.

Ensuring quality control in field-welded repairs requires guidelines which include all those applicable to shop welding, plus additional ones that derive from the conditions in the field. Basic concerns that field welding and shop welding share in common are personnel, procedures, and inspection. The contractor verifies the competence of the personnel welding in the field. To this effect, welders must pass qualification tests. According to "D1.5 Bridge Welding Code," all procedures must be qualified by tests, except for those that employ shielded metal arc welding that utilizes filler metals with a minimum specified yield strength less than or equal to 90,000 lb/in^2.

With regard to inspection, this includes: minimum preheat temperature, minimum interpass temperature, welding current, voltage, travel speed, proper protection of filler metals, and appropriate welding techniques.

In addition:

- Contractors should be selected on the basis of their reputation, having a documented and verifiable experience.
- Allow the possibility of choosing other alternatives when awarding the contract to the lowest bidder, so that the reputation of the contractor should also play a role.
- Include in the specifications all elements that affect the quality of the work, including the weather conditions during the work andthe type of material being used, such as the type of steel to be welded, etc.
- Review with care the proposals from the contractors, assuring that all factors have been included.
- Verify the competence of the inspectors.
- Underline the need for efficient communications, using at times welding engineering consultants.

As previously mentioned, the type of steel to be welded is one of the essential datum to be established. Different steels, graded according to ASTM specifications, have different characteristics of weldability which must be established. Other factors to be taken into consideration are, for instance, the cleanliness of the material being joined, that in the field may pose problems not experienced in the shop. Welding in the field can be affected by thick, flaky, scaling rust which create cavities, pits, and other surface irregularities. Prior to welding, it is necessary that scale and rust be removed using wire brushing, shot or sandblasting, or grinding. When cleanliness is not up to the required standards, porosity may occur in the weld.

Atmospheric conditions on the site may also have a damaging effect on welding that must be avoided at all costs. One of the factors for instance is the effect of precipitation which could keep the metal wet when the requirement is to keep the welding area absolutely dry. Another factor is ambient temperature which is subjected to strict limitations by the code. In fact, the Structural Welding Code requires that below 32°F, the steel must be preheated to a minimum of 70°F. Moreover, if the ambient temperature drops below 0°F, no welding is permitted.

Another atmospheric agent that could have deliterious effects on field welding is the wind. In fact, even light breezes can badly affect gas metal arc and gas-shielded flux cored arc welding. To this effect, the use of self-shielded flux core or shielded metal arc electrodes is recommended (Miller, 1993).

Load capacity upgrading

Many of the bridges in need of rehabilitation through retrofitting were required to be restored due to deterioration through aging or damage by other sources. Yet another important aspect was to upgrade these structures when their load-carrying capacity for live loads did not reflect the present-day standards. To this respect, in working with the upgrading of bridges, reference is made to the AASHTO code, Sec. 3.7 Highway Loads and specifically, Sec 3.7.2, Classes of Loading. In Sec 3.7.2, mention is made about four standard classes of highway loading, which include: H 20; H 15; HS 20; and HS 15.

Specifically, each loading is described in the following sections of the code: Sec. 3.7.5 H Loading indicates that H loadings consist of a two-axle truck and the designation H followed by a number indicates the gross weight in tons of the standard truck. In this category are H 15 and H 20; Sec 3.7.6 HS Loading, indicates that HS loadings consist of a tractor truck with semitrailer and the designation HS followed by a number indicates the gross weight in tons of the tractor truck.

In this category are HS 15 and HS 20. More specifically each load carries a number indicating the year in which that type of load had been specified by AASHTO. For instance, H 15-44 indicated that this load had been designated by AASHTO in 1944. This basic information is furnished to point out the reference source that should be consulted in the preliminary design of bridge upgrading, especially due to the fact that standard loading conditions for highway bridges have changed over the years, creating confusion about the exact meaning of the various designations being used. For more information, see AASHTO (1996).

Preparing prestressed concrete girders

Prestressed reinforced concrete girders for bridges have become very popular since the 1950s and like other reinforced concrete members, have aged through the years, requiring repairs to the concrete and to the tendons. Usually, the damage received by girders is caused either by deterioration due to atmoshpheric conditions or to impact. When the damage is within the concrete itself, conventional methods are applied for removing the damaged concrete and casting new concrete in its place.

When the damage occurs to the tendons, then special methodologies should be used to replace the broken part of the cable of the tendons and replace it by joining the new piece of tendons to the remaining existing tendons and eventually tension the newly added part.

A special methodology that will allow the repair of the tendons as previously described has been developed by Alberta Transportation and Utilities, with special credit given to Reid Crowther & Partners Ltd., Calgary, Alberta, Canada. The patented procedure of replacing the damaged portion of the tendon and stressing it again involves the cutting of the damaged part, replacement of the damaged part, the joining of the new part to the remaining tendons, coupling the new strand to the old one at each end of the new replacing part. At this point the tendon is stressed with the use of a turn buckle device referred to as "splice nut assembly."

The removal of the damaged concrete is first obtained by cutting by saw up to 13 mm (1/2 in) in order to prevent the cutting of tendons and shear reinforcing. After this, the concrete is removed by hand tools and light chipping tools. Then any damaged parts of the tendons and of the shear reinforcement must be removed. To perform the splicing of the new portion of the cable with the existing remaining cable, it is necessary to have at least 1.2 m of space in order to insert the tools required for the splicing. The coupling used to splice the news cable with the old is of the type conventionally used and does not need any special fabrication.

The splice nut assembly, instead, requires special fabrication. This is patented by Alberta Transportation and Utilities and has a tensile capacity of 30 kips. The prestressing operation is executed by turning the nut into the threaded cylinder.

The operation procedure follows the following steps:

- coupling of the new strand with a standard strand coupling device and chucks
- cutting the new strand allowing a gap of 5 cm
- coupling ends of the new strands to the cylinder portion of the splice nut assembly
- connecting the remaining ends of the existing tendons to the nut portion of the splice nut assembly
- attaching the dial elongation gauge to the new strand
- fitting the nut into the cylinder portion of the splice nut assembly

The splice nut assembly can be easily fabricated by any local machine shop following a set of specifications (Lanyi, 1993).

Aluminum decking for retrofitting projects

A new family of metal decks has recently been produced and applied to two projects that verify their performance. The Reynolds Metals Company conducted extensive testing and analyses for the development of an aluminum deck panel system. Such studies eventually resulted in a system consisting of a welded, multivoid aluminum extrusion structural system. This included a wearing surface mounted in the shop, creating lightweight panels ready to be installed in the field. The thin-bonded wearing surfaces to be used constituted a main challenge. After many tests conducted on three different options, the selected one was "Flexolith" manufactured by Tamms. This consists of a 0.375-in-thick, shop-applied, two-part epoxy coating material. A hard silica grit or similar material is embedded in the epoxy for superior skid resistance and low, long-term wear characteristics. Such a material had already been in use by the Virginia Department of Transportation and others to protect the concrete decks of bridges from deicing chemicals.

Several types of these aluminum decks have been produced. Of these, one has orthotropic design weighing only 20 lb/ft^2. Another one weighed 22 lb/ft^2. Practical applications of these decks include the Corbin Suspension Bridge in Huntingdon County, PA, which was retrofitted with the 20 lbs persquare foot system and the Little Buffalo Creek Bridge in Virginia which was retrofitted with the 22 lb/ft^2 system.

Some benefits derived from the retrofitting of bridges with this new kind of deck include the following: extruded aluminum sections include corrosion resistance, high resistance to fatigue, and light weight; the material itself is environmentally safe as well as recycleable; benefits due to the light weight of the deck that increased the live load capacity of the structure; high performance in harsh weather conditions; reduced time for the erection of aluminum decks in comparison to other kinds of decks; particularly advantageous for movable decks due to the light weight; competitiveness in cost with respect to other deck types (Ahlskog, 1997).

Retrofitting timber bridges

The presence of a substantial number of timber bridges in the U.S. is demonstrated by the interest shown in the AASHTO concerning this type of structure. Section 16 deals with timber structures and Section17 deals with preservative treatment of wood. Most of the existing structures are in the range of 40 ft and include short wooden members in which the critical stresses are mostly shear rather than flexural. This is supported by calculations made on the basis of AASHTO design specifications of 1992. Examining failures in timber members, that typical failure due to shear stresses are manifested by longitudinal splitting of the timber along the longitudinal axis, passing through the nuetral axis of the section while bending stresses develop ruptures along the bottom fibers.

To reinforce timber beams against shear failure, special reinforcement has been developed using fiber-reinforced plastic (FRP). Such methods have been tested with positve results, both at Washington State University and at the University of West Virginia. More specifically, patents for high-strength fiber-reinforced glue-laminated beams and aligned fiber reinforcement panels for structural wood members had been awarded to D. A. Tingley. However, this product, manufactured under the name *FiRP* by American Laminators, has been applied mostly to new beams rather than to the retrofitting of old ones.

A new technology has instead been developed for reinforcing timber members, not only new members, but also for retrofitting old members against shear failure. Basically, the method consists of bonding diagonal FRP strips to the sides of beams. Carbon FRPs have not been considered for reasons of economy, but instead E-glass fibers were used for their moderate costs since their strength was sufficient for the task in question. With regard to the resins used in association with the fibers, the following types had been considered: resorcinol (Casein Resins); two-component epoxy (Ciba-Geigy); and vinyl ester (Dow Chemical).

When applying the diagonal FRP strips to the timber face, adequate adhesives are required to transfer stresses. To this effect, tests have been made to determine the interfacial shear strength between wood and several combinations of fiber and matrix. The bonded materials tested included woven fibers; stitched fibers; prepreg; and roving. The adhesives tested included resorcinol; epoxy; and vinyl ester. The results of woven fibers (M-10) impregnated with resorcinol had the highest mean interfacial shear strength. The reason for such a result was that the surface texture of the cloth and the wood allowed the resorcinol to establish a stonger bond. Stitched fibers impregnated with resorcinol also yield satisfactory results. On the other hand, prepreg FRP gave poor results. For further data see Harichandran, Ronald S.; Hatfield, Frank J.; and Kifle, Bruck (1997). Shock transmission units (STU) for bridge substructure strengthening STUs, which been developed and gradually improved are now feasible for application not only to new bridges but also for the retrofitting of existing structures.

This mechanism is a special device that joins together several structures or several members of any given structure. The device allows slow, long-term movements within the structure, but it acts temporarily as a rigid connection between structures or components, transmitting impact forces or other forces of short duration. In other words, structures that need to expand or contract slowly under temperature changes are free to do so when linked together with STUs and at the same time they can share an impact load. The impact load applied to one of the structures is subdivided among all of the structures interconnected with STUs. Thus, the benefit is obvious because the impact force will be shared by all structures rather than the first one being impacted.

Considering the development of the mechanism, originally the first STU types included a device filled with gas/oil that needed regular expensive adjustments and maintenance. Also, the initial purchase costs of those STUs were high. Then in 1970, a new version of the system was developed using silicone putty developed in the United States and used in the space programs. This new substance deformed a readily slow pressure and acted as a rigid body under impact. The new STU device was eventually patented by Chaffe and Mander for application for bridges.

The last STU type had only one moveable part and consisted of a cylinder containing a piston loosely fitted and connected to a transmission rod. The void around the rod was filled with silicone putty. The device is then interposed between the two structures being connected. The cylinder is connected to one structure and the rod is connected to the other structure being joined. The transmission rod passes through the entire length of the cylinder and regardless of the position of the piston the silicone putty filler remains constant. When slow motions occur, as in the case of thermal movements, the putty is slowly squeezed through the gap around the piston or through a small hole. When impact forces are applied, the silicone putty acts as a rigid body as previously mentioned.

The STUs are rated in terms of the capacity of the force being transmitted such as 25 to 50 tons. The STUs are usually mounted horizontally to carry horizontal forces, yet they can also be mounted vertically with special adaptation. The silicone putty is quite stable within a wide range of temperatures, which allows the STU to be used efficiently both in summer and winter conditions.

Practical application of STUs to bridges is to distribute horizontal forces to various bridge components, as in the case of horizontal forces due to seismic loads or horizontal forces due to the breaking of railroad traffic, including any form of rail transportation (Pritchard, 1993).

Passive isolation bearings

A recent technological development applicable to the retrofitting of bridges as well as to new bridges is the so-called "passive isolation bearings." These components come in three different types applicable to zones of low to moderate seismic areas. Their application has been in bridges in the state of Illinois, where many case studies can be found. Such bearings have the ability of reducing the earthquake forces being transferred from one part of the bridge to another. This occurs through three different behaviors: the period shift caused by the lateral stiffness of the bridge; the better distribution of seismic forces to all abutment and pier supports; and the dissipation of seismic energy by plastic deformation or friction of bearing elements.

The flexiblility of the bearings allows the upper structure to deflect horizontally with respect to the abutments and piers within the limits required for the proper serviceability of the bridge. With regard to the longitudinal stiffness of the bridge, it is well known that when conventional bearings are used, this stiffness only involves those bearings purposely fixed in the longitudinal direction. When passive isolation bearings are used, the longitudinal stiffness of the bridge is the sum of the stiffness of the passive isolation bearings.

Another important feature, previously indicated, is that these kind of bearings can dissipate a large amount of kinetic energy being transmitted by seismic forces. The mechanism of the dissipation is just like that of shock absorbers. These bearings act using the elastomeric properties of lead which can deform and, in so doing, absorb energy. In a sense, these bearings are acting very much like the neoprene pad bearings used in the World Trade Center Towers in New York, where the connection between the columns and the trusses at each floor absorb the oscillations of the structure set in motion by wind.

The three types of bearings previously mentioned include elastomeric isolation bearings, friction-type isolation bearings, and FIP bearings. The first type uses a lead core, the second type includes sliding surfaces, and the last type is a combined system including the characteristics of both the elastomeric and friction isolation bearings. Various applications of these bearings in Illinois include the following bridges: Sexton Creek bridge (Alexander County, 1989), Cache River bridge (over the diversion channel, 1991), approaches, mainlines, and the fly-over ramps of the Poplar Street bridge (East St. Louis, MO, late 1960s), and State Route 13 bridge (over railroad, St. Clair County).

More specifically, the elastomeric isolation bearing consists of a force control mechanism capable of dissipating earthquake energy and also providing better distribution of seismic forces among bridge supports. One of the manufacturers is Dynamic Isolation Systems (DIS). The supports connect the upper structure with the substructure, eliminating any possibility of girder unseating that often occurs during earthquakes. When the bearings deform under service loads, their deformation depends upon the rubber and the lead core deformation. Such deformations are elastic under the live loads and are plastic under earthquake forces. Plastic deformations are determined by the plastic deformation in the lead.

These kinds of bearings reduce the horizontal stiffness and increase the structural period, which in turn result in a reduction of the inertial forces during earthquakes.

Friction-type isolation bearings use PTFE sliding surfaces with controlled frictional resistance to connect the upper structure with the substructure, longitudinally and transversely. The Eradiquake brand is made in the United States by R. J. Watson.

The FIP bearing (manufactured by FIP Structural Systems of Italy) has been successfully used both in Europe and the United States. The manufacturer produces two types of FIP bearings. In one type, small thermal movements between the upper structure and substructure are allowed in each direction by the sliding surfaces and elastic deformation of steel pins. Type II bearings instead allow large movements using a shock transmitter and a second sliding surface. During earthquakes, the shock transmitter is locked and the steel pins are engaged (Karshenas, et al., 1997). Another valuable resource on this topic is "Methods of strengthening of highway bridges," report 293 by the National Cooperative Highway Research Program (Transportation Research Board; Washington, DC).

Case studies

The following case studies present factual examples of the many projects that illustrate the present state of affairs, mostly in the United States, concerning the rehabilitation of bridges, including upgrading.

Julien Dubuque Bridge

The Julien Dubuque Bridge over the Mississippi River on U.S. Route 20 had been scheduled for renewal by the Iowa Department of Transportation. The project engineers were Arthur W. Hedgren, Jr., Ph.D., P.E., and William Schmitt, P.E., of HDR Engineering, Inc. The bridge had a length of 5,760 ft. The total length of the spans includes girder approach spans of 1,448 ft on the Iowa side, a 1,539-ft three-span, through-arch truss unit over the river, and 2,773 ft girder span on the Illinois approach. Of particular significance for this bridge built in 1943 is that it received the AISC Annual Award of Merit for Most Beautiful Bridge, Class 1, and is also on the National Historic Register.

The remodeling of the bridge was organized into three phases which included: preparation of the design study reports for deck replacement, strengthening, and structural repairs of the bridge, preparation of contract documents for deck replacement, widening of the deck, member strengthening, and structural repairs, and providing consulting services and construction engineering during the construction phase of the project. Design studies which are part of the first phase includes the following:

- studies of alternatives for the replacement of the bridge deck
- rating of the steel superstructure in connection to the new deck
- design for strengthening and repairs
- cost estimates
- scheduling of construction

The following different types of decks to replace the existing ones, were considered: cast-in-place, precast, orthotropic, exodermic, and steel grid (partially filled). Part of the remodeling of the bridge included the substitution and widening of the existing deck. The existing deck included a roadway of 24 ft and a 5-ft wide concrete sidewalk on one side and a steel walk on the other side. A new roadway structure was designed adding 4 ft to the previous one.

The northside concrete sidewalk remained the same, while the steel safety walk had to be substituted to 4.5 ft wide. Existing trusses had to be reinforced and, similarly, the girders in the approach spans had to also be reinforced. Further work included part of the reconstruction of the Iowa approach spans (Hedgren and Schmitt, 1993).

Red Kill Bridge

The Red Kill Bridge in New York State, carrying Route 23A, had been designated for rehabilitation by the New York State Department of Transportation and will be engineered Ryan-Briggs, Associates, P.C., of Troy, New York. The structure, 114 ft long, was a reinforced concrete twin box-girder bridge having a cross-sectional width of 10 ft and a 14-ft high individual box, with 1.25 ft spacing between the two.

Since its construction in 1953, the structure had to be reinforced in 1958. Yet the bridge continued to deteriorate as verified by inspections in 1981 and 1983. Such a deterioration included spalling of the concrete and exposure of the steel reinforcement, including estimates of potential failure of the vertical wall of the boxes.

In addition, wingwalls were examined and considered to be exposed to potential failure. In conlusion, the reports indicated that structural reinforcements were required to avoid potential failures of the bridge.

On the basis of this report, studies had to be made to decide upon different alternatives to be followed. This included the restructuring of the bridge, the closing of the bridge, alternative routes for diverting the traffic in one direction or the other, and evaluation of economic losses and social inconvenience to the nearby urban centers that used the bridge.

Part of the project included the studies of various alternatives, including: retaining the structure in its present condition, continuing with maintenance and repairs, and casting of additional concrete to reinforce the deteriorating components, insertion of a corregated steel pipe into each barrel of the existing culvert, replacing the existing structure with another of similar type, replacing the existing structure with a conventional one; reduction of load capacity, rehabilitation of the structure to its existing sate, and replacement of bridge. After the selection of the alternative, the NYSDOT awarded the contract to the lowest bidder for $370,000 in 1989 (Rogers and Biggs, 1993).

Berkely Bridge

This project involved the strengthening of the steel girder approach to the Berkely Bridge (Norfolk, VA). The dramatic increase in traffic in the area of Norfolk dictated the strengthening and widening of the existing Berkely Bridge to carry the traffic in one direction only, and the construction of a new parallel bridge to carry the traffic in the other direction.

The existing bridge consisted of an approach span of 1,895 ft and a double-leaf bascule span 150 ft long over a navigable channel. The approach spans include thirteen spans, each having a length of 94 ft plus six other spans having a length of 112 ft each. The deck included four lanes 11.5 ft wide and a walkway of 3.6 ft wide. Each span, simply supported, included steel plate girders which originally carried a noncomposite concrete deck.

The work included the removal of the existing deck; the strengthening of the exterior girders of the existing bridge; and the installation of a new deck, wider than the existing one. The new deck was to be enlarged four feet wider than the existing one, obtaining four lanes 12 ft wide plus shoulders 3 ft wide.

Plans for the widening and strengthening of the existing Berkely Bridge and the construction of the new one were assigned to T.Y. Lin International, Alexandria, Virginia, and Hardesty & Hanover, New York, by the Virginia Department of Transportation. The new bridge was completed in 1990 and the old bridge rehibilitation completed in 1991 (Lawrie and Seim, 1993).

Miller Creek bridges and Black Gore Creek bridges

The Vail Pass rehabilitation project located along I-70 in Eagle County, CO, included four cast-in-place, balanced cantilevered bridges, completed in the fall of 1983. These included the Miller Creek bridges, composed of the east-to-west and west-to-east bound structures, and the Black Gore Creek bridges composed of the east-to-west and west-to-east-bound structures.

The Miller Creek bridges had an upper structure consisting of a single-cell trapezoidal box-girder, 42 ft wide and 8 ft deep. The Black Gore Creek bridges had an upper structure consisting of a single cell trapezoidal box-girder, 42 ft wide and 10 ft deep. A report on this job includes quantities of the materials employed for the rehabilitation, including the epoxy resins, concrete, reinforcing steel, and prestressing steel (Corven, 1993).

The four bridges, consisting of concrete box-girders built in the mid-1950s, developed extensive cracks a short time after their completion. On the basis of evaluation performed on these four structures, plans were prepared for their rehabilitation. Part of such plans included the construction of stressing buttresses at the abutments, stressing blocks and deviation beams throughout the span to accommodate supplemental external posttensioning, the repairing of existing cracks with the injection of epoxy resins, and replacement expansion joint devices.

A main factor in the project of rehabilitation of these structures demanded that the bridges were to remain open to traffic, which required, special scheduling of the work. The lengths of the bridges were: 748.5 ft for Miller Creek bridges (eastbound and westbound), 455.3 ft for the eastbound Black Gore Creek bridge, and 518.3 ft for the westbound portion of the bridge.

Pimmit Run Bridge

The bridge at Pimmit Run is part of the George Washington Memorial Parkway that extends 20 miles through Maryland, Virginia, and the District of Columbia. The project of rehabilitation of this structure included the substitution of the old deck, that had gradually deteriorated, with a new deck completing the operation without closing the bridge to traffic. The challenge of maintaining the bridge operable at peak capacity is a particular feature of this case study.

The key solution adopted for maintaining the traffic consisted in dividing the bridge into two parts longitudinally, closing one of the parts during construction, while leaving the other open and working only on weekends. With amazing speed, the work was concluded in only five weekends. The design for the rehabilitation of this bridge included precast concrete panels for the slab, placed segmentally.

The existing structure of this bridge consisted of two edge girders, transverse floor beams, longitudinal stringers, and a cast-in-place concrete deck slab. The bridge's three spans were 104 ft, 132 ft, and 104 ft. The width of 68 ft before the rehabilitation remains the same. However, the distribution of the roadways required to be changed to 12-ft lanes in each direction, a median barrier dividing the lanes, plus 5.25-ft shoulders.

The precast concrete panel for the new bridge deck consisted of elements reinforced transversely to resist stresses through the handling processes. Then the entire deck length was prestressed longitudinally.

The removal of the existing deck was carried out by cutting the deck longitudinally into two parts, and eventually cutting one of the parts at ten-foot intervals, while the other part would carry the traffic. As each part of the deck was removed by a crane, a new precast panel would be installed in place of the old one, leaving a three-inch gap in between panels. To obtain a level surface, each new panel was lowered down on the existing framework by inserting a level bed of grout. The three-inch gaps between slabs left open were eventually filled with high strength concrete mortar. After the mortar had achieved sufficient strength, the longitudinal prestressing created a monolithic slab for the entire span of the bridge.

Hillsborough Bridge

A project for the expansion of the Hillsborough Bridge in Prince Edward Island, Canada had been published describing the proposal. The existing bridge consists of three spans for a total length of 810 ft. The main span is 297.7 long and the two side spans are 256.5 ft each. The deck carries two lanes of traffic for a total width of 24 ft, plus two 4-ft sidewalks. The expansion of the bridge included the addition of another lane of traffic, thus the study had to verify the feasibility of the project, considering that the bridge could not be closed to traffic during the restructuring.

Part of the project demanded the substitution of the deck with pre-cast pre-stressed concrete deck panels and external post-tensioning. Furthermore, to increase the capacity of the supporting steel trusses, the study called for the pre-stressing of the trusses by adding draped post-tensioning tendons. The tendons applied on the side will be encased in a tube and grouted for corrosion protection.

The new concrete deck would cantilever out from the existing supports and be capable of carrying the additional load. The slab panels were designed to be erected two at the same time in sequence and then eventually prestressed in the longitudinal direction to obtain a monolithic deck. Each of the concrete deck units are secured to the steel structure below by means of steel shear stud connectors. Furthermore, adjacent panels are also post-tensioned transversely to secure monolithic action (Corven, 1993).

Aglio Viaduct

The Aglio Viaduct, between Bologna and Florence in Italy, was completed in 1960 and required major restoration due to a premature aging of the structure. The structure includes two side-by-side reinforced arches having a span between the two supports of 537 ft and a rise of 143 ft. Supporting the deck are 18 vertical struts equally spaced. The layout of the bridge includes a straight section of 1,091 ft and a curved section of 349 ft with radius of 1,148 ft having a grade of 2.9 percent. Since its completion, the bridge required repairs for one pier and a slab. Part of the damage included the following:

- severe deterioration of the piers considering those standing from the ground as well as those resting from the arches
- disentegration of concrete especially around the edges
- rusting of reinforcements
- shear failure of the ties in some of the piers, damage to the portal beams
- rusting of the fixed and moveable bearings
- inadequacy of the expansion joints and the drainage system
- damage due to the effect of chlorides through the joints in the bridge deck
- effects of deterioration due to traffic

Major repairs required by the damage listed above involved the piers, the substructure and the deck. The piers required a 6-in coating of a thick layer of rheoplastic cement mixture made with a high adhesive containing adequate steel reinforcement. The piers of the arches were also covered with a 5-in layer using the same kind of concrete and reinforced with adequate steel reinforcing bars. Portal beams supporting the deck also required similar treatment. Furthermore, repairs were conducted on the damaged parts of the concrete arches and application of epoxy and polyurethane on those parts of the bridge that showed particular vulnerablilty to weathering agents (Medeot and Romagnolo, 1993).

Rip Van Winkle Bridge

The Rip Van Winkle Bridge over the Hudson River, 30 miles south of Albany, is a structure built in the mid-1930s. The bridge is 5,000 ft long, and the spans include a series of fixed, cantilevered, and suspended spans supported by steel towers that bear over concrete piers. Several inspections conducted over the years revealed an unsatisfactory performance of the bridge that included: faulty expansion joints that froze and had not been operable for years, cracks in one of the concrete piers underneath the water level, deterioration of the concrete deck, and the need for enlarging the deck. The assignment for inspection and preparation of plans for the rehabilitation of the bridge were assigned by the New York State Bridge Authority to Parsons, Brinckerhoff, Quade and Douglas, Inc.

Reports in the literature on the rehabilitation of the bridge refer to the bowing of the bridge beams, due to the frozen Pier 3 wind-lock. When the wind-lock was eventually released during the reconstruction work, a loud bang was heard when the bowed beams reacquired the original configuration.

The new cast-in-place concrete deck occupied the entire width between the thru-trusses; and the additional sidewalks on each side were added by cantilevering them outside of the trusses. Probably one of the main features to think about when rehabilitating a deck is that of adding a cantilevered portion on each side of the existing structure. Contract of the restoration and widening of the Rip Van Winkle bridge was awarded in the spring of 1991 and completion of the work was at the end of 1992 (Moreau, et al., 1993).

Bridge over the Kinderhook Creek

The bridge over the Kinderhook Creek is located in Stuyvesant Falls, NY, and was placed on the National Register of Historic Places in 1977. Several inspections by the New York Department of Transportation declared the bridge unsafe, due to structural deterioration, and eventually it was closed to traffic in 1991. This old steel truss bridge had been condemned for demolition and was to be replaced with a modern concrete structure. However, due to historical significance of the bridge and due to political pressure initiated by the local population, plans for a new bridge were changed, and plans for the restoration of the old structure were ordered.

The final solution reached, after a complex series of events involving citizens groups and several state agencies, involved the rehabilitation and upgrading of the bridge using steel arches working in combination with the existing trusses. The original bridge built in 1899 by the Berlin Iron Bridge Company, consisted of so-called "camelback" trusses with 12 panels and had a span of 202.25 ft. The width from curb to curb was 17 ft, plus a 4.8-ft-wide sidewalk cantilevered on one side . The floor deck consisted of an open steel grate, with purlins, stringers, and floor beams supporting it. The trusses were supported by masonry piers.

Alternative solutions for the rehabilitation of this bridge, included first, the substitution of each of the truss members with new members, step-by-step until the entire structure had been restored and upgraded to new standards. The other solution instead, the addition of steel arches, was proposed by studies at Bucknell University. There, model testing validated the effectiveness of this method which was eventually patented and applied to several other bridge restorations in the United States. The method proved to be effective in upgrading the structure for carrying the larger loads of present-day traffic. Altogether, this solution was proved to be much more economical than the original proposal of demolition and rebuilding of a new bridge.

The arch rehabilitation system for old steel truss bridges is a patented method that appears to be much more advantageous than any of the typical rehabilitation systems used before. For the particular bridge in question, after reviewing the reports on the bridge inspections made through the years, and after having assessed the feasibility of the arch rehabilitation system, the project for the rehabilitation was assigned to Ryan-Biggs Associates in Troy, New York. In particular, the design for the arches was assigned to BKLB Incorporated in Lewisburg, Pennsylvania, acting as subconsultants.

Specific data supporting the competitiveness of the arch rehabilitation system over the substitution of the bridge with a new structure, reveal that the costs of the arch rehabilitation system for this project was $750,000 (low bid), versus $1.5 million for a new structure (Kim, et al., 1993).

Stony Creek arch bridge

The Stony Creek arch bridge located in Danville, IL, is an historical bridge built in 1894 and designed by Ira O. Baker, a notable engineer in the state. Because of its age and typical configuration, this arch bridge, made of sandstone, was also historically significant due to the fame of its designer. The structure deteriorated considerably through the course of the years and major rehabilitation work was necessary for restoring the structure and upgrading its strength to carry the new HS20 loads.

Two alternatives were examined. One, included the demolition of the bridge and reconstruction of a new modern one. The other consisted of restoring the existing structure, retaining all the elements of the old bridge. Comparing the costs of the two alternatives, the rehabilitation of the old one came to be higher than the cost of building a new bridge, yet although more costly, the rehabilitation of the old structure was eventually selected.

The existing bridge consisted of a sandstone masonry arch spanning 62 ft. The arch, with its elliptical configuration, had sustained major structural damage at the springlines and at the quarter points, where the stresses were higher.

The solution for its restoration had no precedence, and therefore constituted a unique prototype that could eventually be employed in similar other cases. The concept was based on the fact that patching the original arch would not have reinforced the masonry arch sufficiently, and therefore the arch should have been totally supported by a new arch built underneath it.

Such an arch, made of cast-in-place reinforced concrete, would carry itself as well as the dead load of the masonry and the live load due to traffic. The new arch was calculated to act independently without any composite action with the existing masonry arch. By computer analysis, the new arch was designed to be 18in thick at the crown and gradually reaching 30in at the springlines. Also, a new mat foundation in concrete was designed to tie the arch at the base after diverting the flow of the creek during construction.

This case study has a particular merit for pointing out a system for restoration, which is unique in itself, and could probably be applied to similar other cases concerning arched masonry bridges. Completed in 1990, the cost of the project was $1,370,000. The consultant for the project was Environmental Science & Engineering, Inc. (Bell, 1993).

Manayunk Bridge

The Manayunk Bridge, crossing the Schuylkill River in the northwest area of Philidelphia, Pennsylvania, is a concrete structure that includes nine spans for a total of 1,900 ft. The structure, built originally in 1918, had been closed to passenger railroad in 1960 and eventually totally closed in the late 1980s due to the deterioration of the structure. Because of such deterioration the arch bridge was no longer able to carry the railroad.

The owner of the bridge, the Southeastern Pennsylvania Transportation Authority (SEPTA), commissioned to conduct studies for investigating basic alternatives such as demolition and construction of a new bridge, or rehabilitation of the existing structure. Part of the studies involved site surveys, laboratory tests, computer analyses, and evaluation of alternatives.

Placed in the National Register of Historic Places in 1983, the bridge, in 1989, was totally fenced in for safety and security reasons, including the installation of nets on the concrete arch spans to catch spalling concrete. Visual inspection of the concrete surface revealed cracking, scaling, and spalling.

A team comprised of a transportation specialist, a field evaluation and laboratory agency, and two university research laboratories submitted a proposal to SEPTA which was eventually approved it. The proposal included the following: assess potential future degeneration of the structure; assess overall conditions of the bridge; assess the amount of degeneration in the actual conditions, i.e., without any repair; propose the rehabilitation plan; propose plans for demolition; determine several methods of rehabilitation; determine costs for a full rehabilitation plan.

The team was assessed with the duty of inspecting, testing, and evaluating the structure. Field work included: visual inspection; delamination survey; removal of cores; pachometer survey; and the removal of powder samples for relative humidity tests and non-destructive impact echo tests.

Part of the duties which were performed by the team was the finite structural analysis of the bridge. Results of such analysis indicated that the highest stresses within the structure had never been higher than the allowable stresses. Eventually the assignment to the team was completed and the results had been given to SEPTA for final evaluation and eventual selection of the alternative to be implemented (Kline, et al., 1993).

Russian Gulch Bridge

The Russian Gulch Bridge is in the Russian Gulch State Park on Highway 1 in California. The historic structure is part of the Phase II Seismic Retrofit Program undertaken by the California Department of Transportation (CALTRANS). The seismic evaluation and retrofitting project of the structure was recently completed, and the analysis indicated the inadequacies of the bridge to resist transverse horizontal seismic forces. The bridge length is 526 ft. It includes a concrete deck slab, supported on 29 bents, supported in turn by two columns each. Along its length, the bridge includes a spandrel arch with a span of 240 ft and a rise of 85 ft. The arch in turn carries 16 of the 29 bents previously mentioned. The columns having a rectangular cross-section are fixed in both directions at the bottom and while at the top they are pinned in the longitudinal direction and fixed in the transverse direction. Furthermore, there are four expansion joints, one located over each arch span and two toward the middle of the arch.

The retrofitting strategy aimed to identify alternative paths for the transverse seismic forces applied to the bridge, minimizing the force applied to the arch itself and other parts similarly vulnerable. Large displacements can also occur in the longitudinal direction due to the absence of support at the abutments and lack of continuity within the deck. In recommending retrofit alternatives for this bridge, several limitations had to be respected. These included the preservation of the historical character of the bridge, continuous traffic flow, and structural effectiveness. Therefore, the following measures had been recommended: retrofitting of the abutments; retrofitting of the hinges; and retrofitting of the column bases (Abdel-Karim, et al., 1993).

Strawberry Mansion Bridge

The Strawberry Mansion Bridge, built in 1897 in Philadelphia, spans the West River Drive, Schuylkill River, and Kelly Drive. The total length of the bridge is 1,250 ft. It includes six truss spans and four steel arches of the three-hinged type, having a span of 200 ft each. Originally the bridge carried two lanes of traffic plus two sets of tracks.

The project for the evaluation of the bridge had been commissioned by the City of Philadelphia, Department of Streets, to A. G. Lichtenstein & Associates, Inc. Such an evaluation project included inspection and evaluation; non-destructive testing; seismic analysis; design of bridge strengthening and upgrading the structure to carry the latest AASHTO loadings; and design of a promenade within the area of the original trolley lines. The environmental studies were not included, since they were performed by the City of Philadelphia from 1991 to 1995. Furthermore during the process of completing the design studies, the deterioration of the bridge had been so extensive as to require the closing of the structure to traffic in 1996.

Part of the plans produced by A.G. Lichtenstein & Associates, Inc., included two phases. The first phase included the replacing of some of the back walls and abutment seats and floor beams, the stringers, and the deck. The second phase will include painting of the structure, architectural railing and lighting, construction of the promenade, strengthening the truss elements, and rehabilitation of the substructure units. The first phase will be completed in 1997 while the second phase will be completed in 1998 (Homaee, et al., 1997).

San Mateo-Hayward Bridge

The San Mateo-Hayward Bridge is located between Foster City and Hayward City over the San Francisco Bay in California, with a total length of 7.1 miles carrying six lanes of traffic on State Route 92. The firm of Carter and Burgess had been given the contract for the design of the seismic retrofit of this bridge by the State of California through the Business Transportation and Housing Authority (BT & HA). This project was part of a program including six toll bridges assigned to private consultants to accelerate the pace of rehabilitation.

This bridge, opened to traffic in 1967, was one of the first major bridges in the United States with orthotropic steel decks. The bridge is divided into three sections. The first section includes the approaches consisting of T-beams spanning 220 ft, supported by two column bents at each end. The second part is the main section of the bridge, 1.85 miles long. It includes twin steel box-girders with orthotropic steel decks. The last part includes a precast concrete trestle section consisting of 859 30-ft spans. Of these, the main part is the section undergoing rehabilitation.

The main part includes 37 piers based in the water and one on land. The various groups of piers differ from one another, both in geometry and structural systems. The steel superstructure, instead, includes 37 spans consisting of two steel box girders. The spans vary in length from 186 ft up to 750 ft, which is the span over the navigation channel that has a clearance of 138 ft over the mean sea level.

The spans include 18 single-span anchor units alternating with 19 suspended spans. More precisely the suspension spans are carried by the cantilevered parts of the anchor spans. The hinges at the supporting points are tied with pin angle bars along the longitudinal direction, while they are restrained in the transverse direction by shear locks. During the Loma Prieta earthquake the bridge sustained relatively moderate damage that was eventually repaired.

The major concepts in retrofitting this bridge to strengthen its resistance to seismic forces, depend mostly on a ductility enhancement. The various components demanding focal rehabilitation solutions primarily included, bearing assemblies; steel towers, spandrel beams, and tower anchorage assemblies; and rectangular foundations (Vitek, et al., 1997).

Rock Island Centennial Bridge

The Rock Island Centennial Bridge, spanning the Mississippi River and linking Davenport, IA, and Rock Island, IL, was opened to traffic in 1940. Its length is 3,850 ft, and it includes two viaduct approaches and five spans consisting of steel box-girder tied arches, having lengths of 540 ft and 396 ft. The rehabilitation and upgrading of the structure included the repair of the deck and the open drainage system on one end, and the upgrading of the structure load bearing capacity to accommodate for the larger vehicular loadings required by the latest AASHTO *Standard Specifications for Highway Bridges*. Specifically, the arches' rehabilitation required the use of computer analysis.

The arched portion of the bridge includes two arches 51.5 ft apart carrying cantilevered sidewalks on each side. The arch ribs are plated-steel box girders. The roadway and flooring are supported by vertical hangers with a 14-in-deep H-section. Such hangers connect to the ends of the floor beams. The original deck consisted of a 7.25-in reinforced-concrete slab, supported by cross beams spaced at 4.5 ft. The cross beams are in turn supported by longitudinal girders spanning 35.7 ft. The concrete sidewalks were 3.5-in thick supported by steel bulb angles and channels.

Since the construction of the bridge, the City of Rock Island has hired HNTB Corporation for providing annual inspections of the bridge. Up to 1976, the bridge received some minor alterations and periodic repainting, but in the 1980s a program was started to rehabilitate the steel structure. Specifically, in 1987, an analysis was conducted to prioritize required repairs and develop a long-term maintenance plan.

Part of the rehabilitation program included the replacement of the deck, replacement of steel members that have deteriorated, installation of a closed drainage system, replacement of the link hanger connections, and painting of the steel. In the process of the design for the replacement, the City Council decided to upgrade the structure to carry only HS15 loads, rather than HS20, and approved a program that will allow the toll bridge to be in operation during the restoration work with at least one lane of traffic constantly open in each direction. It also approved a construction time limit within 16 months from the beginning of the work.

The sequence of work for the replacement of the deck indicated that, along the approaches, one side of the original deck could be removed and replaced, while the other side would remain untouched and open to traffic. Then eventually the traffic would be switched to the newly installed side, and the other side would undergo similar replacement. However, such a scheduling could not be accomplished for the arched section of the bridge, due to the uncertainties of the assymetrical loading conditions.

In order to design the schedule of reconstruction on the arched section, a computer analysis using GT STRUDL II had to be done, which indicated that the reconstruction process of the deck should follow the same schedule used during the orginal erection of the bridge. Eventually the construction scheduling for the reconstruction of the new deck was modified at the request of the contractors to whom the job had been awarded. The heavy existing concrete deck was eventually replaced by the new orthotropic steel deck of galvanized steel, and the new sidewalks were constructed with a lighter weight system that weigh one-third less than the original sidewalk (Tessiatore and Nusekabel, 1997).

Chelsea Street Bridge

The Chelsea Street Bridge in Boston, MA, over the Chelsea River is a steel bascule-type bridge with an operating span of 135 ft. The structure was severely damaged in 1982 by the collision of an oil freighter. After the collision the bridge was kept open and the evaluation of the damage was assigned to Hardesty & Hanover.

The results from the inspection included the following findings: two truss members had been severed; the bridge remained standing, but the overall geometry of the bridge remained unacceptable; many members of the trusses and floor system were severely damaged and needed replacement; and the rest of the structure remained intact.

The rehabilitation program included two options. The first was to completely dismantle and rebuild the bascule span. The other option consisted of restoring the geometry of the span without removing the span and then replacing the damaged members using temporary supports. H&H adopted the second option. In restoring the bridge geometry, a jack system was employed. The system included three jacking points and the use of cables which have been proved to be particularly effective in the restoration of the geometry distorted by collision (Coates, 1997).

Summit Bridge

The Summit Bridge crosses the Chesapeake and Delaware Canal in Delaware, carrying four lanes of traffic on State Route 896. The structure was built in 1959 with a length of 2,058 ft. The bridge includes a main span length of 600 ft, which also has a suspended span of 300ft carried by the cantilevered portion of the other spans. The bridge had been inspected and an evaluation report had been prepared by A. G. Lichtenstein & Associates, Inc. under contract by the U.S. Army Corps of Engineers once in 1994 and later in 1995.

The inspection revealed several fractured members and zones of fatigue, as well as an uneven wear at those supports that connected a supported span with the cantilevered portion of the next span. Movements of the 6 in. diameter pin in the eyebar linkage plates could be observed due to the 0.375-in gap being produced by the wear of 36 years of service. In addition, a distinct sound could be clearly heard during the movements of the pin within the eye bar.

A rehabilitation program therefore included the replacement of the pins and members that had suffered damage. The work, of course, had to be scheduled to provide a minimal interruption of traffic. The replacement of the hinge mechanism offered two possible solutions. One consisted in replacing the old mechanism with multirotational bearing devices. The other alternative instead consisted of the replacement of the assemblies with a similar one having a higher resistance to gouging, and an increase in the bearing areas of the pins by using structural sleeves. After careful consideration, the second alternative was selected (Homaee, et al., 1997).

Boulevard Bridge

The Boulevard Bridge in Richmond, VA, built in the 1920's is 1,118 ft long. The bridge, carrying only two lanes of traffic plus two 4 ft sidewalks, spans over the James River, the tracks of CSX and Norfolk & Southern Railroads and the Kanawha Canal. The bridge is composed of 24 spans, of which nine are continuous cantilevered trusses, and the remaining ones are simple truss and riveted girder spans.

The proprietor of the bridge, the Richmond Metropolitan Authority (RMA), reviewed several options for the rehabilitation of the bridge and eventually decided to rehabilitate the bridge to accommodate present-day standards. The criteria for rehabilitation included the following: replacing the deteriorated structural system, replacing missing rivets with high-strength bolts, sandblasting the truss members, repainting, replacing of the deck with a new lightweight concrete type, and redesigning the new approach spans. It is interesting to note that during the analysis of the bridge, due to the lack of documents, the type of steel used for the bridge could not be established, and therefore tests had to be conducted on the material itself to establish the mechanical characteristics of the steel.

HNTB Corporation of Chicago, Illinois conducted the required studies for the evaluation of the bridge and eventually proposed the following two alternatives. One consisted of rehabilitating the bridge in order to accommodate two 11-ft wide lanes of traffic plus the addition of a 5-ft wide sidewalk at a cost of $7 million. The other alternative consisted of replacing the entire structure with a new bridge at a cost of $11 million. The first alternative of rehabilitation was eventually selected for a variety of reasons as follows: the first alternative was judged to be more cost effective, environmental issues suggested the rehabilitation of the structure over the reconstruction of a new bridge, concern of the neighbors about the potential of increased traffic in the event that a new bridge was built, loss of revenue in the case of building a new bridge due to the fact that a neighboring toll bridge would lose some of the traffic diverted to the new bridge. The rehabilitation project was particularly successful due to the contribution of the BEST Center and HNTB (Fu and Jausurawong, 1997).

Macombs Dam Bridge

The Macombs Dam bridge is located in New York City and crosses the Harlem River between Manhattan and the Bronx. Constructed in the mid-1890s, the bridge was designed by Alfred Boller and its identity puts the structure within the classification of an historical bridge. The bridge includes a swing span of 415 ft long and adjacent spans that include deck trusses, camelback truss, beam spans, and a viaduct structure totalling 70 spans in all.

Due to a century of service, the overall conditions of the bridge have aged reaching a state of degradation that required attention. On this basis the firms of Modjeski and Masters, Inc. and Urbitran Associates were retained by the New York City and New York State Departments of Transportation to inspect the structure and prepare plans for its rehabilitation and upgrading for carrying new live loads of HS20. As a result of the studies, the rehabilitation contract of $92 million was awarded to Nab Construction Corporation in 1996.

Included in the process of rehabilitation was the basic criterion to conduct the operation with a minimum disturbance to traffic that also involved events at Yankee Stadium which required a special scheduling of various operations concerning the progression of the work.

Part of the basic operations of the bridge's rehabilitation and included the retrofitting of the structure with additional eyebars. In so doing the new eyebars should be able to work in unison with the existing eyebars, preventing the overloading of one over the other and preventing the possibility of making one slack. To this effect, the new eyebars have to be prestressed to an adequate value by using jacking devices. More specifically this load will be applied when no live loads are acting on the structure. More specifically, the jacking devices consist of hydraulic stud tensioners commonly used for work on prestressed concrete operations. The sequence of the jacking operations in a particular role for the satisfactory completion of the retrofitting operations on the eyebars was accomplished by a trial and error system.

With regard to auxillary trusses, it was found necessary to relieve the loads on some of the existing trusses by introducing new ones along the centerline of the bridge. Some complications were incurred at the end of the new truss for providing adequate supports due to the lack of space caused by the existing supports of the other trusses. Additional retrofitting considerations, besides the structure, included the remodernization of the mechanisms controlling the operation of the swing span. In addition the cables carrying power to the machinery room were required to be substituted by new ones which had to be installed within a 15-ft-deep trench below the bottom to accommodate potential deepening of the channel in the future.

Another part of the rehabilitation process included the removal of some of the original stringers and replacement of them with new ones rather than trying to upgrade the older ones.

With regard to the concrete pier bents, it was observed that a serious deterioration had occurred in the caps. The damage often reached the point where the reinforcement and stirrups were exposed, which required total replacement of the caps rather than trying to repair them by patching. In the replacement caps, were precast without any complication and installed, allowing the new stringers to bear on their top.

Further operations in the rehabilitation of floorbeams and longitudinal members included the installation of coverplates on the flanges that strengthen considerably the bending capacity of those members. Some difficulties in the installation of the new plates over the existing ones had been encountered, yet the specific details of installations may be useful to be ascertained from a more in-depth description of the work.

Furthermore the upgrading of the bridge included the replacement of all of the bearings with new ones that will improve the response of the bridge to seismic forces (see LeRoy and Kauhl, 1997).

Bridge retrofitting for high-velocity railways

High-velocity train systems may include not only new projects starting from scratch but also the retrofitting and improving of existing railways and railroad bridges. A new generation of railroad transportation is no longer a distant goal as was a few decades ago. The reality of new railroads system employing high velocity train is now is a tangible achievement which have been demonstrated in different countries already including Europe, Asia, and North America.

The definition of high-velocity trains is in the order of magnitude of 200 mi/h but it is expected in the near future that such a velocity could reach over 300 mi/h. Therefore the interaction between bridge structures and high-velocity trains will create a new generation of rail poses a new set of design parameters. To this effect different aspects of this problem are briefly discussed for the major influence that these new bridge criteria criteria will have in the future.

United States

For instance, considering Amtrak high-speed trains between New York City and Boston, the the tilt system for high speed trains is one part of the improvements for raising the traffic velocity in such route. More specifically in the section between New Haven and Boston where the line is nonelectrified and the track layout is particularly tortuous, a new system for high velocity trains would include the following basic criteria. This includes the electrification of the line using high power electric ac traction power cars of a light-weight type that will guarantee high acceleration performance.

A major factor in the project in question is the winding of the rail tracks that includes a large number of curves which cannot be eliminated due to the typography or the terrain. To overcome the influence of such curves on traveling speed the tilt system developed to allow the increase of velocity inspite of the tortuosity of the tracks (Hesser, 1997).

Norway

In Norway, a 5-km length of double track was supposed to have been completed by 1994 within Norway's Ostfold region. Such a line constituted the first project to be implemented in Norway for high velocity trains with speeds in excess of 130 km/hr and inclusive of tracts allowing speeds of 200 km/h. The first part of the project will include a double tract of 34 km to be extended equal t between Oslo and Halden.

Begun in 1987, this project by Norwegian State Railways was completed in 1996 after including quite a number of variances throughout the course of the planning and construction. Part of the future extension of this project includes the completion of high velocity rail-line transportation that will connect Oslo to the Swedish border at Kornsjo and further to Goteborg. For more specific data, see Jackson (1994).

Czech Republic

Czech Railways is also interested in upgrading the system in terms of high velocity transport on rail up to the range of 160 km/hr and more. The project involves quite a variety of solutions concerning not only the interaction of high-velocity trains with bridges but also a series of others such as: curve realignments, tunnels, underpasses, improvement at railroads stations, improvement of level crossing, new signal systems, train control by telecommunications, and environmental protection (Malek, 1995).

Japan

The Railway Technical Research Institute in Tokyo is working on Shinkansen bullet train designs that analyze the sound levels of such high-velocity trains with respect to a series of parameters which include also bridges. Bridges and other specific characteristics of such high speed lines contributing to loud sounds include cars and other installations. Specific noises depend on rolling, bridge structure, pantograph spark, pantograph aerodynamic, as well as other aerodynamic noises.

Trying to develop a sound mitigation, a special program based on actual measurements of the level of the noise generated by each of the contributing parameters has been conducted based on testing. In such tests, in fact, the applied methodology consisted in measuring individual noise components by means of microphone arrays and parabola microphones. See Moritoh, et al. (1996) for the contribution of railroad bridges interacting with high speed trains.

Japan's participation in the development of high velocity railways and special bridges required for such new lines are currently being studied.
More specifically a research project is analyzing the combined dynamic behavior of Shinkansen (bullet train) vehicles with irregular rails and with bridges using a finite metal approach. Part of the study involves a mechanical model analyzing the interaction between wheels and irregular rails. Furthermore, bridges in this context are also analyzed by finite element methods. More specifically, part of the project includes also a specific computer program that analyzes the dynamic response of Shinkansen vehicles operating at high speed over irregular rails (Tanabe, et al., 1994).

Further studies on the response of bridge girders and beams subjected to the dynamic loads from high-velocity trains, focuses also on specific flexural structures made out of concrete-steel composite members. Part of such studies include the determination of a new impact formula to be included in the standard specification of Japan National Railways (Asakawa, et al., 1984).

There are still other parts of the research programs that Japan is carrying out on high-velocity railroad systems; important testing on the response of railroad bridges has been recently conducted. Specifically the dynamic responses of different bridges under high-velocity railroad traffic have been conducted on different types of bridge structures. These included three bridge structures consisting respectively of truss girders, composite girders, and preflexed composite girders. Measurements conducted during the testing included impact factors in relationship to stress and deflection (Asakawa, et al., 1983).

The interactions between bridge and railroad vehicles have been specifically studied with respect to the dynamic analysis of bridges under high-speed trains. Part of the study involved the modeling of the vehicle-bridge interaction using an element of particular efficiency and accuracy as well. As reported in the description of the methodology described by Yang and Yau (1997): "In this study, a train is modeled as a series of sprung masses lumped at the bogie positions and a bridge with track irregularities by beam elements."

Other studies on railway bridges interacting with high velocity traffic include also work on problems involving the behavior of bridges along high speed railway lines. A very recent paper by Niemierko (1997) presents an overview of this problem based on the findings of the European Rail Research Institute. Included is a presentation of vehicle, track and foundation, as well as the condition of the maximum track overloading on the high speed traffic. The resonance response during the dynamic loads at high velocity speeds is discussed together with the effects on riding comfort in terms of car body accelerations, and the running stability in terms of decrement factors.

The interaction of high-velocity railroad vehicles with structural bridge components involves the analysis of beam vibrations. Working along these lines, simple beams are specifically analyzed for their behavior under the dynamic loads of high-speed railroad traffic using a special modeling. In such a work the train is modeled using two subsystems. One of these includes the front wheel assemblies and the other the rear assemblies. Considering the wave generated by continuously moving loads on the beam, the suppression of resonance in the beam behavior is a fundamental objective of this work. Specifically, bridges carrying high-velocity railroad system should be designed considering the suppression of resonance in all bridge components by a proper selection of each structural member, especially beams, in terms of length and cross-section. Other specific parameters involved with the designs in question, includes the length of the cars, the distance of the axle, and the speed of the vehicles (Yang, 1997).

Other studies

Further studies on high-velocity railways include also characterization of the traffic on high-speed token-ring networks. A study works specifically with application toward modeling network interconnection and computerization. Moreover the modeling of railroad traffic at any given point is analyzed by using the Markov packet train characterization. Also included is a simulation study of a bridge with a "Lan/Wan interconnection" (Spiegel, et al., 1994).

6

FUNDAMENTALS OF DESIGN FOR STRUCTURAL ENGINEERS

NEW TRENDS

In the design of structures of major relevance, the engineer assumes by right the main role in the design and construction phases, supported by a number of collaborators in various fields that eventually constitutes a supporting team. By structures of major relevance we imply civil engineering projects on land and on water necessary for enclosing spaces for habitation or working conditions: structures on the waterfront delineating the transition from land to water and for accommodation of ships, major industrial facilities, airports, viaducts and bridges, tunnels, etc. In the typology of these projects a large inventory of potential solutions is available by a large number of case studies that can be accumulated throughout the course of recorded history starting from the latest examples and going backward in chronological order.

A general consideration of particular relevance to be added at present is the evolution of some of the basic parameters used in the conceptual design of previous structures. Such consideration refers to aesthetic values that all kinds of structures must implicitly reflect and to the compatibility of any structure with the surrounding environment. While such two considerations may have implicitly been included in the structures of the past, the present demands have been emphatically underlined in recent decades. Therefore, the spectrum of potential solutions of classical structures have to be reexamined under present-day concerns.

These major two considerations, now present in any design program, reflect a philosophy generated by a social conscience that developed at the same time throughout the world. Their acceptance is a basic state of facts that would doubtfully recede with the future, implying a dependable new approach of major stability. In this context, therefore, the traditional role of professional engineers should definitely include an educational background that contains the basics for aesthetics and environmental expertise. At the same time a "team approach" would be required that would include different professional experts to work together with mutual understanding of each other's role. Among the various structures previously mentioned, those that will be affected the most by aesthetics and environmental characteristics would definitely be bridges which constitute the main topic in the present context.

CONCEPTUAL DESIGN: SCHEMATIC PROCESS

In the design of bridges, the structural engineer as the leader of the team should be capable of developing not only the overall design and construction of the structure, but also its original conceptual design phase. This involves a clear definition of the term "conceptual design" without ambiguities in all its fundamental and minute components. The major relevance of this phase of design has not been sufficiently emphasized in the engineering education, and therefore it demands special attention in contrast to the attitude usually adopted in architectural education at the beginning of any design process. Consequently some of the basic principles used to organize a conceptual design in architectural terms could also be used for the conceptual design of engineering structures.

One definition of conceptual design for bridges implies a series of operations in the design process which are implemented at the very beginning. The first operation consists of gathering all possible information concerning the requirement that a bridge has to satisfy. In so doing a major program is outlined, making it possible to begin the process of design. In this phase there are many operations which eventually can be expressed in a long list that will follow the same pattern for all cases.

The list may be longer or shorter, according to the data that are required in the specific case and according to the data that could be available. In the first stage, therefore, there is no effort dedicated to design and creativeness is not part of it. It is rather a laborious operation which may involve a great amount of hours for research, and it is usually developed at the very preliminary stage of conceiving the idea of the project.

For example, included among the many data are economic factors addressing the feasibility of the structure and the various advantages and benefits that the structure will provide. Also part of the data may include the different types of beneficiaries that will gain from the use of the structures, the social impact on the region, and the monetary return out of the investment.

The list may be particularly long when considering how may factors affect the structure in other fields such as geographical considerations, topographical limitations, weather factors, and building code requirements, etc. In conclusion there are an extremely high number of factors to deal with individually just on the preliminary level, allowing for a more in depth analysis later on in the process of design.

In the preliminary stage of conceptual design, the second phase is a more structured one that consists of creating a tangible series of proposals by sketching all kinds of potential solutions that may apply to this case. An inventory of structural systems has to be gathered either from the individual experience of the structural engineers involved or systematically organized through a research process that each engineering firm could put together and keep in its archives.

Examples of different bridge typologies may include, for instance, suspension, cable stayed, arch, box-girders, frames, struts and ties, combinations of various types. These can eventually be schematized from a large number of case studies that could be found in the specialized literature. In this second phase of design, the preparation of sketch drawings is attained by using a topography of the site enlarged to an appropriate scale.

From this plan view a cross section is eventually drawn using the same scale. Such a cross section will show the configuration of the land throughout the section over which the bridge will span. A number of copies of such cross section will then be prepared, and these will constitute the basis for sketching all kinds of possible bridge configurations drawn as front elevations.

With such sketches on the table, a brain storm session will allow the team of designers to have something tangible to discuss. Each team participant will have an opportunity to express comments and judgments for each proposal so that eventually each sketch will be complemented with a series of statements that will lead to a grading evaluation. During such a period, which could extend through more than one session, some proposals will eventually be eliminated restricting gradually to a few the number of final proposals to be studied in greater depth.

The participants of the team may vary; including structural engineers, architects, cost estimators, field engineers familiar with construction methodologies, landscape architects, ecologists, and transportation experts,. These participants may have plenary sessions in which they are all present or they may meet forming smaller individual groups to keep the discussion more focused on spatial aspects.

A third phase in the conceptual design consists of consulting the long list of restrictions gathered in the first phase and applied then to the various proposals sketched in the second phase. Among these, for instance, are include: geological conditions of the site; building code requirements, moving loading conditions, lateral wind force effects, seismic forces; climatic conditions, and traffic engineering requirements.

The fourth phase involves a practical consideration of construction methodologies available for the structure in question. These include contacting local builders who may eventually be the bidders on the construction of the bridge and asking their input on questions such as: their specific experience in projects of this type; the type of machinery that they may assume would be required; the materials that in their experience would be more suitable; the availability of a skilled labor force in the area; etc.

On the basis of the information gathered at this point it may be practical to make a quick evaluation of the original proposals concentrating on some and eventually adding additional proposals; that have not been already formulated. Phase number four will usually consist of preparing a three-dimensional model of the most probable solution from a two-dimensional sketch; which can bring out site considerations and overall configurations of the structure that were not possible to see in a drawing format. With a tangible model to look at, this is a stage in which other nontechnical judges may express opinions including observers from the public in general whose opinions will have a final impact in many cases concerning approval or rejection of the entire project.

At the end of this fourth phase; the process stops with a selection of structure represented by the model. Yet this initial process can eventually be repeated if a general feeling suggests to try again for a better solution. Therefore; there is still time to reconsider the entire project before proceeding with a detailed analysis and design of the final project. At this point; the overall effects of the various parameters are well known; and those parameters, which have the highest level of relevance, are analyzed more closely to be sure that nothing has been left out. Each project; in fact may have different parameters that dominate the design. For instance construction limitations derived from the topography of the site may be one of the dominant factors. Similarly, other dominant factors would be special local conditions for wind loads or seismic. It is difficult to generalize, since each project is different.

The four phases that have been described represent only one way to organize the various stages in which the preliminary process of design may be visualized. As we have shown the individual parameters are so numerous and their interrelationship is definitely compile so that by all means it is possible to reorganize the entire process in a different format. This process is not as rigid as it may appear, but is rather fluid so that each individual designing group is probably expected to have individual formats that may be proven to be more efficient.

Individual parameters

As previously indicated, a preliminary design process subdivided into various phases includes the organization of several factors, each subdivided into a number of parameters. At this point let us present the most significant factors and parameters in their individual roles. Eventually it will be necessary to show the interaction among them, which at the end will create one structural entity.

The number of factors and the individual parameters create a difficult task in trying to present them with a certain coherence, so that the format followed in this context is not necessarily the only one to use. Design objectives for a bridge structure usually include: control, safety, serviceability, economy, and elegance.Design standards reflect the current state of knowledge of the technical aspects of bridge design (control, safety, serviceability, and safety), so that these objectives met without ambiguities from the beginning and do not require any creative interpretations. On the other hand, the last two objectives, economy and elegance, are indeed the main subject of the creative process of design that escape rigid definitions and depend mostly on the experience and creativeness of the designer. In fact, the successful interpretation of these two objectives would constitute eventually the basis for assessing the value of a project in a design competition.

The term elegance often used in this context implies the achievement of three-dimensional forms which possess intrinsic aesthetic qualities perceived by observers. Aesthetics is usually judged on a subjective basis; however, there are some basic characteristics which are objectively recognized. Unfortunately, typical engineering education has not included studies on aesthetics as in architectural schools. This subject, particularly when applied to structural forms, leaves some ambiguities. Without delving into a long discussions on aesthetics, there are some basic components in evaluating the aesthetics as they relate to structures which are primarily intuitive and generally undestood.

Among these is the concept that elegance and economy hand in hand, on the basis that economy usually implies simplicity and simplicity is often indicative of elegance. To this effect, simplicity involves the minimization of components, eliminating redundancies, thus limiting the artifacts to the extreme minimum is inductive to elegance and aesthetic results. These are concepts of a general nature, thus; when applied to real cases it is often necessary to take some allowance.

For instance, the rigid application of reducing the structure to a minimum to obtain elegance is excessive, and often an elegant rendition implies some additional cost. To this effect, some sources propose a cost increase between 5 and 20 percent to be allowed for aesthetic modifications to the basic minimum cost. Although such values are just indicative and do not have any explicit validity, they are an attempt to quantify the value of aesthetic considerations.

Design requirements in turn constitute a long list that is common to all structures. Some of these requirements, however, include specific data that vary from project to project. These, for instance, when related to the site conditions would include: geology, topography, and hydrology. On the other hand, functional requirements may include: roadway alignment, class of highway, roadway cross section, clearance, construction schedule, and detours.

Structural concepts, can be distinguished into two categories: (a) the main components which will eventually constitute the structure as a whole, and (b) structural details.

Of course, the components in category (a) played the major role in the final evaluation of the project. Those in category (b) have also a relevant role not to be underestimated, and therefore proper attention should be given to them right within the development of the conceptual design. Structural details may include substantial characteristics such as the configuration of the cross-section and, the details of the supports.

Specified limits stated in the present-day design standards evaluate the static and dynamic response of structures to a great degree of dependability, yet the analysis is particularly involved and usually is not completed within the preliminary conceptual design.

Among the design factors is the concept of durability of the whole structure. To this effect, it is convenient to consider the following parameters:

- accessibility,
- replaceability,
- redundancy,
- protection against deicing chemicals,
- adequate concrete cover of the reinforcing bars,
- ease of concrete placement in the formwork,
- dependability of drainage,
- effectiveness of expansion joints, and
- proper connection of steel components to concrete members.

STRUCTURAL ART

Structural art as special term, often used in engineering literature was introduced by Billington (1983). Coming up with a practical term that anybody could use with a clear understanding of its meaning was something long overdue for facilitating the expression of a state of facts present for approximately two centuries in the field of engineering. This form of art, purely engineering in nature, developed spontaneously from the knowledge and creative fantasy of new professionals that had come to interpret a new role in the art of building. We are talking about the civil engineers educated by the polytechnic schools that emerged throughout Europe in the early 1800s. This period coincides with the development of the industrial revolution that swept Europe after the development of science at the end of the eighteenth century.

Structural art, in its basic definition, developed independently of architecture, progressing on a parallel track with architecture itself. This, in fact, is the major explanation that justifies the existing schism that still exists between the two professions. In addition to being totally independent of architecture, it was also independent from any art movement as Billington indicates. Yet, this position is contradicted by the art movement referred to as "constructivism," born in Russia in approximately 1920 and expanding later worldwide with the advent of the Bauhaus.

In fact, artistic expressions of structures in this movement are at the basis of constructivism itself, so that a clear connection between structures and art was definitely established since the earlier part of this century. In historical terms, the first mention of structural art in 1812 is to be found in the work of a British civil engineer, Thomas Telford (cited by Billington, 1983). Fine arts and industrial arts have been traditionally separated in a hierarchial order expressed by "arts and crafts." Yet, the new world of structures wedges itself in between the two, creating a logical bond. Therefore, structural design as a new art form is aligned with painting, sculpture, and architecture. A useful artifact can be at the same time a work of art and vice versa. Art is not limited to abstract renditions of ideas, but it can be found in human-made objects as well as in the works of nature.

This is probably one of the most important points, for one cannot dispute the intrinsic aesthetic values that nature displays in its creations. For instance, the aesthetics of the human body and other creatures in the animal world, as well as the aesthetics of natural forms in the landscape, are vivid examples of beauty and function merged into forms revered by artists in the visual as well as in the literary art worlds.

By the same token, the products produced by craftsmen and structural engineers could also have intrinsic aesthetic values, as demonstrated by many artifacts including machines, jewelry, and clothing fashions. Each one of these broad fields of application of utilitarian arts can be seen in practical terms to touch most people deeply in their feelings.

In the present day, many technological products are impressive examples of how industrial arts touch the aesthetic sensitivity of the population at large. Automotive products for instance are typical examples of an industrial art form generating astonishing appreciation as reflected by the price of some prototypes. What may have been at one time the prerogative of only some fine art works in painting, sculpture, and architecture, nowadays industrial art works may demand more attention and better appreciation as demonstrated by the economic values of society.

Structural engineering, with its intrinsic potential for creating works of art, is therefore an accepted reality that has been evolving gradually for two centuries, acquiring more and more recognition, not only in artistic circles but also among the public at large.

Yet in its state of evolution, structural art has not yet conquered a unanimous recognition in the art world, in architecture, and the general public, which may lack a special sensitivity for it due to lack of knowledge and insufficient exposure to it. Thus, it becomes evident that structural art appreciation needs to be promoted not only outside of engineering but even more so among engineers themselves.

Recently a more specific terminology emerged in this field: "structural morphology," which refers to the expression of forms as a final representation of scientific concepts, technology, and aesthetics combined. An engineering structure, whether a building, bridge, dam, or industrial plant. Becomes an accomplished entity in itself that requires a proper identification within its overall personality that includes its functions. In other words, structural forms stand out on their own within the overall configuration of the whole.

SEARCHING FOR ENGINEERING AESTHETICS

Well known engineering projects, in their typical programming format and their educational realm, have traditionally excluded concepts of aesthetics *per se*, leaving a vacuum in the education of young engineers. With the emphasis given to environmental concerns in the last few decades, any engineering structure is required to fit within the surrounding environment, both in the landscapes of the countryside as well as in the cityscapes of urban centers.

To fill this vacuum, architects and environmentalists gradually have become involved with engineering projects. The team approach that results from this collaborative work, however, requires that engineers become knowledgeable to a certain extent in the overall art of designing so that efficient communication may be established between the team workers, starting with the usage of a common language that all understand. Naturally, it is impossible for major results to be accomplished in this context by a concentrated short course in basic design. However, the following are a few presentations made with the purpose of introducing a basic vocabulary concerning design in general.

To begin, one could first introduce the idea that any structure built by man can acquire aesthetic values in itself and become harmonious with its surroundings when a designer succeeds in doing so either through intuition or through the application of basic design principles. The success of a structure is then assessed both by knowledgeable critics as well as by the public in general following different criteria. Many structures that are fashionable in one period may become totally obsolete and anachronistic years later when the social climate is changed. On the other hand, structures may survive the critics through long spans of time, retaining their original aesthetic values after centuries of service.

Aesthetic values within a structure must be pursued by the designer knowledgeable with some fundamentals that apply to any art form. For instance, any artifact may have a practical use or it may be intended only to appeal to the human senses. Buildings are an example of the former, while musical compositions, paintings, and sculptures are examples of the latter. Yet all works must satisfy a basic logic that justifies their entity. All artifacts acquire their individuality to become a unit. This unit can be seen as a continuum or as the assembly of parts or components that holistically form one unit.

Especially for units resulting from the assembly of components, a clear definition of unity is often required. In fact, a unity of parts is not necessarily achieved by an illogical assembly. The components require a rationale that unites them and also the satisfaction of other requirements such as harmony and coherence of the specific qualities that they may have. Thus, with such a brief introduction other fundamentals of design will be presented followed by basic notions that may help to evaluate how structures such as bridges can be made to fit harmoniously within their surroundings.

Human intervention, within the realm of nature, has been observed historically from a small scale to interventions that generate a sense of incredulity. Within the specifics of engineering works, human ingenuity has covered quite a spectrum of projects interacting with nature, varying from the procurement of potable water to the astounding excavation of the long canals of ancient times for navigation and irrigation. Continuing through modern times other major engineering works have interacted with natural formations at a large scale, including the perforation of long tunnels and the spanning of long bridges over water and across mountain slopes.

Parallel, however, to the scientific and technological advances that man has accumulated in this area is his basic ingenuity which has manifested itself in the creative component that is part of any human endeavor. In its tangible works, the human intellect expresses itself through a process of intuition, creativeness, and analysis which, when combined, constitute the essentials of design.

At this point, therefore, it is assumed necessary to cover a few basic fundamentals of design criteria which are usually part of any project. The following topics have been selected primarily for the basic role they play in the overall design procedures and do not constitute a comprehensive approach to a topic that is way too broad to be covered within the context of this work.

Balancing

Intuitive perception of balancing corresponds directly to a simple series of relationships that can be clearly defined without ambiguity and can simply be quantifiable in numerical terms. In the abstract terminology of balancing, one can come directly to specifics and recognize that this concept may apply to various quantities, such as forces and other vector quantities, as well as to scalar quantities that could be geometrical, such as area or volume, or could be of any other nature provided that the quantities can be quantified in numerical terms.

The mathematical definition of balancing (or equilibrium), is clearly expressed by the two conditions that have to be satisfied: the total sum of the quantities has to be equal to zero, and the sum of the moments of the individual quantities also has to be equal to zero. By the first of the two conditions it is implied that all quantities cannot be of equal quality, but they must be subdivided into opposite groups having opposite signs.

We must have opposite quantities as a precondition for balance or equilibrium, because if quantities are only of one kind their sum will always be different from zero. This brings out some sort of paradoxical conclusion on the fact that uniformity does not conduce to equilibrium, while on the other hand, the pretense of contrasting quantities is necessary. The second condition implies that a second group of quantities (moments) must also be in equilibrium, and that these quantities are the products of two factors: the original ones and their respective distances from a line or point.

When dealing with geometrical quantities such as area or volume, or dealing with forces, the result of balancing or reaching equilibrium is probably more intuitive and easier to visualize. Practical examples of balancing can be referred to axes or points, as previously mentioned, and different orders of balance can eventually be identified, originating more complex and interesting concepts which can better be understood by examples.

Let us consider two groups of geometrical elements: one light, the other dark, calling the light negative and the dark positive. Representing each element with a vector, the two groups are differentiated, using vertical upward vectors for the dark and vertical downward vectors for the light. Now consider the equilibrium of these two groups forming one system.

The upward resultant of the dark and the downward resultant of the light elements must be equal so that their sum will be equal to zero. However, their moments about any point must also be zero, which implies that the two resultants have to be applied at the same point. For the second condition to occur, the distribution of the dark and light elements has to be such that the centroids of each of the two groups must coincide. This could be visualized when the distribution of the elements is symmetrical with respect to the common centroid.

To balance the resultants of various groups of elements, therefore, we must verify that their sum is zero and that the centroids of the individual groups also coincide, allowing considerable freedom in the organization of the various elements (composition).

In the design composition, it is possible to organize elements within a set, finding the centroids of such a set. Then more sets are grouped and the centroid of this group is found. Other groups can also be formed, and eventually the various groups can be organized so that their common centroids will coincide with the centroid of another system, creating a series of conditions of equilibrium in which each part is always in a condition of balance, attaining an order that includes more than one condition of equilibrium. When dealing with geometric forms, whether in plan or in space, the mathematical balance is also perceived visually by the observer. However, rather than working at an intuitive level as most artists do, it makes much more sense to work analytically according to the basic rules of equilibrium contained within the discipline of statics.

Axes of symmetry

Symmetry is part of those intuitive concepts which within this context do not require major speculation. Axes of symmetry are intuitively understood in the representation of visual images, including geometrical forms as well as free ones. Symmetry can imply one, two, or more axes as well as an infinite number; and the point of intersection is referred to as the centroid. To find the centroid on the basis of such a property, at least two axes of symmetry are required to identify it without any additional operation.

To define centroids, one can also employ different definitions. For instance, for a figure subdivided into elements, it is possible to say that its centroid is the point of application of the resultant of all the individual components. On this basis, the centroid is easily found algebraically using the theorem of Varignon that asserts that the moment of the resultant of a system about any arbitrary point is equal to the sum of the moments of all the individual elements about the same point. Therefore, knowing the numerical values of each element and the numerical value of the resultant, the location of the resultant is determined twice: once, assuming that the vectors representing the elements are oriented all in one direction; and secondly, assuming another orientation. The intersection of the two resultants locates the centroids.

Through the relationship among centroids of different systems, it is possible to establish meaningful relationships between systems. Through this it is possible to explore the concept of overall equilibrium using mathematical logic more than basic intuition. In fact, designers working intuitively with forms have often reached meaningful solutions by intuition alone, yet mathematical logic offers a much easier methodology.

Axial reference

In any reference system, axes are used starting from a point of origin and extending linearly using any arbitrary system of units. The linearity of the axes is not limited to rectilinear ones as in the case of Cartesian systems in the plane (x, y) or in space (x, y, z).

An axis could also be curvilinear if so desired, providing that the configuration is rational. Examples of that could be the adoption of an arc of a circle or of an entire circle. Such a reference system could be used to identify the relative location of specific geometrical points that one wants to coordinate to one another and locate in their geometrical organization.

In visual terms, once the curvilinear circular arc is established, and a similitude is made with a clock, the relative position of different objects could easily be surveyed considering the radial distance from the center of the circle and also the orientation, as on a radar screen which will visually organize the recording of significant points without ambiguities.

In the specific case of a circle, the polar reference systems are conventionally used with the same accuracy as the typical Cartesian coordinate system with rectilinear axes. In fact, any rational curve can create a linear system of reference could be used, such as an elliptical configuration, a parabolic one, a spiralling one, or any other type.

Rectilinear longitudinal axes used for recording linear arrangements can be used individually. For instance, just using one axis, objects can be identified as being at a given longitudinal distance from the origin and placed at a certain distance to the left or right of such axis.

Recording a free-form on a plane, will be much easier by superimposing a x-y grid and then reproducing the contour in the individual segments as they appear in each quadrant of the grid. Similarly, to put such a free-form into perspective, the process becomes extremely simplified if the grid itself is placed in perspective and the individual segments are proportionally reproduced in the respective quadrant. The same process applies to the three-dimensional representation.

The semantics of design

Whether entering the realm of design as an architect, engineer, or planner, one remains essentially a designer and as such, there are basic commonalities that they all share including the tools that the communal language offers through the semantics of the spoken word. The language of design requires a certain enunciation, as the terms used in this special context assume a significance that transcends the objective definitions of lexical notation.

In this respect, confusion can arise when terms become unclear in the way they are used subjectively. Yet, a glossary can hardly be established with certainty as an official standard has not really been established in practice. Exact definitions, such as are found in other disciplines, are nonexistent in design; however, the predilection of certain terms and expressions in common use constitute a small assembly which have been tentatively gathered hereafter in an attempt to furnish just a sample of what one could expect to find.

Special meanings that evolve through usage can add additional attributes to a specific term or an expression. These attributes can mostly be understood in a relative sense by comparing them to other similar ones in the same context, by reinforcement or by contrast.

To this effect, it is helpful to select some particular terms or expressions and include other specific terms related to them directly or by association of ideas. The development of such a topic, far from being exhaustive, is attempted with caution and reservation just for initiating a process that could lead to a practical use.

Let us start from two basic words that are quite dominant in the glossary of design: *form* and *shape.*

Form and shape

Before entering into the illustration and the significance of the concept of form as presently intended, it is beneficial to introduce such a concept first from the historical perspective since, like most other concepts of human knowledge, their meaning is the result of a gradual evolution. To this effect, form in itself, with its basic implications, can be seen in a more comprehensive light as it had been the subject of several philosophical speculations.

Since early times, philosophy has tried to define the concept of form with various answers throughout the course of history. Democritus (born c. 460 BC), the major exponent of the atomistic school, defined form as the assembly of atoms which were microscopic particles of different size and space whose different arrangement produced different configurations. For Plato (c. 429-347 BC), the Socratic concept of form (idea) became something different, i.e. an entity that transcends space and time. The Platonic ideas exist independently of the human mind. Men can understand these forms because his immortal soul in the periods between incarnations has been exposed to them and therefore can remember them. Knowledge of forms and ideas is a process of recollection.

But the most authoritative interpretation of the concept of form in the ancient world is derived from Aristotle (385-322 BC). Form is inherent in matter and thus is not transcendent as in Plato's philosophy. Things have an immanent tendency to develop their own established forms, such as in the case of embryos that develop into a particular adult body or a seed that becomes a certain plant. Things exist at two different levels: the potential state that identifies matter and at the actual state that is form.

In the Christian philosophy of the Middle Ages, the scholastics reinterpreted the concept of form going back to Plato and Aristotle. St. Thomas Aquinas (1225 - 1274) considered two types of forms: those that transcend matter (God and angels) and those inherent in matter itself *"forma est actus per quem res actu existunt."* With Immanual Kant (1724-1804), the concept of form acquires its generally accepted modern significance. Pure forms of sensitivity affect the perception of things being perceived. Such forms, space and time, are produced by the action of the mind that coordinated the sensations; thus, in this process, the mind does not acquire knowledge by abstracting information from the senses directly.

"Shape" and "form" are often differentiated from one another by artists who try to use words included in dictionaries of different languages without a sound basis for clear and objective definitions. Especially in the pseudoculture of some art critics, the ambiguity of terms contributes to lack of communication, confusion, and state of frustration which is detrimental to any cultural expression. In the English language, the first term is recorded as derived originally from the Middle English, which in turn derived from the Old English, while the second term derives directly from the Latin *"forma."*

To try to find such distinctions between two terms is misleading unless two separate definitions are purposely structured as the basis to establish different concepts. For instance, one may choose to define "shape" as an approximate form partially vague without sharp details which could also be associated to the term "delineation." The term "form" could be defined distinguished from the previous one, using the same logic, and be assumed to include an unambiguous shape having more clarity and more specific details, giving an overall sharper image. Words as we normally use them are part of a cultural patrimony that should carry instant recognition without the use of a dictionary if they accomplish their role for which they have been created (i.e., unequivocal communication among humans).

PRIMARY FUNDAMENTALS

Design

- construing
- planning
- scheme
- construction
- structuring
- creating

- resolution
- operation
- ordering
- outlining
- layout
- programming

Aesthetics

- composition
- philosophical aim
- human ambition
- lifting the spirit
- beauty in itself
- structure of beauty
- inspiration

Style

- identity
- methodology
- manner
- connotation
- mode
- basic character
- essential description
- uniqueness

Composition

- assembly
- organization
- arrangement
- correlation
- construction
- formation
- structure
- unification

Association

- correlation
- interrelation
- reference
- compound
- bond

Unity

- articulated/connected
- unity/synthesis
- articulated/interacting
- assembly
- composition

Enframement

- definition
- assessing
- referencing
- structuring
- within perspective
- organizing

Reference framing

Axial bearing

- facing the rising sun
- frame of reference
- cardinal points
- axial pivoting
- beaming
- cross beaming
- directional axis

- geometrical guidance
- fundamental order
- foundational order

Axial arrangement

- linear
- longitudinal
- processional
- sequential

Distance

- remoteness
- distance
- span
- longitudinal space
- evoking the horizon
- unstimulating

Visual kinetics sequence

- motion
- repetitiveness
- developable
- rhythmic
- unrolling
- predictable
- periodic
- through time and space

Scale magnitude

- macro/micro
- global
- regional
- local

Finite

- limited
- defined
- measurable
- realistic
- factual

Graphic rendering

- sketching
- drawing
- photographic

Documentation

Perspective

- field of view
- viewpoint
- mechanics of vision
- human subjectivity
- judgemental factors

Dominance

- competitiveness
- control
- supremacy
- authority
- emphasis

Intensity

- quantitative
- measurement
- comparative mean
- assessment of degree
- climax

Texture

- tactile
- fabric
- structure
- structurally spaced

Scent

- sensorial
- highly descriptive
- evocator
- stimulating
- reminiscent
- emotionally powerful
- deeply touching
- colorful
- fragrant conveyance
- aromatic
- flowery
- earthy
- biologically active

Shade

- protective
- diffusion
- abating
- damping
- dimming

Language of images

Figuration

- graphical
- representational
- diagrammatic
- pictorial
- illustrative
- schematic
- visualization

Silhouette

- delineation
- essential imagery
- visual synthesis
- statement of elegance
- youthful fragrance
- graphic concision

Object morphology

Perception

- intuition
- general grasping
- comprehension
- emotive reception

Surprise

- anticipation
- curiosity
- expectation
- stimulating
- flavoring
- spicy stimulant

Delight

- exhilaration
- joyfulness

- brightness
- objective of aesthetics
- design goal

Illusions

- mirage
- delusion
- beyond physical vision
- transcending reality
- mental vision
- visual construction
- trompe l'oeil

Sterility

- negative fertility
- bareness
- aseptic state
- blending

Symbolism

- representative
- emblematic
- implication

Circulation factors

Confluence

- nodes
- convergence
- networks
- grids
- ramifications
- interlacing

Space fluidity

- spatial mobility
- kinematic mutation
- planar plasticity
- evolutions
- amorphic state
- boundless kinetics

Looping

- circuiting
- networks
- path layout
- linear fluidity
- linear mapping
- static trajectory
- sculptural ballistics
- kinetic paths
- fluid lines
- linear fluidity

Paths

- representational
- schematic
- logical thinking
- organizing
- assessing
- strategic

Naturalistic dimension

Sight

- view
- appearance

- manifestation
- representational
- visually imaginative
- vista
- pictorial
- chromatic

Visual representation

- visual identity
- morphology
- geometry
- iconographic
- minutely descriptive
- schematically descriptive
- conceptually descriptive

Naturalistic

- simplistic
- spontaneous
- unadulterated
- unsophisticated

Seasonal

- naturalistic
- evocative
- reminiscent
- cyclic
- mutability
- anticipatory

Naturalistic

- bucolic
- placid
- naturalistic
- soothing
- pictorial
- mythological
- literary imagery
- classical revival
- arcadian
- idyllic

Water basin

- blue area vs. green
- liquid spreads
- lakes
- reservoir
- ponds

Land-Water demarcation

- shores
- waters edge
- coasts
- coastlines
- littoral borders
- waterfront

Articulation

- cohesion
- unity
- composition
- geometry
- constructivism
- architectonics
- reticular

Balance

- equilibrium

- stability
- symmetry
- rest
- equalizing
- levelling

Circulation

- flow
- traffic
- motion
- diffusion
- propagation
- transmission
- transition
- accessibility
- blockage
- diffusion
- fluidity
- expansion
- processional
- occlusion
- diversion
- channelling
- conduction
- exiting
- entering

Structuring

- order
- organization
- organic assembly
- grouping
- interactive assembly
- correlating
- structure of logical paths
- diagramming
- structural assembly

Rhythm

- reiteration
- repetition
- periodicity
- regularity
- frequency
- modular
- harmonic
- articulation
- mathematically defined
- metrical

Pace

- component of rhythm
- kinetic factor
- rhythm
- frozen motion
- measurable with kinetic

Scales

- factor of space and time
- visualization of space

Monotony

- free of contrast
- predictable
- homogeneous
- without accents
- unimaginative

Metrological perspective

Size

- measurable
- comparative
- relatively assessable
- factual
- tangible
- perceivable

Quality of scale

- absolute
- relative
- conceptual
- meteorological
- comparative
- objective
- subjective
- human
- theoretical
- practical

Climax

- crescendo
- apex
- escalation
- summit
- zenith
- culmination
- acme
- paroxysm

Contrast

- opposition
- comparative
- dissimilarity
- visually revealing
- comparative evidence
- pictorial sharpness
- visual dissonance
- comparative dominance
- referential reflectivity
- sensory resolution
- meteorological parameter

Transparency

- clarity
- visually communicative
- crystal clarity
- penetrative
- revealing

Occultism

- hidden
- covered
- concealed
- secretive
- unrevealed
- unclear
- obscure
- mysterious

Color

- hue
- tint
- tincture
- dye

- coloration
- pigment
- variegation
- chromatic
- polychromatic
- intensity
- brilliance

Object morphology

- language of form
- language of vision
- analytical viewing
- dictionary of images
- viewing and understanding
- realization through form
- mute language of materials
- when timber speaks

Materiality

- natural matter
- basic substances
- essential elements
- gifts of nature
- first ingredients
- tangible entity
- products of creation

Animating

- vivifying
- energizing
- coloring
- kinematic excitation

Geometric rationalization

Polygonal

- angular
- segmental
- geometric

Polyhedric

- multifaceted
- crystallographic
- spatial assembly
- high-tech framing

Volumetric

- spatial
- spatially articulated
- polyhedric
- prismatic
- cubiform

Subjective responses

Impression

- subjective view
- perception
- unfocused image
- conceptual grasping
- feeling
- sensation

Enclosure

- bounded
- enveloped
- delimited
- confined

- circumscribed
- impounded

Entrance

- ingress
- access
- influx
- incoming
- influence
- penetrating
- insertion

Exit

- egress
- effusion
- emanation
- leakage
- run out
- percolation
- exudation

3-D abstraction

Curvature

- Gaussian
- positive/negative
- zero Gaussian
- convex/concave
- synclastic/anticlastic
- a factor of strength
- constant/variable
- circles and spirals
- celestial curvature
- ellipsis

Curvilinear

- spatial flow
- evoluted
- voluted
- harmonious development
- spheroidal
- spatial tracing
- looping

Scenery

- representational
- theatrical
- imaginative
- picturesque
- colorful
- detailed background
- pictorial landscapes
- panoramic

Pictorial

- representational
- imaginative
- artistically vivid
- colorful
- iconographic
- subjective view
- tableaux vivant (living pictures)

Visual reflection

- specular
- rebounding
- secondary image

- crepuscular
- twilight
- indirect
- echoing

Nature

- omnipresent source
- quintessence of motherhood
- unavoidable entity
- amicable/merciless
- mythologically descriptive
- timeless
- anomalous/perennial
- incumbent

Landscape

- natural scenery
- land formation
- natural background
- natural view
- God's creation

Site

- land-based
- land bound
- naturalistic

Flow

- current
- run
- trickle
- steaming
- circulate

Turbulence

- vorticity
- unsteadiness

Whirlpool

- spinning
- eddy
- swirl
- vortex
- maelstrom
- rotational

VISUAL DYNAMICS

Visual values in the landscape

In architectural composition and in the planning of the micro and macro environment in any creative manifestation expressed in visual terms, a dialogue emerges between the observer and the object. In such a dialogue one finds structure and a meaning that constitutes an area of study that has practical applications as major objectives. Such a dialogue of images is briefly illustrated in Figure 6-1.

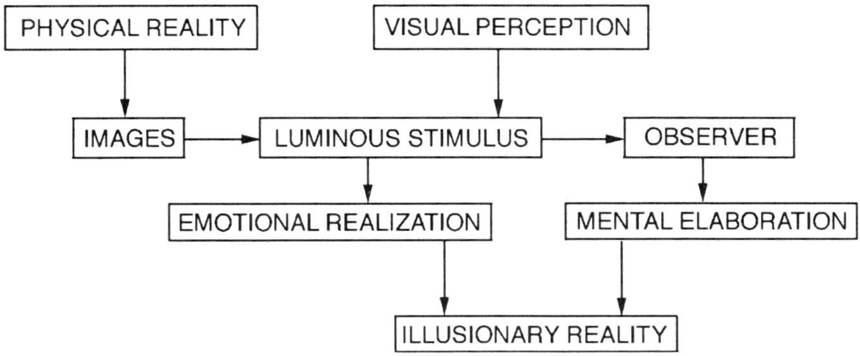

Figure 6-1 Dialog of images diagram.

The image of a physical reality (object, environment, etc.) or of an illusionary reality (created probably by the observer himself on an emotional basis) expresses itself through the fundamental components of form and color which transmit the visual stimulus to the observer through luminous impulses. After receiving a message, the observer analyzes it mentally and reacts with sensations. A positive influence on such sensations is one of the main purposes of design, both in architecture and in the environmental area. The need for understanding the mechanics of the connection between the object and the guidance system for design is evident. It is important to note that methodologies with practical applications in such areas will never be able to substitute the role of subjective criticism due to the complexity and subjectivity of the creative process in design.

Perception and memory of images

While facing a sensorial world of images where color and form manifest themselves in a constant mutation, it is of particular interest to understand the mechanics of the physical process by distinguishing the physiological action that occurs within the observer from the objective situations that exist between the object struck by light and the luminous energy which is reflected by it. In considering the physiological action, one could observe that the luminous energy which derives from the object stimulates the retina of the observer and transfers the impulses to the brain.

The latter analyzes such impulses and stores the results within the memory. Recent studies distinguish the memory action: visual memory and linguistic memory. The former consists only of images which are saved indefinitely, while the latter consists of ideas. Naturally, both forms can operate on the same object, each in its own way. The visual memory has a capacity that is practically inexhaustible; in fact, the observer is usually capable of recognizing any image he previously has seen without any stress. In this type of memory, however, the images are not catalogued in terms of symbols, words or in accordance with any other index; therefore, it is impossible to recall the images to mind using reference indices.

The mechanics of linguistic memory is undoubtedly more complicated and hypothetical within the different proposed stages. The scheme in the following layout indicates that the image on the retina, whether an image of symbols or of objects, is stored for a fraction of a second in visual terms; and in that time span is analyzed with the component elements identified. Once meaning has been extracted from the images, this is then expressed with verbal descriptions in the form of ideas, and is then stored in a memory of short duration having a very limited capacity (between four and six initial images). From here, the ideas pass to another form of memory of long duration which is not very well identified.

The visual image by itself is not capable of leaving a conscious record in the memory, without the aid of mental elaboration. The latter analyzes, distinguishes, accepts, or rejects the individual components and the whole of the image by exploring its significance and associating to it ideas that are expressed through a language. The elements of such a language constitute the key that connects the visual world and humans who perceive it through logic.

In the aesthetic critique of the environment (built-up environment, natural environment, landscape, panoramas, etc.), it is essential in design to structure objective analytical systems which identify and describe qualitatively and quantitatively the visual image of the work underway. The formulation of a descriptive analytical language is therefore at the base from which the system can be structured. The observation of the visual process that we have just discussed brings into perspective another factor of extreme significance in the design: the action of seeing is resolved in the action of interpreting the significance of the stimuli that the retina perceives.

The objective of observation of the physical world and the subjective characteristics of the observer (analytical capacity, memory, culture, emotional components, etc.) are compounded in the action of vision. The images then speak through the observer, either in a direct form with physical sensation that derives from the perception of space (form, color, etc.), or in an indirect form which derives from cultural experiences and associations of ideas. The symbolism of images that derive from cultural experiences can add deep and complex meanings to the images and also exceptional richness of detail.

Visual language

The analyses of the visual reality of the environment through the experiences of individuals have their own conceptual validity, because on the psychological level, man is the measure of the world that surrounds him. Many experiments in such a sector place the subject in the environment itself or in simulated spaces. Often the subject is developed in space through reduced-scale models or in a simulated environment through photographic projections. The major objective is the creation of a language that describes with clarity and without ambiguity the sensation experienced by the observer who is surrounded by the physical space. The terminology that is used varies according to the researcher. For instance, Litton, in describing the visual characteristics of the landscape, uses a terminology which includes the following descriptive elements:

- *Panoramas* — without limits (with foreground, middle ground), horizontal planes, fields, water surfaces.
- *Features* — rocky formations (vegetation) structures.
- *Enclosures* — spaces enclosed by vertical walls (green barriers of trees), enclosed by horizontal surfaces (lakes, valleys).
- *Focal* — points at the horizon.
- *Canopied* — shaded spaces covered by trees (promenades covered by dense trees).
- *Detail* — with numerous elements observable in the near ground (form of the terrain, vegetation, etc.).
- *Ephemeral* — changes due to lighting (sun, shade), of the reflected images due to sun and shade reflecting the seasonal weather, etc.

Jones and Jones (1974) introduced the following descriptive factors: color ingredients, intactness, unity, vividness. Roy Mann Associates proposes shoreline complexity, topographic complexity, vegetative integrity, and vegetative diversity. A work published in 1973 by the U. S. Department of Agriculture Forest Service presents concepts for the management of the landscape, emphasizing the language used for it. In structuring the fundamental components of the natural landscape, this publication uses the following assumptions: (a) the character of the landscape is controlled by the following elements of dominance: form, shape, chromatic values, texture; (b) variable factors that act on the elements of dominance: motion, light, atmosphere, seasonal changes, distance, point of view, scale, time; and (c) principles of influence that act on the elements of dominance: contrast, sequence, alignment, convergence, multivalence, framing.

Application systems

Among recent practical applications is the classification of the visual characteristics of the landscape on a national scale (continental). With the constant increase of various production systems on a larger and larger scale, the development of ecological changes which influence the visual component of the landscape is noticeable. The demographic increase and the increase of the socioeconomic level of life of recent decades have increased the radius of the areas of operation in agriculture, ranching, urban expansion, highways, open-pit mining, etc.

The consequences on the visual appearance of the land have, for quite some time, reached critical proportions and imposed drastic measures. The methodologies proposed in such sectors include the objective breakdown of the landscape into its component elements which must be examined and classified qualitatively, and quantitatively according to a common logic in order to operate it by means of rational systems that maximize the desired results. This includes the defense of the aesthetic values of the landscape and optimization of the product and protection of the source.

Of particular significance has been the application of the system in the management and administration of forests, through the operations of reforestation and logging. The extension and configuration of the logging areas are determined as a function of the visual values of the landscape, in terms of production of timber and preservation of forests.

It is useful to observe schematically the practical example of application in the Visual Management System, the structure of which can usefully be applied to other problems of similar nature. The first operation is the subdivision of the territory under study into areas which can be of three different character types: (A) particularly variable, (B) moderately variable, and (C) basically uniform. These classification apply to: topography of the territory, rock formations, vegetation, and hydrography (lakes, rivers). These character types are:

A. Landscape particularly variable

- mountainous topography with slopes steeper than 60%, with profile, crests, jagged peaks and ridges.
- rock formations of exceptional configuration, with steep surfaces.
- vegetation that does not cover the ground continuously but delineates a layout with open areas and groves of trees; large variety of plants and wooded areas.
- large lakes with water surfaces larger than 50 acres or lakes with smaller surface areas but having particular characteristics in the configuration of the shores and in the type of vegetation, as well as the inclusion of small islands, etc.
- watercourses with particular configuration of their path that form cascades, rapids, lakes, meanders, etc.

B. Landscape moderately variable

- with a topography that includes slopes variable between 30 percent and 60 percent with jagged peaks and rolling profiles.
- rock formations of minor character with moderate mass and boulders.
- vegetation uniformly covering the soil with eventual tree groves that form a layout with no particular demarcation.
- lakes with surface area of 5 to 50 acres with shores having a certain irregularity inscribed by vegetation of limited relevance.
- watercourses with regular paths.

C. Landscape basically uniform

- topography with slopes less than 30 percent.
- rock formations almost nonexistent.
- uniform vegetation that covers the ground without forming any pattern, absence of undergrowth and tall trees.
- lakes with a surface less than 5 acres; absence of fragmentation along the shores; absence of reflections.
- watercourses with intermittent discharge or having a continuous flux of small capacity. Absence of rapids, cascades and other elements.

In conclusion of this first stage, maps are prepared. These maps will include not only topographical features but also areas represented by the symbols A, B, C (Variety Classes). The second stage consists of locating the different points of observation (i.e., roads, paths, camping areas, lakes, beaches, fishing areas, etc.), and in classifying these, whether belonging to first or second order of importance according to whether they are of national or local scale.

Then the territory is divided into different zones which can be of three different types: 1, 2, 3 (sensitivity levels) in terms of the observer's reactions, in terms of the number of observers, and in terms of the observation site. The criterion for the determination of type is as follows:

1. Landscapes observed from points of the first order, when 3/4 of the number of the observers express noticeable interest in the scenic values. Zones of second order used by visitors, including at least 3/4 of the total number who express interest for the visual value.
2. Landscapes observed from points of the first order, when less than 3/4 of the number of observers express considerable interest in the

scenic values. Zone of second order, when at least 1/4 of the total visitors of whom express considerable interest in the visual value.

3. Zone of second order observed by visitors, of whom at least 3/4 express interest in the scenic value.

Then topographic maps are completed with the identification of the areas viewed from different points of observation with the aid of aerial photographs. Each of such zones is identified by a number (1,2,3) that indicates the sensitivity level.

The observed landscapes are then classified in terms of the distance of observation which are indicated as follows: fg, foreground; mg, middle ground; and bg, background. If a zone can be classified under two different sensitivity levels, the one that is adopted is that indicated in the table that follows:

bg2fg1	mg1	bg1	fg2	mg2	bg2	
mg2	fg1	mg1	bg1	fg2	mg2	bg2
fg2 fg1	mg1	mg2	fg2	mg2		
bg1fg1	mg1	fg2	fg2			
mg1	fg1	mg1	bg1			
fg1 fg1						

For instance, between mg1 and fg2, the choice is mg1.

The third and last stage of operation consists of indicating the type of intervention for each of the zones that have been previously classified. The possible types of intervention are indicated by their respective symbols, as follows:

P: preservation - keeping intact the natural configuration, allowing only recreational activities, sports, and tourism that have minimum impact.

C: conservation - permission for alteration of form, profile, chromatic value, and structure which are not noticeable, i.e., that do not change the general character of the landscape.

PC: partial conservation - in addition to the values already expressed in C, the inclusion of new forms, shapes, chromatic values and texture that have a minor intensity and that do not compete in visual power with the essential character of the landscape is allowed.

A: alterations - alteration of the dominant character is allowed. Changes in the form of vegetation and topography must be inspired by the character of the surrounding territory. Structures (roads, buildings, signs, etc.) are allowed, but must be in agreement in form, shape, color, and texture of the natural landscape.

MA: maximum alteration - adding to what has been said in the previous case, form and layout of the structures which alter the landscape are of the same type that can be observed in the surrounding territory but on a higher scale.

It is possible to select the type of intervention which is suggested as a function of variety classes (A, B, C) and as a function of the distance of observation (eg, mg, bg). In this manner, the maps are completed by adding for each zone the symbols of intervention operations (P, C, PC, A, MA).

AESTHETICS

In any type of work, whether purely inspirational or aiming projects of utilitarian nature, the designer cannot escape basic fundamental questions concerning aesthetics in his objective of defining and understanding the meaning of beauty in itself and its actual expression in art. The term, of Greek origin, may imply that "aesthetics" was typically used from ancient times, yet it did not appear in use until the 18th century with Alexander Gottleib Baumgarten (1714-1762).

Beauty and art are topics that relate to one another through a variety of logical connections. Although clearly distinct from one another, together they constitute the essence of aesthetics. This branch of knowledge, while probing the essence of beauty, also investigates the laws that control the way in which art expresses beauty through a variety of media.

The meaning of beauty brings out the hypothesis that such an entity is really intrinsic within some works of art, leaving to the observer the role of discovering and understanding it through spontaneous reactions or through a sensitivity that the individual can acquire through education.

On the other hand, a totally different approach is that the art object is only the catalyst generating a reaction by the observer, who discovers beauty within himself through his sensitivity and feeling of appreciation. In the former case, the process is objective (i.e., aside from the observer, while in the latter case, the process is purely subjective and depends only upon the observer himself). Such a disparity of definitions brings out the extreme limits, while in practical terms many positions of compromise could include both opinions in various degrees.

In latest years, speculation on the meaning of art in general and specifically on the end products that artists produce brings out the nature of the art work interpreted as virtual reality. The product, therefore, has an identity of its own rather than being representational of an hypothetical prototype, therefore the very basic assumptions of Plato's theory on this subject (Langer, *Feeling and form* , 1953).

In the view of modern critics through the work of Dewey's *Art as Experience* (1934) beauty is not controlled by specific rules and cannot be judged objectively. Beauty can be perceived in art through aesthetic experiences which are identified by strong appreciation of the product by the observer who responds with positive feelings of fulfillment and sense of completion.

Such experiences are the result of an interaction between the artistic product and humans who respond to the nature of the art product as expressed in its intrinsic structured order. As the object of critics, art cannot be judged but described in its intent and in its quality.

In *Aesthetic* (1913), Croce addresses the topic specifically. Aesthetic perception is a form of knowledge. Knowledge itself is distinguishable into an intuitive and a logical type of which the first involved the perception of aesthetic objects with their individuality and uniqueness. Aesthetic knowledge is immediate, that is, it cannot be attained through concepts.

The overall meaning of an art object can only be understood by observing the product as a whole and not through the evaluation of the individual components. This supports the intuitive perception of art. Knowledge of art is unattainable through feelings and sensations.

In *The Critique of Judgement* (1790), Kant applies to aesthetics the same basic questions he addressed to many other topics in his work, i.e., the universality of judgment. Any judgmental statement on art products is not the reflection of the intrinsic quality of the product per se, but reflects instead the emotional feeling of the observer.

Therefore, beauty in itself is not an objective quality intrinsic in art waiting to be discovered. Beauty does not posses in itself superior qualities that make it stand separately, but is perceived by the observer through his own emotional response. Although subjective in nature, art judgments can be universally communicated. Furthermore, the nature of beauty involves order at its basis without being structured by rules, and more so, such order harmonizes mental perceptions and understanding.

In the rich tradition of Greek culture, Plotinus (204-270 B.C.) conceived of art as an expression of divine spiritual unity, while Aristotle (384-322 B.C.) in his *Poetics* supported the concept that symmetry, order, and definiteness are the dominant forms in which beauty expresses itself, while throughout the whole Greek period the dominant theme underlining beauty included unity, proportion and variety. Plato (429-347 B.C.) in the *Republic* had specific ideas about art and its content, assuming any artistic production to be a second-hand copy of prototypical ideas existing in the metaphysical world; this is explained since art is a poor replica of the physical world, which is in turn a replica of the metaphysical one.

Order in mathematics and nature

Whether man is the inventor of mathematics, or such a discipline transcends humanity as the prototypical ideas in the Metaphysics of Plato, still remains a concept of wonder. Yet such a world is there to be used and enjoyed by the human spirit searching for serenity of the mind and for a spiritual aesthetic that produces tranquility. Mathematical order in its harmonious state is soothing spiritually and can usually be pursued in design. Similarly, the world of nature follows its own laws of order, which quite frequently merge with the mathematical order in question. A few examples follow.

Phyllotaxis

Within the biological world, the mathematical order pursued in this context shows an example in the phenomenon of phyllotaxis. As the stem of a plant grows and new leaves sprout, their spacing along the stem and their radial progression show a precise geometrical order referred to as "phyllotaxis" (or phyllotaxy). The name derives from the Greek "*phyllon*" (leaf) and "*taxis*" (arrangement). Leaves are arranged along helical patterns (genetic spirals) and the angle between the radial direction of two consecutive leaves is constant for a given plant.

These angles are usually expressed as a fraction of 360 degrees. The values of these fractions usually coincide with those of the Fibonacci series: 1/2, 1/3, 2/5, 5/13, 8/21,... The arrangement of the leaves as expressed in terms of phyllotaxis is indicated by a fraction; for example, 5/13, which means that there are 13 leaves arranged along a spiral that makes 5 turns around the stem between one leaf and the next one that is aligned on the same vertical line.

Such numerical relationships are found in nature the following examples: in the spirals in sunflowers; in pine cones; in the arrangement of leaf buds on stems of tobacco plants; in animal horns; in the spiral of snail's shells; and in the genealogy of male bees.

The Fibonacci Series

In the nineteenth century, French mathematician Edouard Lucas originated the term "Fibonacci Series" after Leonardo Fibonacci, (ca. 1170), born in Pisa, the son of a merchant from Pisa appointed consul over a community of Pisan merchants in Bugia, North Africa (now Bejaia, Algeria).

After studying mathematics in Algeria, he travelled and continued his studies in Egypt, Syria, Greece, Sicily and Provence, completing his major work *Liber abaci* in 1202. With his reputation as the most distinguished mathematician of the Middle Ages, he introduced in the culture of the West the Hindu-Arabic numeral system and the famous numerical series. The series: (1, 1, 2, 3, 5, 8, 13, 21, 34, 55,...) consists of numbers which correspond to the sum of the previous two.

This constitutes the first series that has appeared historically in the mathematics of Europe. It was first expressed by the French mathematician Albert Girard in 1634 as: $u_{n+2} = u_{n+1} + u_n$, where u is the term and n is its place in the sequence. Of fundamental importance in the Fibonacci series are the findings of Scottish mathematician Robert Simson from the University of Glasgow who, in 1753, discovered that as the numbers increased in magnitude, the ratio between succeeding numbers approached phi, indicating the golden section 1.6180. Such a number is also expressed by:

$$\frac{(1 + \sqrt{5})}{2}$$

Golden section

The golden section can easily be visualized in geometrical terms by considering a line of unit length and cutting it down into a longer and shorter segment. The length of the longer member divided by the length of the shorter section is the golden ratio = 1.618. One can also say that the longer member is 0.618 units and the shorter 0.382 units because their sum is equal to 1. The golden rectangle has the longer and the shorter side such that the ratio of their length is equal to 1.618. We should also notice that a golden rectangle has certain properties, such as that it can be divided into a square and into a rectangle. The square will have the size of the short side and the remaining rectangle is in itself a golden rectangle.

The product of the total length (S + L) times the short span (S) is equal to the square of the long side (L^2). Other geometrical applications of the golden section include the pentagon. The golden rectangle was often used for sizing canvasses in the Renaissance, while the pentagram, derived from the geometry of the pentagon, was used in the Middle Ages as a magic symbol. In the Renaissance, the golden section inspired a major interest in the visual arts. Mathematician Lucas Pacioli in his writing, "*Divina proportione*" (1509) illustrated by Leonardo da Vinci, emphasized the aesthetic value of the golden section.

Besides the golden section, it is relevant to mention other similar ratios which have been singled out as to have equal properties of balance and appeal to the human eye. Included among them are the following: Plato proposed, for instance, a geometrical construction in which an equilateral triangle is divided into two right triangles, which are in turn combined to form a rectangle.

The sides of this rectangle are in a ratio equal to 1 divided by the square root of 3. Similarly, another rectangle of pleasing proportion mentioned since ancient times is one that has sides in the proportion of 1 to the square root of 2. Reinforcing these rules of aesthetic appreciation in art are the findings of present-day science through the experimental work of Gustav Theodor Fechner in 1876 who concluded that rectangles within the proportion close to that of the golden section generate positive feelings in the observer.

The silent language of geometry

In the structuring of this ancient field of scientific knowledge (geometry), the first scientists in ancient civilizations were practically motivated by the measuring of land, from which the very same name of the discipline derived. With its gradual development, however, this knowledge grew independently of its original purpose, creating a language that stands at the very root of logical thinking that has enticed one civilization after another.

The forms that eventually emerged from the rationalization of concepts have built an enormous accumulation of visual constructs in the plane and in space which are a constant inspiration of forms in the design of objects, including large-scale artifacts such as buildings, bridges, highways, etc.

Simple statements, containing in a synthetic form the seed for visual representations, intrigue the mind and stimulate creativity. The example of the circle speaks for itself: the focus of points equally distant from one called the center. With a statement of this sort, a person on the other side of the globe speaking on a telephone can cause another person to draw precisely a geometrical pattern that by magic appears as the mind of the person receiving the command performs the task. Similar statements, of course, with extremely more complex logic can make a machine produce intricate drawings by means of rational statements which translate into graphical impulses.

In contraposition to such a language, there are others of different nature which respond to different criteria. The uncoordinated splashing of paint on a surface may communicate visual messages to the observer without a logical coherence depending merely on fortuitous events, yet some sort of stimuli is produced even if outside of any control. In between two such opposite polarities there is room for compromise, since many artists find a comfortable position for self-expression in many intermediate positions.

Still another language eventually emerges in the world that surrounds humanity in its daily experience (i.e., the world of nature through inanimate and living forms). Both the results of forces, the morphology of the two worlds speaks to the human mind-inspiring other visual forms, and suggests subjective interpretations.

The morphologies of the territory, of land formation, of water bodies impose themselves for their impressive magnitude that overcomes any human observer. Similarly, the observation of natural biological growth in the flora and fauna suggests such a complex intricate series of relationships that have caused the generation of the final forms. In all of these aspects of the physical world experienced by man, the human creator in his projects eventually arrives to design configurations under the many stimuli to which he is sensitive.

Constructivism

Of major relevance in contemporary design are the contributions that *constructivism* brought among the general theories of composition, which apply not only to the visual arts and architecture but also to design in general. It is of extreme importance to include at least a short mention about such an inspiring point of view that could offer major guidance and a sense of reference to architects and planners, engineers, and anybody engaged in working in the field in question.

What is constructivism? With the new political equilibrium produced by World War I, this new art movement had gradually taken shape by 1920. Although this movement includes the basic visual arts, its association with architecture is very close. Initially referred to as "constructive plastic art," the new movement is described by the French artist, Jean Gorin, in 1936 as follows: "Purely constructive plastic art is fundamentally architecture; it is the higher function of all genuine architecture.

In and through architecture it attains its fullest expression, forming a 'unity' with her." Constructivism can be defined as a type of composition of geometrical shapes attained from working with industrial materials that produce abstract shapes. It later included some sort of mechanical connotation that tied the elements of the composition to a more refined unity. This philosophy of design still has an enormous impact at present, and one can assume it will never fade away, as it has already influenced the design of the modern world for almost a century. With reference to other design theories, cubism introduced high relief collage using new material in art such as cardboard, glass, hair, etc. It is from this background, in 1913, that a new art movement began in Russia by the efforts of Vladimir Tatlin, El Lissitzky, and Alexander Rodchenko. This movement, now referred to as constructivism, used the same techniques of the cubist and new materials from the industrial production of the time. The first exhibition was in 1915.

The work by Tatlin could be described as free geometric constructions. Two years later in 1917, Naum Pevsner (who later changed his name to Gabo) and his brother Antoine Pevsner joined the movement and in 1920 issued a famous manifesto that announced the philosophy of the new school. After a short period, constructivism expanded in Europe but it gradually vanished in Russia, where its abstract was not grasped by the masses. Pevsner moved to Berlin in 1922 and to Paris a year later. Gabo moved to Berlin in 1922 and to Paris in 1932. Several other artists also moved from Russia to Germany. There at Weimar and later at Dessau, the Bauhaus was the vehicle to expand constructivism that, since 1920, had been imported from Russia through the influence of the pioneering artists. Later, when Moholy-Nagy left the school in Germany to found the New Bauhaus in Chicago, the constructivist idea spread to the United States. From this early period to the present, constructivism has expanded in Europe and America, inspiring a number of artists and architects.

As in any other art movement, constructivism did not grow in a vacuum and was affected by several other contemporary schools. Picasso and cubism influenced constructivism from its inception, and later Piet Mondrian with de Stijl and neoplasticism came in contact with it and contributed to its evolution. To summarize the evolutionary states of constructivism, the following schematic notes, derived from Steven Bann's *The Tradition of Construction*, help to project the development of this movement in a synthetic historical chronology.

1915 First construction by Gabo; Tatlin's "counterreliefs" exhibit in St. Petersburg.

1917 Beginning of "de Stijl" in Leiden, Holland (Mondrian, van Doesburg).

1919 Tatlin's appointment by Free Studio in St. Petersburg; founding of Bauhaus at Weimar, Germany, by Gropius.

1920 "The Realistic Manifesto" by Gabo and Antoine Pevesner (brothers) at Moscow open-air exhibition; Tatlin's second exhibition of the model of "Monument for the Third International."

1921 Speech "On Constructivism" by Varvara Stepanova, Moscow Institute of Artistic Culture.

1922 Migration of Lissitzky and Gabo from Russia to Berlin; founding of International Faction of Constructivist by van Doesburg, Lissitzky and Hans Richter in Dusseldorf, Germany; Russian art exhibition at van Diemen Gallery, Berlin.

1923 Joining of Maholoy-Nagy in the Bauhaus; architectural model exhibit in Paris by van Eesteren and van Doesburg.

1924 Publication of *Gosplan for Literature* by Literary Center of Constructivists.

1925 Soviet Pavilion by Melinikov at Paris International Exposition of Modern, Decorative, and Industrial Arts.

1928 Departure of Gropius and Moholy-Nagy from Bauhaus, leaving Hannes Meyer in charge.

1930 Founding of *Cercle et Caree* by Seuphor and Torres-Garcia in Paris; publication of *Art Concret* by van Doesburg.

1931 Rejection of proposed plans by Gabo and le Corbusier for the Palace of the Soviets.

1932 Dissolution of artistic movements in Russia; founding of *Abstraction-Creation* in Paris; exhibition in Moscow of Tatlin's Letalin flying machine.

1933 Closing of Bauhaus by Nazis; strong criticism against Tatlin and other Russian artists in Russia.

1936 Editing of *Circle* anthology by Gabo, et al.

1946 Founding of *Salon del Realities Nouvelles* in Paris.

1948 Publishing of *Art as the Evolution of Visual Knowledge* by Biederman.

1955 "Yellow Manifesto" by Vasarely at Denise Rene Gallery, Paris.

1957 Completion of *Bijenkorf Construction* in Rotterdam.

1958 Founding of *Structure* magazine by Joost Baljeu and Eli Bornstein.

1961 Erection of Luminodynamic Tower by Schoeffer in Leige, Belgium.

Although briefly presented in such a skeletal format in this context, constructivism requires much more space than presently allowable for its practicality of applications in contemporary design. The reader is referred to numerous specific sources of information covering the large production of works within this philosophy of design. The large spectrum of individual works by numerous artists in various fields can still offer guidance for the understanding of the constructivistic order that underlines the character of this movement.

Environmental aesthetics

In the term "environmental aesthetics," reference is made to the morphological expressions of nature that create instinctive feelings of attraction in the typical observer. Without any further elaboration, it is easy to identify in this term so many forms of the physical environment that stimulate primordial reactions within typical human sensitivity. Thus, selecting water as the focal environmental element, it is possible to identify some such basic elements that inspire feelings of interest and delight in humans.

With specific reference to the marine background along the sea shores, one notices the effects of some physical factors that include the spaciousness of the open water; the color and the mobility of the sea surface; the sky and the horizon considered as the line of confluence between the two; the land in its various configurations; the shoreline where land and water interact with each other through the surf; the granular sand along the beaches; shells, algae, and driftwood at the water's edge; the smell and taste of the salty water; the breezes with their moisture and scent; the sun; the tonality of colors; and the flight of sea birds.

Well-known catalysts for emotional feelings, these factors are essential parameters in the scheme of basic design that must be kept on the active list of the fundamental resources that any professional can use. Each of these factors, primordial in nature, will never lose efficiency and magnetic power regardless of the repetitive usage that any single project may employ. Naturally, similar lists can also be obtained with some overlapping when the setting changes from the scenario of the marine coastline to that of other natural settings involving water such as lakes and rivers.

As basic elements of a matrix, all of these factors in the list like the one above will require individual elaborations and a logic interrelating one another so that multiple interactions can occur successfully. Since the very beginning of the design process, different hypotheses (based on the selection of these factors) can be tested and developed to attain various solutions verifying the major benefits of the usage that such elements can provide.

Dynamic symmetry

Design fundamentals applied to the architecture of buildings, can also be applied to the design of other structures including bridges. For instance, concepts of unity, balance and equilibrium of areas and masses; axial, biaxial, and polar symmetry, orientation, rhythm, etc., can obviously be applied to any kind of design. Such concepts are not new and they have been used not only in architecture but in all expressions of art. However, among such design factors of ancient date, there is a specific new theory of design referred to as "dynamic symmetry".

Promoter of such theory was Jay Hambidge (1867-1924). Born in Canada, Hambidge was basically an artist and scholar on classical Greek art. His scholarly work in fact based primarily on Greek culture was the basis on which he constructed his theoretical structure of design applicable to any form of art, and consequently applicable also to the design of bridges which was the focal point in this contest.

Introducing his writings on dynamic symmetry, Hambidge assumes that the basic principle underling the greatest art is based on the proportions of the human figure and the growing plant. Furthermore, such design theories have been recognized in the practical use of artists according to the writer. More specifically a given for dynamic symmetry by Hambidge explains that such a terminology can be expressed as a fundamental of design related to the architecture of the humans and plants. In observing the so-called plan schemes, a distinction is made about two different types. One includes the art of several civilizations, such as the Saracenic, Mahomedan, Chinese, Japanese, Persian, Hindu, Asyrian, Coptic, Byzantine, and Gothic, while the second type of scheme refers to Greek and Egyptian art. Comparing the two, the author concludes that the latter is superior to the former.

More precisely, such plan schemes can be considered symmetries, i.e. express the relationship between individual components of the design and the design itself considered as a whole. There are basically two types of symmetry in nature which are in turn named static and dynamic. Although distinct from one another, it is possible to say that static symmetries can be considered also to be a special case of dynamic symmetry therefore, whether static or dynamic are at the bases of any art form, yet the dynamic symmetry is definitely superior to the static counterpart.

In applying both types of symmetry, the process may include a conscious part as well as a subconscious part. Static symmetry is more specifically defined from examples such as crystals and observing the cross section of certain fruits. Other examples of static symmetry in the works of nature, include diatoms and radiolaria. Static symmetry can also be seen applied by several cultures such as the Copts, Byzantines, Saracens, Islamic, and the Gothic and Renaissance designers, using the properties of patterns of regular planar figures such as the square and the equilateral triangle. Vice versa, Egyptian and Greek culture used static symmetry was based on an area divided into a number of equal parts, such as a square and a half, three quarters, one quarter, one third, two thirds, etc.

An example of the application of static symmetry applied to a Greek statue such as the Tenea Apollo explains that such a sculpture could be enclosed in a rectangular area composed of three squares and two thirds of a square. Such a Greek sculpture however it followed a previous Egyptian scheme. On the other hand the Egyptian statue of Aminophis can be defined as related to dynamic symmetry. Considering dynamic symmetry, in nature, can be defined as an orderly assembly of parts such as sea shells and the arrangement of leaves on a plant.

The tentative to organize art on a rational basis has a long tradition that starts in the early stages of most civilizations. Particular qualities describing form, assembly of forms is specific characteristics of material in other qualities have usually produced no fruits. Terms have been defined so that there will be an objective understanding through which artists could communicate. For instance, examples of such quantities used in design can include, order, balance of areas, balance of masses, equilibrium, schematic organization of parts, unity, duality, composition, proportioning, etc. More precisely, such qualities previously mentioned, have a general application in all types of design, including visual arts as well as architecture.

Along these basic definition of qualities in the art world, there are also several attempts to qualify design into basic criteria proven in a subjective way to be universal and applicable to any design application. For instance, one could recall the rule of The Golden Rectangle used from the Greek civilization up to the Renaissance, when artists used the ratio of the two sides of the rectangle in many compositions including the overall dimensions of paintings. Significant contribution to the making of a recently universal rule of design can be found in the writings of Jay Hambidge (1867-1924). Born in Canada, he spent his life in the world of the arts, specializing on Greek culture after studying in Greece for a number of years.

Hambidge's new theory of design is referred to as "Dynamic Symmetry," which distinguishes from the counterpart referred to as "Static Symmetry." More precisely, the former may at times be part of the later. The term dynamic symmetry includes two words which have been purposely chosen from the Greek roots, form and measurements. In the modern sense we find basically three types of symmetries: biological, spherical and radial. In the biological sense symmetry implies the repetition of parts in a plant or animal in an orderly fashion. Spherical symmetry instead is based on a nucleus with a spherical shape around which components are arranged concentrically or radiating from the center. Radial symmetry is based upon a generic form that includes an axis such as a cylinder or a bowl form along which the parts of the body radiate.

Following the writings of Hambidge, he explains in the introduction to *The Elements of Dynamic Symmetry*, the basic criteria that stands at the basis of his rules of design prior to specifics of individual topics discussed in the book. Focusing on the classical expressions of art, Hambidge assumes that the fundamental approach to design is based upon the relationship of the human components as well as that of the growing plant.

Furthermore, he recognizes that such a rule of proportioning have been used in all fields of human creativity involving art in general and the production of utilitarian artifacts. Specifically then, the author defines the term "dynamic symmetry" as the principle of design to be found in the architecture of man and plants.

Quoting from Hambidge, "The determination of form in a specific example of design means, in a sense, the elimination of the personal element. With this element removed the residue represents merely the planning knowledge possessed by the artist. However, the concepts of personal element and residue are not easily perceived even if a series of examples tries to interpret such terminology to the typical reader."

Plan schemes (a term used by Hambidge) can be found in the works of art of many civilizations, which include: Saracenic, Mahomedan, Chinese, Japanese, Persian, Hindu, Asyrian, Coptic , Byzantine, and Gothic. More specifically, he explains that all these plan schemes belong to one type. However, another type of plan scheme can be found in Egyptian and Greek art. Of the two plan schemes, the latter is much more valuable than the former, as demonstrated in nature.

Such plan schemes are then identified as *symmetries* with the strict meaning derived from the Greek origins of the word. Further quoting from Hambidge he explains the meaning of symmetry as: ". . .literally it signifies the relationship which the composing elements of form in design, or in an organism in nature, bear to the whole."

The author explains the existence of two types. One of these is referred to as *static symmetry* and the other *dynamic symmetry.* Static symmetry is further defined as : "the orderly arrangement of units of form about a center of plane." Examples of such a definition can be found in snow crystals and in the cross sections of certain types of fruits as well as in the organism of diatoms and radiolaria.

Such a static symmetry can usually be found in most forms of art either at the conscious or unconscious level. Among Coptics, Byzantines, Saracens, Mahomedans, and the Gothic and Renaissance, Hambidge indicates that static symmetry was based on the properties of planer figures, including squares and equilateral triangles. But, a series of examples of static symmetry is eventually described in turn contraposition to dynamic symmetry.

Summarizing the analysis of Hambidge's *The Elements of Dynamic Symmetry*, this work definitely had an impact in the world of design as reflected by the three editions previously mentioned. With further references on Jay Hambidge and his work on dynamic symmetry, the following publications are hereby furnished for the readers who may want to get to know the topic in question and it's fine details.

Such publications include: *The Parthenon and Other Greek Temples: Their Dynamic Symmetry* (Hambidge, 1924), *The Diagonal* (Hambidge, November, 1919 to October, 1920), *Geometry of Vases* (Caskey, 1922), "A Dissertation on Dynamic Symmetry" (Bragdon, *Architectural Review,* October, 1924), "The Artist's Year Book," 1905-06 (Hosking, *Art News,* January 26, 1924). Illustrations help to visualize the details of this theory. Among these are the logarithmic or equiangular spiral and several others that can be easily distinguished from the context of the book itself.

APPENDIX 1: AN INVENTORY OF NOTABLE BRIDGES

Bridges are particular structures that constitute a public patrimony for various countries worldwide. Civic pride is often associated with bridges that become famous for one reason or another. Other great engineering structures such as bridges are admired as examples of human ingenuity; along tall buildings, canals, and other megastructures that have been built throughout history.

A list of prominent bridges in the United States was compiled by the Survey of State Highway Engineers and is reprinted here from *The World Almanac and Book of Facts* (1996; copyright 1995, with permission of K-III Reference Corporation, all rights reserved).

This list provides a general overview of structures that represent existing prototypes under different categories that could be used as basic references. Such references are assumed to be of practical use to any bridge designer, either at the preliminary stage of design, or later when more detailed data would be of help. Each one of these bridges is a case study that speaks for itself, and although they are spread geographically around the world, they exist for anyone to look at them in their physical reality.

SUSPENSION

Year	Bridge	Location	Longest span (ft)
1964	Verrazano-Narrows	New York, NY	4,260
1937	Golden Gate	San Francisco Bay, CA	4,200
1957	Mackinac	Straits of Mackinac, MI	3,800
1931	George Washington	Hudson River, NY-NJ	3,500
1950	Tacoma Narrows	Tacoma, Washington	2,800
1936	Transbay	San Francisco Bay, CA	2,310
1939	Bronx-Whitestone	East River, New York, NY	2,300
1970	Pierre Laporte	Quebec	2,190
1951	Delaware Memorial	Wilmington, DE	2,150
1968	Delaware Memorial (new)	Wilmington, DE	2,150
1957	Walt Whitman	Philadelphia, PA	2,000
1929	Ambassador	Detroit-Canada	1,850
1961	Throgs Neck	Long Island Sound, NY	1,800
1926	Benjamin Franklin	Philadelphia, PA	1,750
1924	Bear Mountain, NY	Hudson River	1,632
1952	Wm. P. Lane Memorial	Sandy Point, MD	1,600
1903	Williamsburg	East River, New York, NY	1,600
1969	Newport	Narragansett Bay, RI	1,600
1883	Brooklyn	East River, New York, NY	1,595
1939	Lion's Gate	Burrad Inlet, B.C.	1,550
1930	Mid-Hudson	Poughkeepsie, NY	1,500
1963	Vincent Thomas	Los Angeles Harbor	1,500
1909	Manhattan	East River, New York, NY	1,470
1936	Triborough	East River, New York, NY	1,380
1931	St. Johns	Portland, OR	1,207
1929	Mount Hope	Rhode Island	1,200
1960	Ogdensburg, NY	St. Lawrence River	1,150
1939	Deer Isle	Maine	1,080
1931	Simon Kenton Mem.	Ohio River, Kentucky	1,060
1867	John A. Roebling	Ohio River, Cincinnati, OH	1,057
1971	Dent	Clearwater County, ID	1,050
1900	Miampimi	Mexico	1,030
1849	Wheeling, WV	Ohio River, W. Virginia - Ohio	1,010

CANTILEVER

1917	Quebec	Quebec	1,800
1974	Commodore Barry	Chester, PA	1,622
1958	Mississippi River	New Orleans, LA	1,575
1988	Mississippi River	New Orleans, LA	1,575
1995	Mississippi River	Gramercy, LA	1,460
1936	Transbay	San Francisco Bay	1,400
1968	Mississippi River	Baton Rouge, LA	1,235
1955	Tappan Zee	Hudson River	1,212
1930	Lewis and Clark	Longview, WA	1,200
1909	Queensboro	East River, New York, NY	1,182
1927	Carquinez Strait	California	1,100
1958	Parallel Span	California	1,100
1930	Jacques Cartier	Montreal, Canada	1,097
1968	Isaiah D. Hart	Jacksonville, FL	1,088
1956	Richmond	San Francisco Bay, CA	1,070
1929	Grace Memorial	Charleston, SC	1,050
1980	Newburgh-Beacon	Hudson River, NY	1,000
1963	Newburgh-Beacon	Hudson River, NY	1,000
1949	Martin Luther King	St. Louis, MO	963
1975	Caruthersville, MO	Mississippi River	920
1977	Saint Mary's	Saint Mary's, W. Virginia - Ohio	900
1969	Silver Memorial	Pt. Pleaseant, W. Virginia - Ohio	900
1981	Ravenswood	West Virginia	900
1987	Carl Perkins	Ohio River, Kentucky	900
1986	Mississippi R	Natchez, MS	875
1940	Mississippi R	Natchez, MS	875
1938	Blue Water	Pt. Huron, MI	871
1972	Mississippi R	Vicksburg, MS	870
1972	N. Fork American R	Auburn, CA	862
1940	Baton Rouge	Mississippi River	848
1899	Cornwall	St. Lawrence River	843
1940	Mississippi River	Greenville, MS	840
1961	Helena, AR	Mississippi River	840
1963	Brent Spence	Covington, KY	831
1963	Cincinnati	Ohio River	830
1963	Mississippi River	Donaldsville, LA	825
1940	Mississippi River	Vicksburg, MS	825
1929	Clark Memorial	Ohio R, KY	820
1961	Campbellton-Cross Pt.	New Brunswick - Quebec	815

1935	Rip Van Winkle	Catskill, NY	800
1938	Cairo	Ohio R., Illinois - Kentucky	800
1932	Washington Mem.	Seattle, WA	800
1936	McCullough	Coos Bay, OR	793
1935	Huey P. Long	New Orleans	790
1916	Memphis (Harahan)	Mississippi River	790
1892	Memphis	Mississippi River	790
1949	Memphis-Arkansas	Mississippi River	790
1904	Mingo Jct., OH	Ohio River	769
1910	P&LE RR Bridge	Ohio River, Beaver, PA	750
1932	Bi-State Vietnam Gold Star	Henderson, KY	720
1992	Jamestown-Verranzzano	Jamestown, RI	636
1943	Pit River	Redding, CA	620
1941	Columbia River	Kettle Falls, WA	600
1954	Columbia River	Umatilla, OR	600
1954	Columbia River	The Dalles, OR	573
1968	W. 17th St.	Huntington, WV	562

SIMPLE TRUSS

1976	Chester	Chester, WV	745
1917	Metropolis	Ohio River	720
1929	Irvin S. Cobb	Ohio River - Illinois - Kentucky	716
1922	Tanana River	Nenana, Alaska	700
1933	Henderson	Ohio River -Indiana - Kentucky	665
1967	I-77, Ohio River	Williamstown, WV	650
1917	MacArthur, IL-MO	St. Louis	647
1919	Louisville	Ohio River	644
1992	St. Charles	Missouri River, MO	625
1933	Atchafalaya	Morgan City, LA	608
1924	Castleton	Hudson River, NY	598
1937	Deleware R	Easton, PA	550
1930	Swindell Bridge	Pittsburg, PA	545
1889	Cincinnati	Ohio River	542
1952	Allegheny River, Topeka	Pittsburgh, PA	534
1930	Martinez	California	528
1951	Rankin	Pittsburgh, PA	525
1914	Old Brownsville	Brownsville, PA	520
1906	Donora-Webster	Donora-Webster, PA	515
1909	Hulton	Pittsburgh, PA	505
1967	Tanana River	Alaska	500

STEEL TRUSS

1988	Glade Creek	Raleigh County, WV	784
1973	Atchafalaya R	Krotz Springs, LA	780
1972	Piscataqua R	New Hampshire - Maine	756
1972	Atchafalaya R	Simmesport, LA	720
1957	SR-3, Rappanhannock R	Middlesex County, VA	648
1978	Atchafalaya R	Morgan City, LA	607
1938	US-22	Deleware River, NJ	540
1955	Interstate (I-5)	Columbia R., Oregon -Washington	531
1910	McKinley, St. Louis	Mississippi River	517
1972	Mississippi River	Muscatine, IA	512
1896	Newport	Ohio River, KY	511
1989	Atchafalaya River, US 190	Krotz Springs, LA	506
1970	Lake Koocanusa	Lincoln County, MT	500
1931	Lucy Jefferson Lewis	Cumberland River, Kentucky	500
1958	Lake Oahe	Mobridge, SD	500
1958	Lake Oahe	Gettysburg, SD	500

CONTINUOUS TRUSS

1966	Columbia River (Astoria)	Oregon -Washington	1,232
1977	Francis Scott Key	Baltimore, MD	1,200
1995	Central	Ohio River, Kentucky - Ohio	850
1943	Dubuque, IA	Mississippi River	845
1966	Charles Braga	Fall River, MA	840
1956	Earl C. Clements	Ohio River, Illinois - Kentucky	825
1953	John E. Matthews	Jacksonville, FL	810
1950	Maurice J. Tobin	Boston, MA	801
1940	Governor Nice Memorial	Potomic River, MD	800
1957	Kingston-Rhinecliff	Hudson River, NY	800
1986	Rochester-Monaca	Rochester-Monaca, Pennsylvania	780
1918	Sciotoville	Ohio River	775
1976	Carroll L. Cropper	Ohio River, IN-KY	750
1981	Sewickley	Sewickley, PA	750
1984	13th St. Bridge, Ohio River	Ashland, KY	740
1959	Monaca-E. Rochester	Monaca-E. Rochester, PA	730
1976	Betsy Ross	Philadelphia, PA	729
1929	Madison-Milton	Ohio River, IN-KY	727
1966	Matthew E. Welsh	Mauckport, IN	725

1994	Sixth Street	Huntington, WV	720
1977	Bert T. Combs-Lambert	Ohio River	720
1970	Vanport	Vanport, PA	715
1962	Champlain	Montreal, Quebec	707
1964	John F. Kennedy	Louisville, KY	700
1973	Girard Point	Philadelphia, PA	700
1954	Pennsylvania Turnpike	Delaware R., Philadelphia, PA	682
1949	George Platt	Philadelphia, PA	680
1938	Port Arthur-Orange	Texas	680
1926	Cape Girardeau, MO	Mississippi River	677
1929	Cincinnati	Ohio River	675
1946	Chester, IL	Mississippi River	670
1970	Gulfgate	Port Arthur, TX	664
1994	Williamstown-Marietta	Williamstown, WV	650
1955	Jefferson City	Missouri River	640
1930	Quincy, IL	Mississippi River	628
1961	Shippingport	Shippingport, PA	620
1959	U.S. 181, over harbor	Corpus Christi, TX	620
1935	Bourne		616
1935	Sagamore	Cape Cod Canal, MA	616
1965	Clarion River (I-80)	Clarion, PA	612
1975	Donora-Monesson	Donora-Monesson, PA	608
1991	Hoffstadt Creek	Mt. St. Helens, WA	600
1957	Blatnik	Duluth, MN	600
1965	Rio Grande Gorge	Taos, NM	600
1991	Jefferson City	Missouri River	596
1962	W. Branch Feather River	Oroville, CA	576
1967	Glenwood	Pittsburgh, PA	567
1936	Meredosia	Illinois R.	567
1936	Mark Twain Mem.	Hannibal, MO	562
1957	Mackinac	Mackinac Straits, MI	560
1932	Pulaski Skyway	Passaic R, New Jersey	550
1966	Emlenton	Emlenton, PA	540
1936	Homestead High Level	Pittsburgh, PA	534
1960	Brownsville High Level	Brownsville, PA	518
1945	Mansfield-Dravosburg	Pittsburgh, PA	500

CONTINUOUS BOX AND PLATE GIRDER

1982	Houston Ship Chan	Houston, TX	750
1967	San Mateo-Hayward #2	San Fran, Bay, CA	750
1977	Intracoastal Canal	Gibbstown, LA	750
1976	Intracoastal Canal	Forked Island, LA	750
1969	San Diego-Coronado	San Diego Bay, CA	660
1987	Columbia River	Umatilla, Oregon - Washington	660
1994	Acosta	Jacksonville, FL.	630
1981	Douglas	Juneau, AK	620
1976	Wax L. Outlet	Calumet, LA	618
1981	Glenn Jackson (I-205)	Coulmbia River, OR-WA	600
1963	Poplar St.	St. Louis, MO	600
1982	Illonis River	Pekin, IL	550
1982	I-440	Arkansas River	540
1980	US-64, Tennessee River	Savannah, TN	525
1988	Mon City	Monongahela, PA	520
1965	McDonald-Cartier	Ottawa, Ontario	520
1984	Columbia River	Richland, WA	450
1987	Veterans	Pittsburgh, PA	440
1987	SR 76, Cumberland River	Dover, TN	440
1987	SR 20, Tennessee River	Perryville, TN	440
1970	Willamette River, I-205	West Linn, OR	430
1970	I-430	Arkansas River	430
1984	FAU 3456, TN River	Chattanooga, TN	420
1965	I-24 , Tennessee River	Marion Co. TN	420
1978	Snake River	Clarkston, WA	420
1975	36th St.	Charleston, WV	420
1974	Dunbar-S. Charleston	S. Charleston, WV	420

CONTINUOUS PLATE

1973	Ship Channel (I-610)	Houston, TX	630
1971	W. Atchafalaya	Henderson, LA	573
1981	Illinois 23	Illinois River, Illinois	510
1968	Trinity River	Dallas, TX	480
1978	San Joaquin River	Antioch, CA	460
1977	Thomas Johnson Mem.	Solomons, MD	451
1979	Lewis	St. Louis, MO	450
1975	I-129	Missouri River, Iowa	450

1967	Mississippi River	La Crosse, WI	450
1972	Whiskey Bay Pilot	Ramah, LA	425
1966	I-480	Missouri River, Iowa - Nebraska	425
1972	I-435	Missouri River, MO	425
1972	I-80	Missouri River, Iowa - Nebraska	425
1987	I-435	Missouri River, Kansas - Missouri	425
1983	US-36	Missouri River, Kansas - Missouri	425
1972	I-635, Kansas	Missouri River, Kansas - Missouri	425
1978	I-24	Cumberland River, Kentucky	420

CABLE-STAYED

1988	Dames Point	Jacksonville, FL	1,300
1995	Houston Ship Channel	Baytowne-LaPonte, Texas	1,250
1983	Mississippi River, I-310	Luling, LA	1,222
1987	Sunshine Skyway	Tampa Bay, Florida	1,200
1991	Talmadge Mem	Savannah, GA	1,100
1979	Columbia River	Pasco-Kennewick, WA	970
1985	E. Huntington	Huntington, WV	900
1985	Mississippi River	Quincy, IL	900
1980	Veterans Memorial	West Virginia - Ohio	820
1991	Cochrane/Africatown	Mobile, AL	780
1995	Chesapeake & Delaware	Canal Bridge Dover-Wilmingtion, DE	750
1994	Great River Bridge	Burlingtion, IA	660
1991	Neches River	Port Arthur-Orange, TX	640
1990	James R.	Henrico Co., VA	630
1972	Sitka Harbor	Sitka, AK	450

I-BEAM ARCH

1980	Interstate 20	hreveport, LA	438
1988	Route 18	Weston's Mill Rond, NJ	276

STEEL ARCH

1977	New R. Gorge	Fayetteville, WV	1,817
1931	Kill Van Kull	Bayonne, NJ	1,652
1973	Fremont	Portland, OR	1,255

1964	Port Mann	British Columbia	1,200
1916	Hell Gate	East River, New York, NY	1,038
1959	Glen Canyon	Colorado River, Page AZ	1,028
1967	Trois-Rivieres	St. Lawrence River, Quebec	1,100
1990	Roosevelt Lake	Arizona	1,080
1962	Lewiston-Queenston	Niagara River, Ontario	1,000
1976	Perrine	Twin Falls, ID	993
1941	Rainbow	Twin Falls, ID	984
1986	Moundsville	Ohio River, West Virginia	912
1985	I-255	Mississippi River, MO	910
1972	I-40, Mississippi River	Memphis, TN	900
1936	Henry Hudson	Harlem River, New York, NY	840
1967	Linclon Trail	Ohio R., Indiana - Kentucky	825
1978	I-57, Cairo, Il	Mississippi River	821
1980	I-65 Mobile River	Mobile, AL	800
1961	Sherman Minton	New Albany, IN	800
1978	I-470 Bridge, Ohio River	Wheeling, WV	780
1930	West End	Pittsburgh, PA	780

CONCRETE ARCH

1993	Lake Street Bridge	St. Paul, MN	556
1971	Selah Creek (twin)	Selah, WA	549
1968	Cowlitz River	Mossyrock, WA	520
1931	Westinghouse	Pittsburgh, PA	460
1923	Cappelen	Minneapolis, MN	435

SEGMENTAL CONCRETE

1988	Zilwaukee Bridge (twin)	Zilwaukee, MI	392
1985	Red River Bridge	Coushatta, LA	370

TWIN CONCRETE TRESTLE

1979	I-55/1-10	Manchac, LA	181 (157)
1969	L. Pontchartrain Causeway	Mandeville, LA	126 (720)
1972	Atchafalaya Flwy	Baton Rouge, LA	93 (984)
1963	Lake Pontchartrain	Slidell, LA	28 (547)
1983	Interstate 310	Kenner, LA	25 (925)

CONCRETE SLAB DAM

1927	Conowingo Dam	Maryland	4,611
1952	SR-4, Roanoke River	Mecklenburg Co., VA	2,785
1936	Hoover Dam	Lake Mead, Arizona - Nevada	1,324

DRAWBRIDGES

Vertical lift

1959	Arthur Kill	NY-NJ	558
1965	Pennsylvania Railroad	Kirkwood -Mt. Pleasant, DE	548
1935	Cape Cod Canal	Massachusetts	544
1960	Delair, NJ	Delaware River	540
1937	Marine Parkway	Jamaica Bay, New York City	542
1931	Burlington, NJ	Delaware River	534
1908	Willamette River	Porland, OR	521
1968	Second Narrows	Vancouver, B.C.	493
1912	A-S-B Fratt	Kansas City	428
1945	Harry S. Truman	Kansas City	427
1955	Roosevelt Island	East River, New York, NY	418
1980	US-17, James R.	Isle of Wright Co., VA	415
1932	M-K-T-R River	Missouri River	414
1969	Cape Fear Memorial	Wilmington, NC	408
1930	Aerial	Duluth, MN	386
1941	Main St.	Jacksonville, FL	386
1962	Burlington	Ontario	370
1922	Cincinnati	Ohio River	365
1967	SR-156, James R.	Prince George Co., VA	364
1957	Industrial Canal	New Orleans, LA	360
1950	Red R.	Moncla, LA	360
1936	Tribo	Harlem River, New York, NY	344
1961	Corpus Christi Harbor	Corpus Christi, TX	344
1939	U.S. 1&9, Passaic R.	Newark, NJ	333
1930	Martinez	California	328
1960	St. Andrews Bay	Panama City, FL	327
1929	Penn-Lehigh	Newark Bay	322
1987	Industrial Canal	New Orleans, LA	320
1920	Chattanooga	Tennessee River	310

Bascule

1969	East Pearl R.	Slidell, LA	482
1940	Black R.	Lorain, OH	333
1917	SR-8, Tennessee	Chattanooga, TN	306
1956	Duwarnish River	Seattle, WA	300
1955	Chehalis R.	Aberdeen, WA	288
1968	Elizabeth R.	Chesapeake, VA	280
1913	Broadway	Portland, OR	278
1954	Fuller Warren	Jacksonville, FL	224

Swing bridges

1927	Fort Madison	Slidell, LA	545
1991	SW Spokane St.	Seattle, WA	480
1930	Rigolets Pass	New Orleans, LA	400
1950	Douglass Memorial	Washington, DC	386
1945	Lord Delaware	Mattapone River, VA	252
1957	Eltham	Pamunkey River, VA	237

FLOATING PONTOON

1963	Evergreen Pt.	Seattle, WA	7,518
1961	Hood Canal	Pt. Gamble, WA	6,471
1993	Lacey V. Murrow	LakeWashington, Seattle, WA	6,543
1989	3rd Lake Washingtion	Seattle, WA	6,130

SWING SPAN

1903	East Omaha	Missour River	519
1952	US-17	York River, VA	500
1897	Duluth, MN	St. Louis Bay	486
1899	C.M.& N. R. River	Chicago	474
1913	Rte. 82, Conn-River	E. Haddam, CT	465
1914	Coos Bay	Oregon	458

NOTABLE INTERNATIONAL BRIDGES

SUSPENSION

1998	Akashi Kaikyo	Japan	6,529
1996	Store Daelt (East Bridge)	Denmark	5,328
1981	Humber	England	4,626
1997	Tsing Ma	Hong Kong	4,518
1988	Minami Bisan-Seto	Japan	3,576
1988	Bosphorus I	Turkey	3,609
1973	Bosphorus II	Turkey	3,576
1966	Pont 25 de Abril	Portugal	3,323
1964	Forth (road bridge)	Scotland	3,301
1966	Severn	England	3,241

CANTILEVER

1917	Quebec	Canada	1,801
1890	Forth (rail)	Scotland	1,710
1974	Nanko	Japan	1,673

STEEL ARCH

1932	Sydney Harbour	Australia	1,650

CONCRETE ARCH

1980	Krk I	Croatia	1,280
1964	Gladesville	Australia	1,001

STEEL PLATE AND BOX GIRDER

1974	President Costa e Silva	Brazil	984
1956	Sava I	Yugoslavia	856
1966	Zoobruke	Germany	850

CABLE-STAYED

1999	Tatara	Japan	2,920
1995	Pont de Normandie	France	2,808
1993	Yangpu	China	1,975
— —	Meiko Chuo	Japan	1,936
1991	Skarnsundet	Norway	1,739
1986	Alex Fraser	Canada	1,526
1985	Yokohama	Japan	1,509
1987	Hooghly River	Calcutta	1,499
1996	Severn	Japan	1,496

Note that the enclosed list is presently being updated for the *1998 World Almanac and Book of Facts*. The reader is advised to check this source when it is published.

APPENDIX 2: BRIDGE DESIGN STANDARDS

The design and construction of bridge structures all around the world is controlled by a variety of standards that vary from one country to another. Specifically in the United States the first official publication was prepared in 1931 by the American Association of State Highway Officials (AASHTO). This was recognized as the first document of this sort in the United States. Prior to this highway bridges were designed and built under the direction of the chief bridge engineer in the highway departments in individual states. Through the effort of such highway engineers, this first document was eventually produced. The document referred to as *Standard Specifications for Highway Bridges and Incidental Structures,* was eventually adopted not only by the highway departments of different states, but also by other agencies having jurisdiction over the design of bridges, in the United States and eventually in other countries. This document has been updated a number of times, approximately every four years with interim updating to include the progress gradually accomplished over the course of time. Its name was eventually changed to *Standard Specifications for Highway Bridges*.

A major updating of this document took place in 1987 when the committee in charge of the *Standard Specifications for Highway Bridges* initiated in 1983 a study evaluating general design philosophies on bridges in the United States and in several countries abroad. As a result, the 1987 new standards reflected a substantial change. This consisted mostly in adding a new working stress design methodology referred to as "load factor design," which had been broadly accepted in other countries.

The final document was referred to as *AASHTO LRFD Bridge Design Specifications*. These specifications were developed by a team of 50 members under the scrutiny of the AASHTO Highway Subcommittee on Bridges and Structures and twenty other technical committees. Currently bridge designers can follow either one of the following specifications: the long-standing *AASHTO Standard Specifications for Highway Bridges*, and the alternative newly-adopted *AASHTO LFRD Bridge Design Specifications*.

The last new edition of *Standard Specifications For Highway Bridges* (i.e., the sixteenth edition published in 1996) is briefly introduced here for the convenience of the reader. Part of the table of contents is summarized as follows with the permission of the AASHTO:

Section 3—Loads

Part A—Types of Loads

Part B—Combinations of Loads

Part C—Distribution of Loads

3.26 Distribution of Wheel Loads and Design of Composite Wood—Concrete Members
3.27 Distribution of Wheel Loads on Steel Grid Floors
3.28 Distribution of Loads for Bending Moment in Spread Box Girders
3.29 Moments, Shears, and Reactions
3.30 Tire Contact Area

Section 4—Foundations

Part A—General Requirements and Materials

4.1 General
4.2 Foundation Type and Capacity
4.3 Subsurface Exploration and Testing Programs

Part B—Service Load Design Method Allowable Stress Design

4.4 Spread Footings
4.5 Driven Piles
4.6 Drilled Shafts
4.7 Note: Article Number Intentionally Not Used

Part C—Strength Design Method Load Factor Design

4.8 Scope
4.9 Definitions
4.10 Limit States, Load Factors, and Resistance Factors
4.11 Spread Footings
4.12 Driven Piles
4.13 Drilled Shafts

Section 5—Retaining Walls

Part A—General Requirements and Materials

5.1 General
5.2 Wall Type and Capacity

7.4 Tubular Piers
7.5 Abutments

Part C—Strength Design Method Load Factor Design

7.6 General

Section 8—Reinforced Concrete

Part A—General Requirements and Materials

8.1 Application
8.2 Concrete
8.3 Reinforcement

Part B—Analysis

8.4 General
8.5 Expansion and Contraction
8.6 Stiffness
8.7 Modules of Elasticity and Poisson's Ratio
8.8 Span Length
8.9 Control of Deflections
8.10 Compression Flange Width
8.11 Slab and Web Thickness
8.12 Diaphragms
8.13 Computation of Deflections

Part C—Design

8.14 General
8.15 Service Loads Design Method (Allowable Stress Design)
8.16 Strength Design Method

Part D—Reinforcement

8.17 Reinforcement of Flexural Members
8.18 Reinforcement of Compression Members
8.19 Limits for Shear Reinforcement

8.20 Shrinkage and Temperature Reinforcement
8.21 Spacing Limits for Reinforcement
8.22 Protection Against Corrosion
8.23 Hooks and Bends
8.24 Development of Flexural Reinforcement
8.25 Development of Deformed Bars and Deformed Wire in Tension
8.26 Development of Deformed Bars in Compression
8.27 Development of Shear Reinforcement
8.28 Development of Bundled Bars
8.29 Development of Standard Hooks in Tension
8.30 Development of Welded Wire Fabric in Tension
8.31 Mechanical Anchorage
8.32 Splices of Reinforcement

Section 9—Prestressed Concrete

Part A—General Requirement and Materials

9.1 Application
9.2 Concrete
9.3 Reinforcement

Part B—Analysis

9.4 General
9.5 Expansion and Contraction
9.6 Span Length
9.7 Frames and Continuous Construction
9.8 Effective Flange Width
9.9 Flange and Web Thickness—Box Girders
9.10 Diaphragms
9.11 Deflections
9.12 Deck Panels

Part C—Design

9.13 General

10.11 Expansion and Contraction
10.12 Flexural Members
10.13 Cover Plates
10.14 Camber
10.15 Heat-Curved Rolled Beams and Welded-Plate Girders
10.16 Trusses
10.17 Bents and Towers
10.18 Splices
10.19 Strength of Connections
10.20 Diaphragms and Cross Frames
10.21 Lateral Bracing
10.22 Closed Sections and Pockets
10.23 Welding
10.24 Fasteners (Rivets and Bolts)
10.25 Links and Hangers
10.26 Upset Ends
10.27 Eyebars
10.28 Forked Ends
10.29 Fixed and Expansion Bearings
10.30 Floor System

Part C—Service Load Design Method Allowable Stress Design

10.31 Scope
10.32 Allowable Stresses
10.33 Rolled Beams
10.34 Plate Girders
10.35 Trusses
10.36 Combined Stresses
10.37 Solid-Rib Arches
10.38 Composite Girders
10.39 Composite Box Girders
10.40 Hybrid Girders
10.41 Orthotropic—Deck Superstructures

Part D—Strength Design Method Load Factor Design

10.42 Scope
10.43 Loads
10.44 Design Theory
10.45 Assumptions
10.46 Design Stress for Structural Steel
10.47 Maximum Design Loads
10.48 Symmetrical Beams and Girders
10.49 Unsymmetrical Beams and Girders
10.50 Composite Beams and Girders
10.51 Composite Box Girders
10.52 Shear Connections
10.53 Hybrid Girders
10.54 Compression Members
10.55 Solid Rib Arches
10.56 Splices, Connections, and Details
10.57 Overload
10.58 Fatigue
10.59 Deflection
10.60 Orthotropic Superstructures

Section 11—Aluminum Design

11.1 General
11.2 Bridges
11.3 Soil-Metal Plate Interaction Systems
11.4 Structural Supports for Highway Signs, Luminaires, and Traffic Signals
11.5 Bridge Railing

Section 12—Soil-Corrugated Metal Structure Interaction Systems

12.1 General
12.2 Service Load Design
12.3 Load Factor Design

12.4 Corrugated Metal Pipe
12.5 Spiral Rib Metal Pipe
12.6 Structural Plate Pipe Structures
12.7 Long-Span Structural Plate Structures
12.8 Structural Plate Box Culverts

Section 13—Wood Structures

13.1 General and Notations
13.2 Materials
13.3 Preservative Treatment
13.4 Deflection
13.5 Design Values
13.6 Bending Members
13.7 Compression Members
13.8 Tension Members
13.9 Mechanical Connections

Section 14—Elastomeric Bearings

14.1 General
14.2 Definitions
14.3 Material Properties
14.4 Bearing Design Methods
14.5 Anchorage
14.6 Design Forces for Supporting Structure
14.7 Stiffeners for Steel Beams and Girders
14.8 Provisions for Installation Effects

Section 15—TFE Bearing Surface

15.1 General
15.2 Design

Section 16—Steel Tunnel Linear Plates

16.1 General and Notations

16.2 Loads
16.3 Design
16.4 Chemical and Mechanical Requirements
16.5 Section Properties
16.6 Coatings
16.7 Bolts
16.8 Safety Factors

Section 17—Soil Reinforced-Concrete Structure Interaction Systems

17.1 General
17.2 Service Load Design
17.3 Load Factor Design
17.4 Reinforced-Concrete Pipe
17.5 Reinforced-Concrete Arch, Cast-in-Place
17.6 Reinforced-Concrete Box, Cast-in-Place
17.7 Reinforced-Concrete Box, Precast
17.8 Precast Reinforced-Concrete Three-Sided Structures

Section 18—Soil-Thermoplastic Pipe Interaction Systems

18.1 General
18.2 Service Load Design
18.3 Load Factor Design
18.4 Plastic Pipe

Section 19—Pot Bearings

19.1 General
19.2 Design

Section 20—Disc Bearings

20.1 General
20.2 Design

Division I—A

1. Introduction
2. Symbols and Definitions
3. General Requirements
4. Analysis Requirements
5. Design Requirements for Bridges in Seismic Performance Category A
6. Design Requirements for Bridges in Seismic Performance Category B
7. Design Requirements for Bridges in Seismic Performance Categories C and D

Division II

1. Structure Excavation and Backfill
2. Removal of Existing Structures
3. Temporary Works
4. Driven Foundation Piles
5. Drilled Piles and Shafts
6. Ground Anchors
7. Earth Retaining Systems
8. Concrete Structures
9. Reinforcing Steel
10. Prestressing
11. Steel Structures
12. Steel Grid Flooring
13. Painting
14. Stone Masonry
15. Concrete Block and Brick Masonry
16. Timber Structures
17. Preservative Treatment of Wood
18. Bearing Devices
19. Bridge Deck Joint Seals
20. Railings
21. Waterproofing
22. Slope Protection

23. Miscellaneous Metal
24. Pneumatically Applied Mortar
25. Steel and Concrete Tunnel Liners
26. Metal Culverts
27. Concrete Culverts
28. Wearing Surfaces
29. Embedment Anchors

APPENDIX 3: HIGHWAY LIVE LOADS

Special attention is given in this section to live loads applied to State Highway System Bridges, specifically to vehicular loads. Such loads are divided into three systems: *H loads, Alternative loads,* and *P loads.*

H LOADS

The H and HS trucks are used as typical loads for determining the live load capacity of bridges. The H trucks and respective loads include a number of different types. They were originally established in 1920 by the AASHTO and have been periodically revised up to present. Lane load is defined as a live load consisting of a 15-ton truck load followed by a 20-ton load, which in turn is followed by another 15-ton truck load. For calculating the maximum positive moments, the loads are applied as indicated in (a); for the determination of maximum negative moments, the loads are applied as indicated in (b).

(a) For single spans less than 145 ft, the H load consists of one truck. For single spans larger than 145 ft, the H load to be used is the lane load. Such loading conditions must be used in determining the maximum positive moment.

(b) For continuous spans smaller than 45 ft, one HS truck load should be applied. For continuous spans larger than 45 ft, one lane load must be applied. Such loading conditions must be used in determining the maximum negative moment.

ALTERNATIVE LOADS

These loads were approved in 1956 to take into account special military loads. Such loads must be applied to any bridge on the Interstate Highway System. These loads consist of two axles of 24 kips spaced four feet apart. Notice that when spans are less than forty feet, these loads produce larger movements than those produced by H loads. For bridge decks transversely reinforced, the alternative load consists of one axle, including two wheels six feet apart, each one carrying 16 kips.

P LOADS

P loads, or permit design loads, are special loads that require special permits adopted by the State of California. Such loads represent idealized vehicles to be used for testing the capacity of bridges. Highway loads are assumed to be 10-feet wide and traffic lanes are assumed to be 12-ft wide. The ten-foot-wide traffic load can be applied anywhere within the twelve-foot wide-lane.

APPENDIX 4: STANDARD SPECIFICATIONS FOR RAILROAD BRIDGES BY AREA

A major reference in the design of bridges subjected to railroad loading is the *Manual for Railway Engineering* (1996) published by American Railway Engineering Association (AREA), which is consistently updated and republished periodically with different frequency. AREA, founded in 1899, groups together railway officers, engineers, and supervisors engaging in engineering design and construction and maintenance of railroads.

The latest publications of the *Manual* include Volume 2, *Structures* and Volume 4, *Systems Management*. In the organization of this manual, the material is subdivided into different chapters, which in turn contain various parts also subdivided into different sections.

Volume 2 contains the following of particular interest: Chapter 2, "Track Measuring Systems;" Chapter 7, "Timber Structures;" Chapter 8, "Concrete Structures and Foundations;" Chapter 9, "Seismic Design for Railway Structures;" Chapter 11, "Engineering Records and Property Accounting;" Chapter 15, "Steel Structures;" Chapter 16, "Economics of Railway Engineering and Operations;" Chapter 19, "Bridge Bearings;" Chapter 28, "Clearances;" Chapter 29, Waterproofing;" and Chapter 32, "Engineering Management Systems."

APPENDIX 5: COMPUTER SOFTWARE SOURCES

The following are the sources of the software presented in the text. Since the information included is just indicative, it is necessary to contact the sources to obtain more information about the individual software.

American Iron and Steel Institute
1101 17th Street NW, Suite 1300
Washington, DC 20036-4700
Phone: (202) 452-7100; fax: (202) 463-6573

California Department of Transportation
Engineering Service Center
MS #9, Box 942874
Sacramento, CA 94274-0001
Phone: 916-227-8495; fax: 916-227-8174
E-Mail: trfmpl.cabrown@trmx3.dot.ca.gov

Portland Cement Association
5420 Old Orchard Road
Skokie, IL 60077-1083
Phone: (847) 966-6200; fax: (847) 966-9781

American Institute of Steel Construction
One East Wacker Drive, Suite 3100
Chicago, IL 60601-5403
Phone: (312) 670-2400; Fax: (312) 670-5403

ADAPT
1733 Woodside Road, Suite 220
Redwood City, CA 84061
Phone: (415) 306-2400; fax: (415) 364-4678
E-Mail: 102166.3101@compuserve.com

BEST Center
Department of Civil Engineering
University of Maryland
College Park, MD 20742
Phone: (301) 405-2011; fax: (301) 314-9129
Website: http://www.ence.umd.edu/research.d/best.html

Engineering Designing Programmers, Inc.
MDX software
Phone: 573-446-3221; Fax: 573-446-3278
BBS: 573-446-3399
Website: www.mdsoftware.com
E-Mail: mdx@sprynet.com

Skilling Ward Magnusson Barkshire, Inc.
1301 5th Avenue #3200
Seattle, WA 98101-2314
Phone: (206) 292-1200; fax: (206) 292-1201
Website: http://skilling.com E-Mail: rdu449@aol.com

Florida Department of Transportation
Structures Design Office
606 Suwannee Street, MS 33
Tallahassee, FL 32399
Fax: (904) 488-6352

LEAP Software, Inc.
P.O. Box 16827
Tampa, FL 33687-6827
Phone: (813) 985-9170; fax: (813) 980-3642

APPENDIX 6: BRIDGE ORGANIZATIONS

International Bridge, Tunnel and Turnpike Association (IBTTA)
2120 L St. NW, Ste. 305
Washington, D.C. 20037
Tel: (202) 659-4620

American Railway Bridge and Building Association (ARBBA)
18154 Harwood Ave.
Homewood, IL 60430
Tel: (708) 799-4650

Association For Bridge Construction And Design (ABCD)
P.O. Box 23264
Pittsburgh, PA 15222
Tel: (412) 281-9900

National Society for the Preservation of Covered Bridges (NSPCB)
44 Cleveland Ave.
Worcester, MA 01603-1405
Tel: (508) 756-4516

International Association of Bridge, Structural
and Ornamental Iron Works (IABSOIW)
1750 New York Ave. NW
Washington, D.C. 20006
Tel: (202) 383-4810
(202) 383-4800

The Bridge Foundation (TBF)
9 John Armstrong Rd.
St. Thomas Towns Post
Bangalore 560 084, Karnataka, India
Tel: 80 5586485

International Association for Bridge and Structural Engineering (IABSE)
ETH-Honggerberg
CH-8093 Zurich, Switzerland
Tel: 1 3772647

International Association of Shell and Spatial Structures (IASS)
IASS Secretariat
CEDEX: Laboratorio Central
de Estructuras y Materiales
c/ Alfonso XII, 3
28014 Madrid
Spain

Special notice is made with regard to IASS, which at first glance may seem independent of bridges. Yet this organization, founded in the late 1950s by Eduardo Torroja in Spain, groups together the leaders in structural engineering as well as a few architects interested in this topic. Among these professionals are prominent engineers who specialize in bridges and who publish in this field. The organization is extremely active by sponsoring an international conference each year in different cities across the world and publishing the proceedings. Furthermore, special subgroups meet at the international level at different dates and in different cities. It is also important to note a bulletin published every few months by IASS, which circulates among the hundreds of the association's members.

APPENDIX 7: BRIDGE PROPOSAL ANNOUNCEMENTS

When an agency, private or public, is announcing the bids for the construction of a bridge, usually the contract documents, including plans, have already been provided. Announcements are usually placed in various publications such as *Development Business* (published biweekly by the United Nations), or *ENR Journal*, etc. Occasionally, the contract may include not only the construction of the structure but also design proposals. While the format may vary from case to case, it is useful to consider an example of such an advertisement.

The following example has been extracted from one of these publications and concerns the bid for the construction of a bridge. The names and specific reference data have been changed.

Sample announcement

The Government of the State of Atlantica proposes to retain the services of a design/build firm for the design and construction of an 1800 ft toll-bridge spanning Coral Harbor, linking Hope Town to Sun Island.

Firms interested in prequalifing for this project are invited to submit comprehensive details of their experience and past performance in undertaking similar design/build projects. Prequalification documents, including questionnaires, may be obtained from the Project Coordination Unit at the Ministry of Public Works not later than 1600 hours on February 16, 1997. Every application for the prequalification document must be accompanied by a nonrefundable fee of fifty U.S. dollars (true value), if applying from outside of the state, and thirtyfive dollars (true value) if applying from within the state. This fee is to cover shipping, handling, and reproduction costs.

After evaluation of the information submitted, contractors which prequalify shall be invited to tender for construction of the works. Tenders for design/construction of the works are expected to be let in April 1997 and the estimated period of construction of the entire project is 14 months.

The state will not be bound to assign any reason for not prequalifing any applicant and will not accept responsibility for any costs incurred by any applicant.

Completed questionnaires must be returned not later than 1600 hours on March 9, 1997 to:

Mr. John Smith
Acting Director of Public Works
Mainstreet
P.O. Box 71477
Coral Beach, Atlantica 00000

APPENDIX 8: ANNOUNCEMENTS OF CONSULTING DESIGN CONTRACTS

When public agencies announce the need for engineering design contracts related to bridges they follow some general formats that may vary from agency to agency. A typical example of an announcement by North Carolina Department of Transportation Highway Design Branch (advertising through the North Carolina Purchase Directory) follows:

EXAMPLE DOCUMENT

The North Carolina Department of Transportation desires to engage a private engineering firm for structure design at three bridge sites on the following project: I-2402AA Guilford County; I-85 Greensboro bypass from I-85 South of Greensboro to East of US 220, Site 1 Flyover Bridge on I-85 over I-85 Bypass, 180 meters long, 33 degrees skew, 21 meters—curve over curve; Site 4 I-85 Collector Distributor Bridge over I-85 Ramp B, 75 meters long- 144 degree skew-9 meters clear roadway- curve over curve, and Site 7 I-40 Collector Distributor Bridge over I-85 and I-85 Collector Distributor—120 meters long, 105 degrees skew, 10.2 meters clear roadway—curve over tangent.

Scope of Work: Structure and Foundation Design and Geotechnical Investigations

The method of payment for this project will be lump sum.

One firm will be selected for this work.

The engineers performing the work and in responsible charge of the work must be registered Professional Engineers in the State of North Carolina and must have good ethical and professional standing. Any firm wishing to be considered must be properly registered with the Office for the Secretary of State, and if required, with the North Carolina Board of Registration for Professional Engineers and Land Surveyors. Any firm proposing to use corporate subsidiaries or subcontractors must include a statement that these companies are properly registered with the NC Board of Licensing of Geologists. It will be the responsibility of the prime firm to verify the registration of any corporate subsidiary or subcontractor prior to undertaking the work and assuming the liability. The selected firm will be required to furnish proof of Professional Liability insurance coverage in the minimum amount of $1,000,000. The firm must have an adequate accounting system to identify costs chargeable to the project.

The Department of Transportation is committed to an annual goal of 10% for Disadvantaged Business Enterprise participation in federally funded projects and, for state-funded projects, annual goals of 10% for minority participation, and 5% for Women's Business Enterprises Participation.

North Carolina firms qualified to do the required work will be given priority consideration. A North Carolina firm is a firm that maintains an office in North Carolina staffed with an adequate number of employees judged by the Department to be capable of performing a majority of the work required.

The evaluation of firms submitting letters of interest for this will be based on the following considerations and their respective weights:

1. The evaluation of the performance on any previous contracts with the North Carolina Department of Transportation: 30%
2. The firm's experience and staff to perform the type of work required, to include any designated subconsultants: 25%
3. The firm's ability to meet the time schedule established for the work: 15%
4. The firm's outstanding workload with the branch: 15%
5. Percentage of MBE/WBE participation: 5%
6. Percentage of the work to be performed in North Carolina: 10%

The highway design branch maintains a file on each qualified firm that has expressed an interest in preparing designs for the branch. Included in this file is a company brochure of Form PEFQUAL-1 listing personnel and their qualifications for performing desired work, the company's present activities, and financial qualifications. At the time this initial information is submitted, a sample of recent work plans (roadway design, structure design, geotechnical, hydraulic, photogrammetry, route surveys, etc.) will be needed for evaluation. The firm must have a Private Consulting Firm Questionnaire (current conflict of interest assessment) on file. If you have not submitted this data or if it needs to be updated, please send the new data to the State Design Engineer prior to, or along with, your letter of interest. Having this data on file in the Design Branch eliminates the need to resubmit it with each letter of interest.

APPENDIX 9:
BRIDGE DESIGN AWARDS

AMERICAN INSTITUTE OF STEEL CONSTRUCTION BRIDGE DESIGN COMPETITION

Since 1928 bridge awards have been given periodically by AISC to those structures that have been judged by a jury on the basis of aesthetics, economics, and innovation in structural design.

1996 AISC Bridge Design Awards

The 1996 competition, held in Chicago on October 15-17 included 24 winners in eight categories constructed between May 1, 1991 and April 30, 1996. From *Modern Steel Construction,* October, 1996, pp. 31-95 the names of the bridges awarded for the various categories are included. For more specific details the reader is referred to the original source.

Long Span—the Clark Bridge over the Mississippi River between Alton, IL, and North St. Charles County, MO.

Judges' comments: "The bridge's double fans create a very dramatic appearance"; "The single pylon reduces obstructions in the waterway and the cables contribute to the torsional stiffness of the roadway."

Long Span—the Fred Hartman Bridge over the Houston Ship Channel.

Judges' comments: "Unique appearance with good details"; "A beautiful structure."

Long Span—the Hoffstadt Creek Bridge of Washington State's Mt. St. Helens.

Judges' comments: "A good looking bridge with good site utilization"; "An attractive and interesting truss design."

Long Span—the I-526 Bridge over the Cooper River.

Judges comments: "A highly innovative structure with excellent details"; "Well proportioned with very clean lines."

Long Span—the Great River Bridge across the Mississippi River at Burlington, IA.

Judges' comments: "The main tower and cables serve as a beautiful gateway across the Mississippi River."

Medium Span—the U. S. Naval Academy Bridge, built as the gateway to Maryland's historic capital.

Judges' comments: "The right bridge for this site"; "An excellent example of form following function"; "A modern design with unique features that complement an historic area."

Medium Span—the Bob Michel Bridge crossing the Illinois River between East Peoria and Peoria.

Judges' comments: "Solves a problem in a clean, permanent way."

Medium Span—the Ramp Y at I-95 Davie Boulevard Interchange in Fort Lauderdale.

Judges comments: "The double curvature and single box design creates a beautiful but economic design."

Short Span—South Station Connection Ramp in downtown Boston.

Judges' comments: "Incredibly complex in design and construction but successfully executed with curved girders and trapezoidal boxes."

Short Span—the Woodlands Water Crossing at Mall Ring Road in Woodlands, TX.

Judges' comments: "The use of tubular trusses is an elegant and beautiful solution".

Short Span—Rainbow Arch Bridge in Redfield, SD.

Judges' comments: "A beautiful, light arch design that provides a very clear roadway."

Grade Separation—Route 19-Interstate 80 Interchange in Patterson, NJ .

Judges' comments: "A very complex ramp design with an innovative use of steel to solve the tight radius"; "Great architectural treatment of the abutments."

Grade Separation—Quivira Road Viaduct route through Lenaxa, KS.

Judges' comments: "A very economical solution - the haunched girders provided longer spans with reduced depth and a better appearance."

Grade Separation—20th Street HOV Structure/Denver Union Terminal Ramp in Colorado.

Judges' comments: "Graceful curves with good attention to detail"; "Good transition from steel to concrete."

Movable Span—8th Street Bridge Over the Sheboygan River in Sheboygan, WI.

Judges' comments: "An innovative operating system coupled with a beautiful design and appearance."

Movable Span—the New Brickell Bridge across the Miami River.

Judges' comments: "An excellent application of state-of-the-art technology."

Railroad—Aviation/Rosecrans Railroad Underpass in the City of Hawthorne, CA.

Judges' comments: "A wonderful design given the difficult site and critical elevations"; "An innovative erection technique."

Railroad—Union Pacific Railroad Joppa Bridges near Marion, IL.

Judges' comments: "An excellent job given the difficult construction constraints"; "An innovative design and construction project."

Railroad— CSX Ninemile Bridge over the James River nine miles downstream of Lynchburg, VA.

Judges' comments: "Very clean horizontal and vertical lines"; "Great use of steel for a clean, good looking bridge."

Special Purpose—Tudor Road Trail Crossing in the state of Alaska.

Judges' comments: "A very beautiful and graceful solution given the site constraints."

Special Purpose—Market Square Walkway in Roanoke, VA.

Judges' comments: "The bridge blends beautifully into its environment at each end and provides pedestrians with a great view of the surrounding city."

Special Purpose—Endicott Causeway Breach in Alaska.

Judges' comments: "A very successful execution given the extremely challenging construction conditions."

Reconstruction—Smithfield Street Bridge in Pittsburgh.

Judges'comments: "A highly successful renovation of a significant bridge"; "New steel and sensitive engineering has given this bridge a new lease on life."

Reconstructed—Pulaski Road Over I-55 Spudi over the Stevenson Expressway in Chicago.

Judges' comments: "An innovative intersection expansion"; "An excellent solution in a tight urban setting."

1993 AISC Bridge Design Awards

The AISC 1993 Prize Bridge Competition, held in Atlanta on November 11-12, judged more than 100 entries. The jury, consisting of four bridge experts awarded nine Prize Bridge Awards and eight Awards of Merit. The list of the winners published by *Modern Steel Construction*, November, 1993, pp. 19-72, includes specific data of the winning bridges. The following list was taken from the above source:

Long Span. I-255 over the Mississippi River.

Medium span, high clearance. Alsea Bay Bridge replacement near Waldport, OR.

Medium span, high clearance. Lawyer's Canyon Bridge in Idaho on US 95.

Medium Span, low clearance. Flaming Geyser Bridge No. 3024 across Washington's Green River.

Grade separation. Florida's Turnpike,I-595 Interchange in Broward County, FL.

Grade separation. American Falls SH39 Over UPRR, State Highway 39, American Falls, ID.

Elevated highway. Mainline approach structure, I-70 Eastbound, Colorado.

Elevated highway. Tropicana Flyover, Las Vegas, NV.

Moveable span. Potato Slough Bridge, San Francisco East Bay, California.

Railroad. Green Bay & Western Railroad Bridge #95, Wisconsin Rapids, WI.

Short span. Pine Street Bridge over Palmetto Canal, New Orleans, LA.

Short span. Bridge No. 13010 over Paulins Kill Creek, Warren County, NJ.

Special purpose. Veterans Administration Skybridge, OR.

Special purpose. South River Corridor pedestrian bridge, Rochester, NY.

Special purpose. Skyway at the St. Pauls Companies Headquaters,St. Paul, MN.

Special purpose. Olmstead Island footbridges, C&O Canal National Historic Park, MD.

Reconstructed. Belle Vernon Bridge, Belle Vernon, PA.

1991 AISC Bridge Design Awards

The AISC Prize Bridge Competition for 1991 was held in St. Louis, MO on September 16. The jury composed of five members judged over 130 entries awarding eight prize bridge awards and nine awards of merit. The list of the winning projects with full description of the projects was published in *Modern Steel Construction,* September, 1991, pp. 13-58. Limiting the information just to the category, bridge denomination, and location, the following list is herein presented, referring the reader to the original scource for more detailed information.

Long span—Roosevelt Lake Bridge, Roosevelt Lake, AZ.

Medium span, low clearance—Ramp B - Turkey Lake Road over I-4, Orlando, FL.

Medium span, low clearance—Ramp X Roadway and Ramp R roadway, Danbury, CT.

Medium span, high clearance—Southern Boulevard Route 9W bridge over Normanskill, Albany, NY.

Medium span, high clearance—McCarran Boulevard Bridge, Reno, NV.

Short span—Susquehanna Street Bridge, Allentown, PA.

Short span—St. Peter Street Bridge, St. Paul, MN.

Railroad—Soo Line Railroad Bridge over Marshall & Snelling Avenues, St. Paul, MN.

Elevated highway or viaduct—SR-386, I-65, CSX R.R., Two-Mile Pike, Nashville, TN.

Grade separation—I-10—I-17 Interchange, Phoenix, AZ.

Grade separation—Bridge "B" over MCO entrance road, Orlando, FL.

Moveable span.—US 10 over St. Croix River, Prescott, WS.

Moveable span—Greenpoint Avenue Bridge, New York City, NY.

Special purpose—The Link, Kansas City, MO.

Special purpose—Geneva Spur pedestrian bridge, West Chicago, IL.

Reconstructed—Paducah Tied Arches, Metropolis, IL and Paducah, KY.

Reconstructed—Martin Luther King Bridge, St. Louis, MO.

PORTLAND CEMENT ASSOCIATION CONCRETE BRIDGE DESIGN COMPETION

The Portland Cement Association has also offered bridge competitions on a two-year basis starting in 1988. The association granted awards on the basis of excellence and design of concrete bridges.

1996 PCA Bridge Biennial Design Awards

This competition included 56 entries in various categories. The following nine winners were named and received the awards at the American Concrete Institute Awards Breakfast on April 1997, in Seattle, WA:

Big Qualicum Bridge—Vancouver Island, BC, 856 ft long.

Deport Road Bridge—Coventry, CT, clear span of 20 ft and a rise of 10 ft.

Kellogg CBD Viaduct—Wichita, KS, 2,908 ft long.

Log Farm Access Bridge— Ottawa, ONT, 243 ft long.

North Santiam River Bridge—Gates, OR, three spans of 95 ft, 137 ft, and 95 ft.

Los Angeles River Bridge—Los Angeles, CA, 1,414 ft long.

Reconstruction of the Mendota Bridge—Mendota, MN, 4,113 ft long.

Rehabilitation of Portage Avenue Culvert at Omand's Creek—Winnipeg, Manitoba, 154 ft long.

Rockcut Bridge—Stevens and Ferry Counties, WA, 190.5 ft span.

1994 PCA Bridge Biennial Design Awards

This competition included 63 entries in various categories. The following eight winners recieved awards at the American Concrete Institute Awards Breakfast held on March, 1995 in Salt Lake City, Utah. The list of the winning projects is as follows.

Acosta Bridge over the St. Johns River—Jacksonville, FL, 1,645 ft long.

Interstate Route H-3, Windward Viaduct—Haiku Valley, Oahu, HI, 6,600 ft long.

Natchez Trace Parkway Bridge over Tennessee Route 96. Franklin, TN, 1,648 ft long.

Oak Canyon Bridge and Spring Canyon Undercrossing—San Diego, CA, total length of 720 and 780 ft.

Route 5—14 Separation and Overhead—Los Angeles, CA, the length of the two connectors is 1,040 and 1,582 ft long respectively.

The Scripps Crossing—San Diego, CA, 140 ft long.

Semiahmoo Trail Pedestrian Bridge—Surrey, BC, 80 ft long.

Upper Middle Road- 16 Mile Creek Bridge—Oakville, ONT, 1,099 ft long.

For more information about individual projects including jury comments, photographs, drawings, and other data for the Portland Cement Association Awards, see *Engineered Concrete Structures*, Portland Cement Association, 1996, Vol 9, no. 3, and 1994, Vol. 7, no. 3.

APPENDIX 10: BRIDGE BEAM DESIGN THEORY

The bridge design software available from Leap Software, Inc. was been introduced in Chapter 3 including three example cases. The following is the bridge beam design theory used in that software and the explanation of how to apply these concepts.

THEORY

This section outlines bridge beam design theory and explains how to apply these concepts.

Structural Modeling

Two structural models are used for analysis of the various load components. The particular model used for a given load component depends on whether the load acts on the bare precast (i.e., non-composite) or the composite beam.

The first model is a two-dimensional, simple-span beam with the section properties of the precast beam only (fig. TH-1). The length of the beam of this model depends on which span of the structure is being considered. The model is analyzed for beam self-weight, cast-in-place deck weight, and any other dead loads acting on the precast section alone.

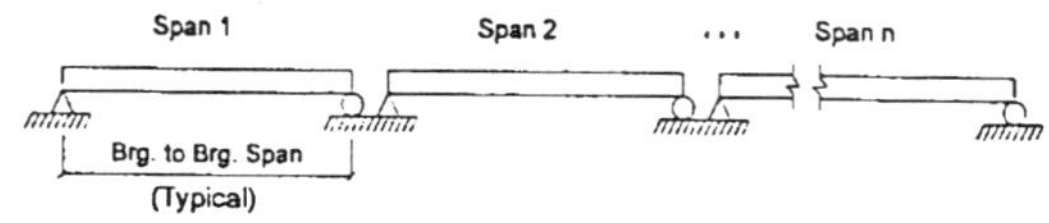

Figure TH-1. Simple Span Models used for Analysis of Beam Self-Weight, Superimposed Dead Load on Precast, and Deck Weight.

The second model is a continuous, two-dimensional, linear structure with knife-edge support at the piers (fig. TH-2). The section properties of the total composite cross-section are used for each span. This model is analyzed for both dead load acting on the composite cross-section and moving live load. Trucks are moved across the entire length of the structure, facing both left and right, with variable wheel spacing. Lane loads are analyzed with special modifications as required by AASHTO for maximum negative moment and shear.

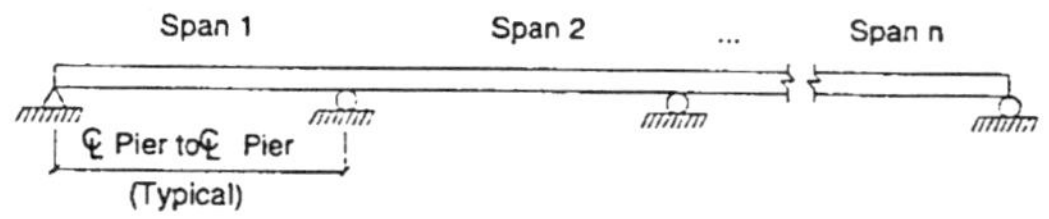

Figure TH-2. Continuous Beam Model Used for Analyzing Superimposed Dead Load on Composite and Live Load

The results of analyses using both models are used to obtain final moment and shear envelopes for which the beams are designed.

Live Load Moment Envelopes and Shear Envelopes

Moment and shear envelopes for live load are developed in two phases: (1) lane load analysis and (2) truck + preceding/trailing uniform load analysis. The truck load comprises the axle loads and preceding/trailing uniform loads acting together. At each section the values of the live load envelopes are obtained by comparing these two envelopes and taking the critical values.

Figure A10-1 Bridge beam design theory (reproduced courtesy of LEAP Software, Inc.).

For purposes of the preceding table, the sections fall into the following categories:

Longitudinal Stringer	AASHTO Girder Rectangular Longitudinal Stringer Special Longitudinal Stringer
Spread Box Girder	Spread Box Girder
Multi-beam	Box Multi-Beam K = 1.0 Channel Beam K = 2.2 Double Tee K = 2.2 Nonvoided Rect Beam K = 0.7 Rect. Beam w/Round Voids K = 0.8
Flat slab	Flat Slab

Figure TH-4 illustrates the model for determining the distribution factor for an exterior beam using the simple span assumption explained above. The first wheel is placed at a distance from the curb equal to 2 ft as indicated in AASHTO fig. 3.7.6A.

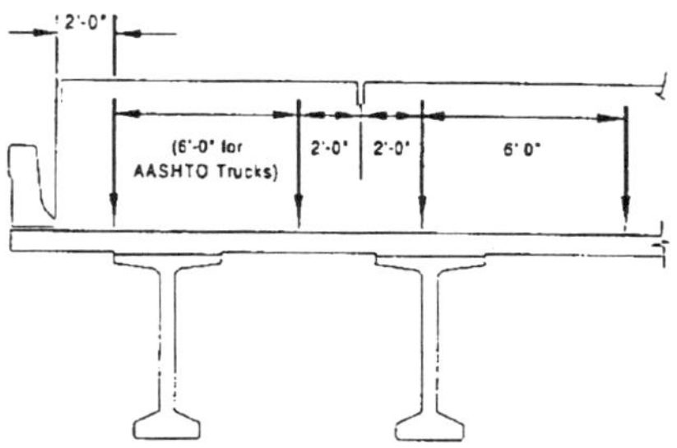

Figure TH-4. Model used for Determining Distribution Factor for Exterior Beam

Kern Points

The upper and lower kern points of a beam cross section provide information on the effect of adding a strand to a pattern. While they do not predict the magnitudes of the effects, they are, nevertheless, helpful when designing a strand pattern.

If a strand were added under the lower kern point, additional compressive stress would be induced in the bottom of the beam and additional tensile stress would be induced in the top of the beam. If a strand were added between the two points, compression would be induced in both the top and the bottom of the beam. Adding a strand above the upper kern point has the opposite effect as that of adding a strand under the lower kern point.

If a strand were placed at the lower kern point, the stress in the top of the beam would be exactly zero. Knowing this,

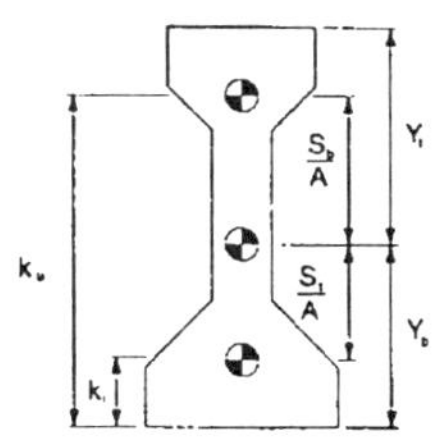

Figure TH-5. Typical Kern Point Locations

Figure A10-1 (continued)

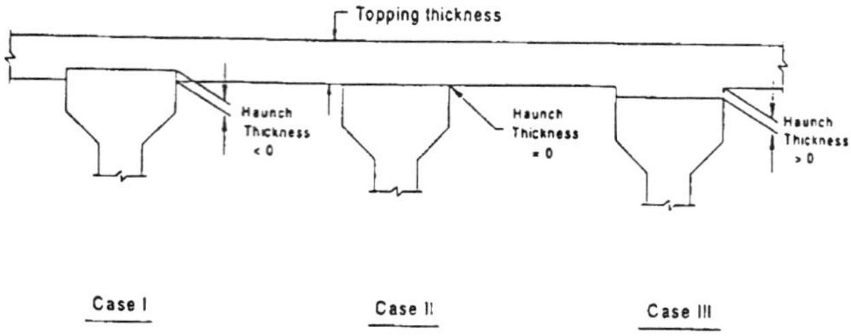

Figure TH-6. The Three Cases Addressed for Computing Effective Width of Compression Flange

Prestress Losses

AASHTO Method

The initial relaxation loss (RET, psi) is computed using the formulas found in the PCI Journal (Ref: "Recommendations for Estimating Prestress Losses," PCI Journal, Volume 20, No. 4, July-August 1975, Article No. JR 162., p.9). Elastic shortening (ES) is computed by a modified closed-form equation, the derivation of which is given in the next section. Other components of loss are computed by the code-specified formulas. The following modified BPR formulas are used to determine the final loss of prestress in pretensioned beams:

Derivation of Closed-Form Solution for Prestress Loss Due to Elastic Shortening, ES.

Given:

A_c = Area of concrete section.

A_s = Total area of strands.

e = Eccentricity of centroid of strands.

E_{ci} = Modulus of elasticity of concrete at release.

E_s = Modulus of elasticity of prestressing steel.

ES = Loss due to elastic shortening (difference between stress in prestressing steel immediately before and immediately after release).

f_{cir} = Net stress immediately after release at the centroid of the strands at the section of maximum moment.

I = Moment of inertia of beam cross-section.

Figure A10-1 (continued)

APPENDIX 11: HOMAGE TO TWO LEGENDARY FIGURES

The author has always felt a great sense of respect and admiration for his two countrymen Pier Luigi Nervi and Riccardo Morandi who have left a legendary memory with their work. He had the opportunity of meeting with each one of them for a short period of time during the fall of 1973. On that occasion the author spent one hour with each of them in their respective offices in Rome conducting a seminar with his students from Clemson University. On these occasions, both Professor Nervi and Professor Morandi kindly dedicated two books to the author which he has treasured with a great sense of respect. Now, on the occasion of writing this book, the author takes the opportunity to preserve the handwritten notations of these two men by enclosing them within the cover of this work.

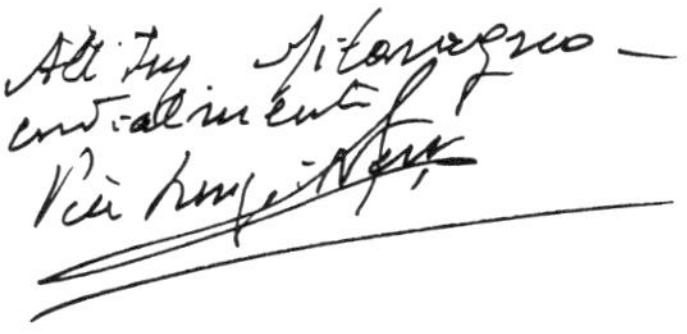

APPENDIX 12: PRELIMINARY TRUSS SELECTION

When, in the preliminary design process, the designer reaches the phase of selecting an appropriate structural system, the choice may include truss systems. In such a case, the designer is confronted with a number of choices concerning truss types. To help select the proper truss type, a series of typical trusses that could be used for bridge structures is herein included. These originally appeared in *Simplified Truss Design*, by M. Melaragno (Van Nostrand Reinhold, 1980, presently out of print) is hereby reproduced by permission of the copyright holder, the author.

This list of different types of trusses with parallel chords includes the most typical truss type configurations. These types could be used for loadings applied to the top chord or to the bottom chord as in the case of through truss bridges. Once the selection is made, the designer has the option of making a quick verification of the selected type by calculating from the tables the axial forces in each member and the exact length of each member as well. The table is divided into several values of n = L, H which is the ratio of the span over the depth of the truss. Having established the value of n, the table includes three factors: K_1, K_2, K. The product K_1P gives the value of the axial force for every member of the truss produced by the load P applied on the top chord. The product K_{2p} gives the value of the axial force for every member of the truss produced by the load p applied on the bottom chord. The product KH gives the exact length of each member of the truss.

Having quickly determined these values, the designer may opt to make a rapid calculation of the required members subjected to axial tension or compression and verifying the feasibility of the selected truss.

Table 1 Flat Howe 4 Panels at Top & Bottom

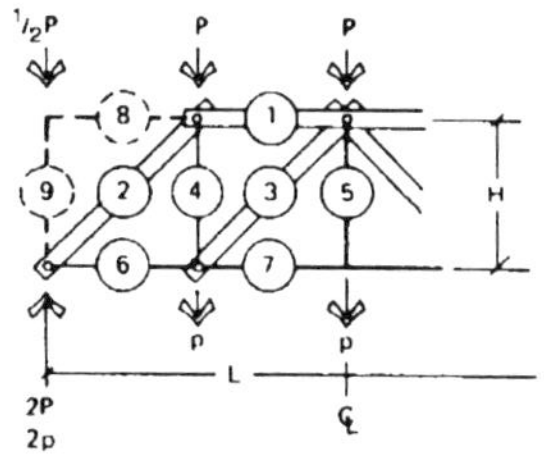

$n = \frac{L}{H}$

$N = \sqrt{n^2 + 16}$

Member	Values of n: 7		8		9		10		11		12		13		General Formulas	
	AXIAL FORCE COEFFICIENTS															
	K_1	K_2	K_1	K_2	K_1	K_2	K_1	K_2	K_1	K_2	K_1	K_2	K_1	K_2	K_1	K_2
1	-2.63	-2.63	-3.00	-3.00	-3.38	-3.38	-3.75	-3.75	-4.13	-4.13	-4.50	-4.50	-4.88	-4.88	$-3/8\,n$	$-3/8\,n$
2	-3.02	-3.02	-3.35	-3.35	-3.69	-3.69	-4.04	-4.04	-4.39	-4.39	-4.74	-4.74	-5.10	-5.10	$-3/8\,N$	$-3/8\,N$
3	-1.01	-1.01	-1.12	-1.12	-1.23	-1.23	-1.35	-1.35	-1.46	-1.46	-1.58	-1.58	-1.70	-1.70	$-1/8\,N$	$-1/8\,N$
4	0.50	1.50	0.50	1.50	0.50	1.50	0.50	1.50	0.50	1.50	0.50	1.50	0.50	1.50	1/2	3/2
5	0	1.00	0	1.00	0	1.00	0	1.00	0	1.00	0	1.00	0	1.00	0	1
6	2.63	2.63	3.00	3.00	3.38	3.38	3.75	3.75	4.13	4.13	4.50	4.50	4.88	4.88	$3/8\,n$	$3/8\,n$
7	3.50	3.50	4.00	4.00	4.50	4.50	5.00	5.00	5.50	5.50	6.00	6.00	6.50	6.50	$1/2\,n$	$1/2\,n$
8	0	0	0	0	0	0	0	0	0	0	0	0	0	0	0	0
9	-0.50	0	-0.50	0	-0.50	0	-0.50	0	-0.50	0	-0.50	0	-0.50	0	$-1/2$	0

	LENGTH COEFFICIENTS							
	K	*K*	*K*	*K*	*K*	*K*	*K*	*K*
Horizontal	1.750000	2.000000	2.250000	2.500000	2.750000	3.000000	3.250000	$1/4\,n$
Diagonal	2.015564	2.236068	2.462214	2.692582	2.926175	3.162278	3.400368	$1/4\,N$
Vertical	1.000000	1.000000	1.000000	1.000000	1.000000	1.000000	1.000000	1
ANGLES (IN DEGREES) BETWEEN MEMBERS								
Between diagonal and horizontal members	30	27	24	22	20	18	17	$4/n = \tan a$
Between diagonal and vertical members	60	63	66	68	70	72	73	$90 - a$
Between vertical and horizontal members	90	90	90	90	90	90	90	90

Table 2 Flat Howe 6 Panels at Top & Bottom

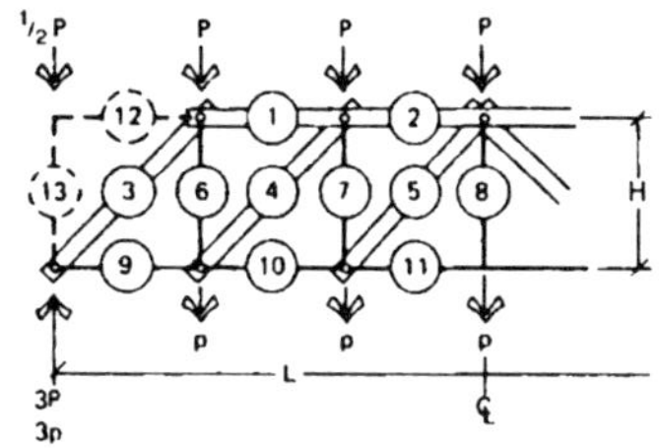

$n = \frac{L}{H}$

$N = \sqrt{n^2 + 36}$

Member	VALUES OF *n*														General Formulas	
	7		8		9		10		11		12		13			
	AXIAL FORCE COEFFICIENTS															
	K_1	K_2	K_1	K_2	K_1	K_2	K_1	K_2	K_1	K_2	K_1	K_2	K_1	K_2	K_1	K_2
1	-2.92	-2.92	-3.33	-3.33	-3.75	-3.75	-4.17	-4.17	-4.58	-4.58	-5.00	-5.00	-5.42	-5.42	- 5/12 *n*	- 5/12 *n*
2	-4.67	-4.67	-5.33	-5.33	-6.00	-6.00	-6.67	-6.67	-7.33	-7.33	-8.00	-8.00	-8.67	-8.67	- 2/3 *n*	- 2/3 *n*
3	-3.84	-3.84	-4.17	-4.17	-4.51	-4.51	-4.86	-4.86	-5.22	-5.22	-5.59	-5.59	-5.97	-5.97	- 5/12 *N*	- 5/12 *N*
4	-2.30	-2.30	-2.50	-2.50	-2.70	-2.70	-2.92	-2.92	-3.13	-3.13	-3.35	-3.35	-3.58	-3.58	- 1/4 *N*	- 1/4 *N*
5	-0.77	-0.77	-0.83	-0.83	-0.90	-0.90	-0.97	-0.97	-1.04	-1.04	-1.12	-1.12	-1.19	-1.19	- 1/12 *N*	- 1/12 *N*
6	1.50	2.50	1.50	2.50	1.50	2.50	1.50	2.50	1.50	2.50	1.50	2.50	1.50	2.50	3/2	5/2
7	0.50	1.50	0.50	1.50	0.50	1.50	0.50	1.50	0.50	1.50	0.50	1.50	0.50	1.50	1/2	3/2
8	0	1.00	0	1.00	0	1.00	0	1.00	0	1.00	0	1.00	0	1.00	0	1
9	2.92	2.92	3.33	3.33	3.75	3.75	4.17	4.17	4.58	4.58	5.00	5.00	5.42	5.42	5/12 *n*	5/12 *n*
10	4.67	4.67	5.33	5.33	6.00	6.00	6.67	6.67	7.33	7.33	8.00	8.00	8.67	8.67	2/3 *n*	2/3 *n*
11	5.25	5.25	6.00	6.00	6.75	6.75	7.50	7.50	8.25	8.25	9.00	9.00	9.75	9.75	3/4 *n*	3/4 *n*
12	0	0	0	0	0	0	0	0	0	0	0	0	0	0	0	0
13	-0.50	0	-0.50	0	-0.50	0	-0.50	0	-0.50	0	-0.50	0	-0.50	0	- 1/2	0

LENGTH COEFFICIENTS								
	K	*K*	*K*	*K*	*K*	*K*	*K*	*K*
Horizontal	1.166667	1.333333	1.500000	1.666667	1.833333	2.000000	2,166667	1/6 n
Diagonal	1.536591	1.666667	1.802776	1.943651	2.008327	2.236068	2,386304	1/6 N
Vertical	1.000000	1.000000	1.000000	1.000000	1.000000	1.000000	1.000000	1
ANGLES (IN DEGREES) BETWEEN MEMBERS								
Between diagonal and horizontal members	41	37	34	31	29	27	25	$6/n = \tan a$
Between diagonal and vertical members	49	53	56	59	61	63	65	$90 - a$
Between vertical and horizontal members	90	90	90	90	90	90	90	90

Table 3 Flat Howe 8 Panels at Top & Bottom

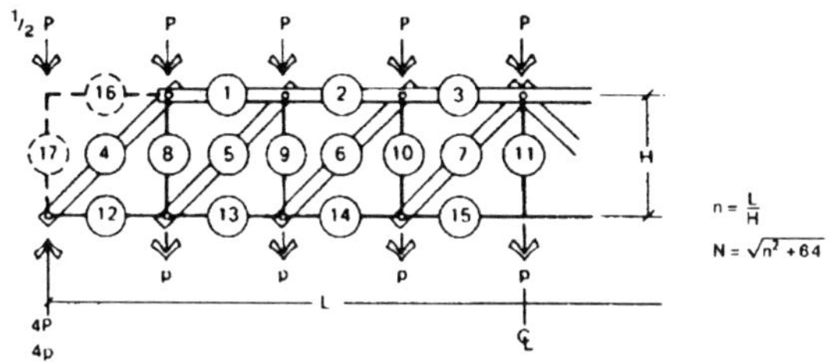

Member	VALUES OF *n*														General Formulas	
	7		8		9		10		11		12		13			
	AXIAL FORCE COEFFICIENTS															
	K_1	K_2	K_1	K_2	K_1	K_2	K_1	K_2	K_1	K_2	K_1	K_2	K_1	K_2	K_1	K_2
1	-3.06	-3.06	-3.50	-3.50	-3.94	-3.94	-4.38	-4.38	-4.81	-4.81	-5.25	-5.25	-5.69	-5.69	$-7/16\,n$	$-7/16\,n$
2	-5.25	-5.25	-6.00	-6.00	-6.75	-6.75	-7.50	-7.50	-8.25	-8.25	-9.00	-9.00	-9.75	-9.75	$-3/4\,n$	$-3/4\,n$
3	-6.56	-6.56	-7.50	-7.50	-8.44	-8.44	-9.38	-9.38	-10.31	-10.31	-11.25	-11.25	-12.19	-12.19	$-15/16\,n$	$-15/16\,n$
4	-4.65	-4.65	-4.95	-4.95	-5.27	-5.27	-5.60	-5.60	-5.95	-5.95	-6.31	-6.31	-6.68	-6.68	$-7/16\,N$	$-7/16\,N$
5	-3.32	-3.32	-3.54	-3.54	-3.76	-3.76	-4.00	-4.00	-4.25	-4.25	-4.51	-4.51	-4.77	-4.77	$-5/16\,N$	$-5/16\,N$
6	-1.99	-1.99	-2.12	-2.12	-2.26	-2.26	-2.40	-2.40	-2.55	-2.55	-2.70	-2.70	-2.86	-2.86	$-3/16\,N$	$-3/16\,N$
7	-0.66	-0.66	-0.71	-0.71	-0.75	-0.75	-0.80	-0.80	-0.85	-0.85	-0.90	-0.90	-0.95	-0.95	$-1/16\,N$	$-1/16\,N$
8	2.50	3.50	2.50	3.50	2.50	3.50	2.50	3.50	2.50	3.50	2.50	3.50	2.50	3.50	5/2	7/2
9	1.50	2.50	1.50	2.50	1.50	2.50	1.50	2.50	1.50	2.50	1.50	2.50	1.50	2.50	3/2	5/2
10	0.50	1.50	0.50	1.50	0.50	1.50	0.50	1.50	0.50	1.50	0.50	1.50	0.50	1.50	1/2	3/2
11	0	1.00	0	1.00	0	1.00	0	1.00	0	1.00	0	1.00	0	1.00	0	1
12	3.06	3.06	3.50	3.50	3.94	3.94	4.38	4.38	4.81	4.81	5.25	5.25	5.69	5.69	$7/16\,n$	$7/16\,n$

13	5.25	5.25	6.00	6.00	6.75	6.75	7.50	7.50	8.25	8.25	9.00	9.00	9.75	9.75	3/4 n	3/4 n
14	6.56	6.56	7.50	7.50	8.44	8.44	9.38	9.38	10.31	10.31	11.25	11.25	12.19	12.19	15/16 n	15/16 n
15	7.00	7.00	8.00	8.00	9.00	9.00	10.00	10.00	11.00	11.00	12.00	12.00	13.00	13.00	n	n
16	0	0	0	0	0	0	0	0	0	0	0	0	0	0	0	0
17	–0.50	0	–0.50	0	–0.50	0	–0.50	0	–0.50	0	–0.50	0	–0.50	0	– 1/2	0
LENGTH COEFFICIENTS																
	K		K		K		K		K		K		K		K	
Horizontal	0.875000		1.000000		1.125000		1.250000		1.375000		1.500000		1.625000		1/8 n	
Diagonal	1.328768		1.414214		1.505199		1.600781		1.700184		1.802776		1.908042		1/8 N	
Vertical	1.000000		1.000000		1.000000		1.000000		1.000000		1.000000		1.000000		1	
ANGLES (IN DEGREES) BETWEEN MEMBERS																
Between diagonal and horizontal members	49		45		42		39		36		34		32		$8/n = \tan a$	
Between diagonal and vertical members	41		45		46		51		54		56		56		$90 - a$	
Between vertical and horizontal members	90		90		90		90		90		90		90		90	

Table 4 Flat Howe 10 Panels at Top & Bottom

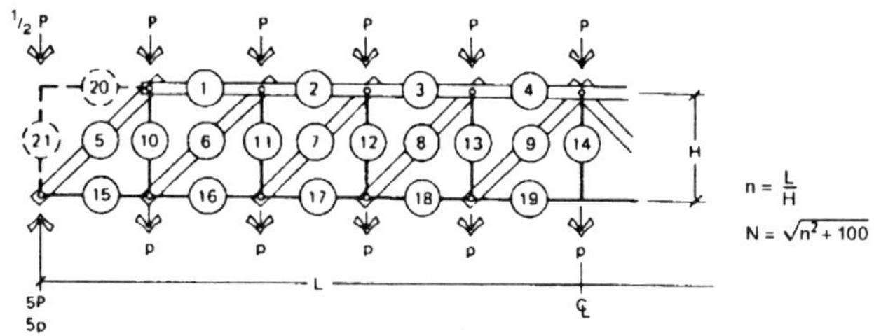

Member	7		8		9		10		11		12		13		General Formulas	
	VALUES OF n															
	AXIAL FORCE COEFFICIENTS															
	K_1	K_2	K_1	K_2	K_1	K_2	K_1	K_2	K_1	K_2	K_1	K_2	K_1	K_2	K_1	K_2
1	-3.15	-3.15	-3.60	-3.60	-4.05	-4.05	-4.50	-4.50	-4.95	-4.95	-5.40	-5.40	-5.85	-5.85	$-9/20\ n$	$-9/20\ n$
2	-5.60	-5.60	-6.40	-6.40	-7.20	-7.20	-8.00	-8.00	-8.80	-8.80	-9.60	-9.60	-10.40	-10.40	$-4/5\ n$	$-4/5\ n$
3	-7.35	-7.35	-8.40	-8.40	-9.45	-9.45	-10.50	-10.50	-11.55	-11.55	-12.60	-12.60	-13.65	-13.65	$-21/20\ n$	$-21/20\ n$
4	-8.40	-8.40	-9.60	-9.60	-10.80	-10.80	-12.00	-12.00	-13.20	-13.20	-14.40	-14.40	-15.60	-15.60	$-6/5\ n$	$-6/5\ n$
5	-5.49	-5.49	-5.76	-5.76	-6.05	-6.05	-6.36	-6.36	-6.69	-6.69	-7.03	-7.03	-7.38	-7.38	$-9/20\ N$	$-9/20\ N$
6	-4.27	-4.27	-4.48	-4.48	-4.71	-4.71	-4.95	-4.95	-5.20	-5.20	-5.47	-5.47	-5.74	-5.74	$-7/20\ N$	$-7/20\ N$
7	-3.05	-3.05	-3.20	-3.20	-3.36	-3.36	-3.54	-3.54	-3.72	-3.72	-3.91	-3.91	-4.10	-4.10	$-1/4\ N$	$-1/4\ N$
8	-1.83	-1.83	-1.92	-1.92	-2.02	-2.02	-2.12	-2.12	-2.23	-2.23	-2.34	-2.34	-2.46	-2.46	$-3/20\ N$	$-3/20\ N$
9	-0.61	-0.61	-0.64	-0.64	-0.67	-0.67	-0.71	-0.71	-0.74	-0.74	-0.78	-0.78	-0.82	-0.82	$-1/20\ N$	$-1/20\ N$
10	3.50	4.50	3.50	4.50	3.50	4.50	3.50	4.50	3.50	4.50	3.50	4.50	3.50	4.50	7/2	9/2
11	2.50	3.50	2.50	3.50	2.50	3.50	2.50	3.50	2.50	3.50	2.50	3.50	2.50	3.50	5/2	7/2
12	1.50	2.50	1.50	2.50	1.50	2.50	1.50	2.50	1.50	2.50	1.50	2.50	1.50	2.50	3/2	5/2
13	0.50	1.50	0.50	1.50	0.50	1.50	0.50	1.50	0.50	1.50	0.50	1.50	0.50	1.50	1/2	3/2

14	0	1.00	0	1.00	0	1.00	0	1.00	0	1.00	0	1.00	0	1.00	0	1
15	3.15	3.15	3.60	3.60	4.05	4.05	4.50	4.50	4.95	4.95	5.40	5.40	5.85	5.85	9/20 *n*	9/20 *n*
16	5.60	5.60	6.40	6.40	7.20	7.20	8.00	8.00	8.80	8.80	9.60	9.60	10.40	10.40	4/5 *n*	4/5 *n*
17	7.35	7.35	8.40	8.40	9.45	9.45	10.50	10.50	11.55	11.55	12.60	12.60	13.65	13.65	21/20 *n*	21/20 *n*
18	8.40	8.40	9.60	9.60	10.80	10.80	12.00	12.00	13.20	13.20	14.40	14.40	15.60	15.60	6/5 *n*	6/5 *n*
19	8.75	8.75	10.00	10.00	11.25	11.25	12.50	12.50	13.75	13.75	15.00	15.00	16.25	16.25	5/4 *n*	5/4 *n*
20	0	0	0	0	0	0	0	0	0	0	0	0	0	0	0	0
21	-0.50	0	-0.50	0	-0.50	0	-0.50	0	-0.50	0	-0.50	0	-0.50	0	- 1/2	0

LENGTH COEFFICIENTS

	K	*K*	*K*	*K*	*K*	*K*	*K*	*K*
Horizontal	0.700000	0.800000	0.900000	1.000000	1.100000	1.200000	1.300000	1/10 *n*
Diagonal	1.220656	1.280625	1.345362	1.414214	1.486607	1.562050	1.640122	1/10 *N*
Vertical	1.000000	1.000000	1.000000	1.000000	1.000000	1.000000	1.000000	1

ANGLES (IN DEGREES) BETWEEN MEMBERS

Between diagonal and horizontal members	56	51	48	45	42	40	38	$10/n = \tan a$
Between diagonal and vertical members	35	39	42	45	48	50	52	$90 - a$
Between vertical and horizontal members	90	90	90	90	90	90	90	90

Table 5 Flat Howe 12 Panels at Top & Bottom

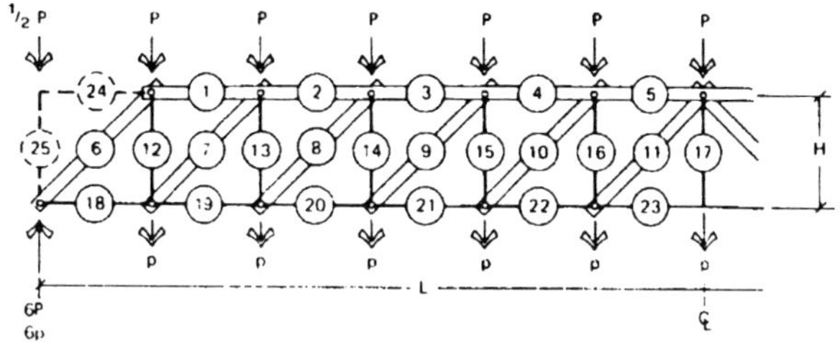

$n = \frac{L}{H}$

$N = \sqrt{n^2 + 144}$

Member	VALUES OF *n*														General Formulas	
	7		8		9		10		11		12		13			
	AXIAL FORCE COEFFICIENTS															
	K_1	K_2	K_1	K_2	K_1	K_2	K_1	K_2	K_1	K_2	K_1	K_2	K_1	K_2	K_1	K_2
1	-3.21	-3.21	-3.67	-3.67	-4.13	-4.13	-4.58	-4.58	-5.04	-5.04	-5.50	-5.50	-5.96	-5.96	- 11/24 *n*	- 11/24 *n*
2	-5.83	-5.83	-6.67	-6.67	-7.50	-7.50	-8.33	-8.33	-9.17	-9.17	-10.00	-10.00	-10.83	-10.83	- 5/6 *n*	- 5/6 *n*
3	-7.88	-7.88	-9.00	-9.00	-10.13	-10.13	-11.25	-11.25	-12.38	-12.38	-13.50	-13.50	-14.63	-14.63	- 9/8 *n*	- 9/8 *n*
4	-9.33	-9.33	-10.67	-10.67	-12.00	-12.00	-13.33	-13.33	-14.67	-14.67	-16.00	-16.00	-17.33	-17.33	- 4/3 *n*	- 4/3 *n*
5	-10.21	-10.21	-11.67	-11.67	-13.13	-13.13	-14.58	-14.58	-16.04	-16.04	-17.50	-17.50	-18.96	-18.96	- 35/24 *n*	- 35/24 *n*
6	-6.37	-6.37	-6.61	-6.61	-6.88	-6.88	-7.16	-7.16	-7.46	-7.46	-7.78	-7.78	-8.11	-8.11	- 11/24 *N*	- 11/24 *N*
7	-5.21	-5.21	-5.41	-5.41	-5.63	-5.63	-5.86	-5.86	-6.10	-6.10	-6.36	-6.36	-6.63	-6.63	- 3/8 *N*	- 3/8 *N*
8	-4.05	-4.05	-4.21	-4.21	-4.38	-4.38	-4.56	-4.56	-4.75	-4.75	-4.95	-4.95	-5.16	-5.16	- 7/24 *N*	- 7/24 *N*
9	-2.89	-2.89	-3.00	-3.00	-3.13	-3.13	-3.25	-3.25	-3.39	-3.39	-3.54	-3.54	-3.69	-3.69	- 5/24 *N*	- 5/24 *N*
10	-1.74	-1.74	-1.80	-1.80	-1.88	-1.88	-1.95	-1.95	-2.03	-2.03	-2.12	-2.12	-2.21	-2.21	- 1/8 *N*	- 1/8 *N*
11	-0.58	-0.58	-0.60	-0.60	-0.63	-0.63	-0.65	-0.65	-0.68	-0.68	-0.71	-0.71	-0.74	-0.74	- 1/24 *N*	- 1/24 *N*
12	4.50	5.50	4.50	5.50	4.50	5.50	4.50	5.50	4.50	5.50	4.50	5.50	4.50	5.50	9/2	11/2

13	3.50	4.50	3.50	4.50	3.50	4.50	3.50	4.50	3.50	4.50	3.50	4.50	3.50	4.50	7/2	9/2
14	2.50	3.50	2.50	3.50	2.50	3.50	2.50	3.50	2.50	3.50	2.50	3.50	2.50	3.50	5/2	7/2
15	1.50	2.50	1.50	2.50	1.50	2.50	1.50	2.50	1.50	2.50	1.50	2.50	1.50	2.50	3/2	5/2
16	0.50	1.50	0.50	1.50	0.50	1.50	0.50	1.50	0.50	1.50	0.50	1.50	0.50	1.50	1/2	3/2
17	0	1.00	0	1.00	0	1.00	0	1.00	0	1.00	0	1.00	0	1.00	0	1
18	3.21	3.21	3.67	3.67	4.13	4.13	4.58	4.58	5.04	5.04	5.50	5.50	5.96	5.96	$11/24\ n$	$11/24\ n$
19	5.83	5.83	6.67	6.67	7.50	7.50	8.33	8.33	9.17	9.17	10.00	10.00	10.83	10.83	$5/6\ n$	$5/6\ n$
20	7.88	7.88	9.00	9.00	10.13	10.13	11.25	11.25	12.38	12.38	13.50	13.50	14.63	14.63	$9/8\ n$	$9/8\ n$
21	9.33	9.33	10.67	10.67	12.00	12.00	13.33	13.33	14.67	14.67	16.00	16.00	17.33	17.33	$4/3\ n$	$4/3\ n$
22	10.21	10.21	11.67	11.67	13.13	13.13	14.58	14.58	16.04	16.04	17.50	17.50	18.96	18.96	$35/24\ n$	$35/24\ n$
23	10.50	10.50	12.00	12.00	13.50	13.50	15.00	15.00	16.50	16.50	18.00	18.00	19.50	19.50	$3/2\ n$	$3/2\ n$
24	0	0	0	0	0	0	0	0	0	0	0	0	0	0	0	0
25	-0.50	0	-0.50	0	-0.50	0	-0.50	0	-0.50	0	-0.50	0	-0.50	0	- 1/2	0

LENGTH COEFFICIENTS

	K	K	K	K	K	K	K	K
Horizontal	0.583333	0.666667	0.750000	0.833333	0.916667	1.000000	1.083333	$1/12\ n$
Diagonal	1.157704	1.201850	1.250000	1.301708	1.356568	1.414214	1.474317	$1/12\ N$
Vertical	1.000000	1.000000	1.000000	1.000000	1.000000	1.000000	1.000000	1

ANGLES (IN DEGREES) BETWEEN MEMBERS

Between diagonal and horizontal members	60	56	53	50	47	45	43	$12/n = \tan a$
Between diagonal and vertical members	30	34	37	40	43	45	47	$90 - a$
Between vertical and horizontal members	90	90	90	90	90	90	90	90

Table 6 Flat Pratt 4 Panels at Top & Bottom

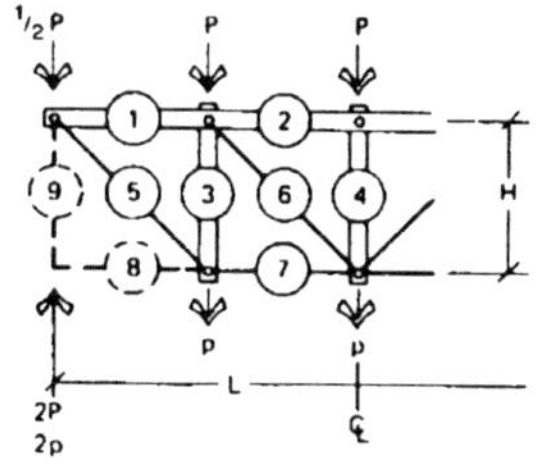

$n = \frac{L}{H}$

$^{1}N = \sqrt{n^2 + 16}$

Member	VALUES OF *n*														General Formulas	
	7		8		9		10		11		12		13			
AXIAL FORCE COEFFICIENTS																
	K_1	K_2	K_1	K_2	K_1	K_2	K_1	K_2	K_1	K_2	K_1	K_2	K_1	K_2	K_1	K_2
1	-2.63	-2.63	-3.00	-3.00	-3.38	-3.38	-3.75	-3.75	-4.13	-4.13	-4.50	-4.50	-4.88	-4.88	- 3/8 *n*	- 3/8 *n*
2	-3.50	-3.50	-4.00	-4.00	-4.50	-4.50	-5.00	-5.00	-5.50	-5.50	-6.00	-6.00	-6.50	-6.50	- 1/2 *n*	- 1/2 *n*
3	-1.50	-0.50	-1.50	-0.50	-1.50	-0.50	-1.50	-0.50	-1.50	-0.50	-1.50	-0.50	-1.50	-0.50	- 3/2	- 1/2
4	-1.00	0	-1.00	0	-1.00	0	-1.00	0	-1.00	0	-1.00	0	-1.00	0	-1	0
5	3.02	3.02	3.35	3.35	3.69	3.69	4.04	4.04	4.39	4.39	4.74	4.74	5.10	5.10	3/8 *N*	3/8 *N*
6	1.01	1.01	1.12	1.12	1.23	1.23	1.35	1.35	1.46	1.46	1.58	1.58	5.10	1.70	1/8 *N*	1/8 *N*
7	2.63	2.63	3.00	3.00	3.38	3.38	3.75	3.75	4.13	4.13	4.50	4.50	4.88	4.88	3/8 *N*	3/8 *N*
8	0	0	0	0	0	0	0	0	0	0	0	0	0	0	0	0
9	-2.00	0	-2.00	0	-2.00	0	-2.00	0	-2.00	0	-2.00	0	-2.00	0	- 2	0
LENGTH COEFFICIENTS																
	K		*K*		*K*		*K*		*K*		*K*		*K*		*K*	
Horizontal	1.750000		2.000000		2.250000		2.500000		2.750000		3.000000		3.250000		1/4 *n*	
Diagonal	2.015564		2.236068		2.462214		2.692582		2.926175		3.162278		3.400368		1/4 *N*	
Vertical	1.000000		1.000000		1.000000		1.000000		1.000000		1.000000		1.000000		1	

ANGLES (IN DEGREES) BETWEEN MEMBERS								
Between horizontal and diagonal members	30	27	24	22	20	18	17	$4/n = \tan a$
Between diagonal and vertical members	60	63	66	68	70	72	73	$90 - a$
Between vertical and hortizontal members	90	90	90	90	90	90	90	90

Table 7 Flat Pratt 6 Panels at Top & Bottom

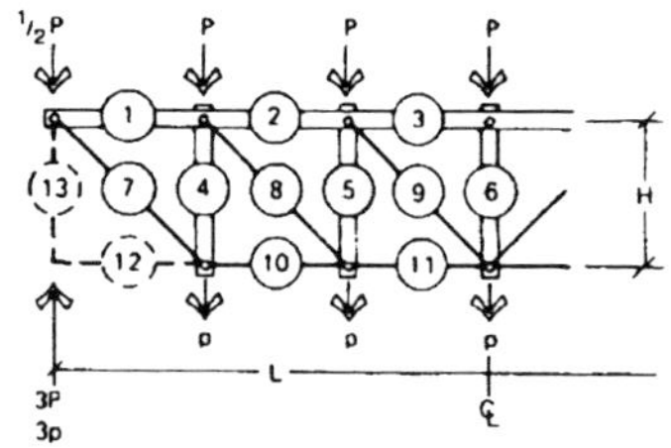

$n = \frac{L}{H}$

$N = \sqrt{n^2 + 36}$

Member	VALUES OF *n*														General Formulas	
	7		8		9		10		11		12		13			
	AXIAL FORCE COEFFICIENTS															
	K_1	K_2	K_1	K_2	K_1	K_2	K_1	K_2	K_1	K_2	K_1	K_2	K_1	K_2	K_1	K_2
1	-2.92	-2.92	-3.33	-3.33	-3.75	-3.75	-4.17	-4.17	-4.58	-4.58	-5.00	-5.00	-5.42	-5.42	- 5/12 *n*	- 5/12 *n*
2	-4.67	-4.67	-5.33	-5.33	-6.00	-6.00	-6.67	-6.67	-7.33	-7.33	-8.00	-8.00	-8.67	-8.67	- 8/12 *n*	- 8/12 *n*
3	-5.25	-5.25	-6.00	-6.00	-6.75	-6.75	-7.50	-7.50	-8.25	-8.25	-9.00	-9.00	-9.75	-9.75	- 9/12 *n*	- 9/12 *n*
4	-2.50	-1.50	-2.50	-1.50	-2.50	-1.50	-2.50	-1.50	-2.50	-1.50	-2.50	-1.50	-25.0	-1.50	- 5/2	- 3/2
5	-1.50	-0.50	-1.50	-0.50	-1.50	-0.50	-1.50	-0.50	-1.50	-0.50	-1.50	-0.50	-1.50	-0.50	- 3/2	- 1/2
6	-1.00	0	-1.00	0	-1.00	0	-1.00	0	-1.00	0	-1.00	0	-1.00	0	-1	0
7	3.84	3.84	4.17	4.17	4.51	4.51	4.86	4.86	5.22	5.22	5.59	5.59	5.97	5.97	5/12 *N*	5/12 *N*
8	2.30	2.30	2.50	2.50	2.70	2.70	2.92	2.92	3.13	3.13	3.35	3.35	3.58	3.58	3/12 *N*	3/12 *N*
9	0.77	0.77	0.83	0.83	0.90	0.90	0.97	0.97	1.04	1.04	1.12	1.12	1.19	1.19	1/12 *N*	1/12 *N*
10	2.92	2.92	3.33	3.33	3.75	3.75	4.17	4.17	4.58	4.58	5.00	5.00	5.42	5.42	5/12 *n*	5/12 *n*
11	4.67	4.67	5.33	5.33	6.00	6.00	6.67	6.67	7.33	7.33	8.00	8.00	8.67	8.67	8/12 *n*	8/12 *n*
12	0	0	0	0	0	0	0	0	0	0	0	0	0	0	0	0
13	-3.00	0	-3.00	0	-3.00	0	-3.00	0	-3.00	0	-3.00	0	-3.00	0	- 3	0

LENGTH COEFFICIENTS								
	K	K	K	K	K	K	K	K
Horizontal	1.166667	1.333333	1.500000	1.666667	1.833333	2.000000	2.166667	$1/6\ n$
Diagonal	1.536591	1.666667	1.802776	1.943651	2.088327	2.236068	2.366304	$1/6\ N$
Vertical	1.000000	1.000000	1.000000	1.000000	1.000000	1.000000	1.000000	1
ANGLES (IN DEGREES) BETWEEN MEMBERS								
Between horizontal and diagonal members	41	37	34	31	29	27	25	$6/n = \tan a$
Between diagonal and vertical members	49	53	56	59	61	63	65	$90 - a$
Between vertical and horizontal members	90	90	90	90	90	90	90	90

Table 8 Flat Pratt 8 Panels at Top & Bottom

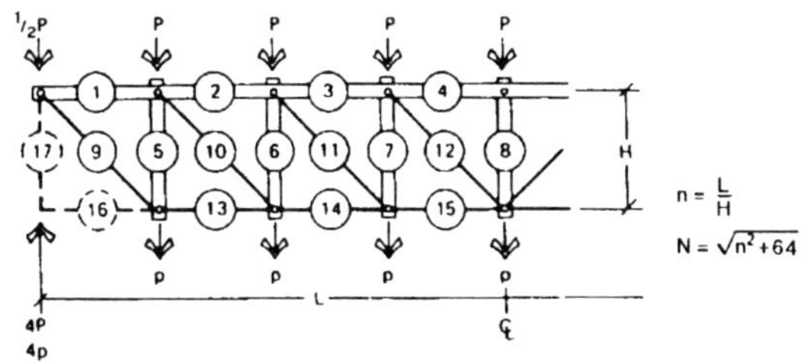

Member	VALUES OF *n*														General Formulas	
	7		8		9		10		11		12		13			
	AXIAL FORCE COEFFICIENTS															
	K_1	K_2	K_1	K_2	K_1	K_2	K_1	K_2	K_1	K_2	K_1	K_2	K_1	K_2	K_1	K_2
1	-3.06	-3.06	-3.50	-3.50	-3.94	-3.94	-4.38	-4.38	-4.81	-4.81	-5.25	-5.25	-5.69	-5.69	- 7/16 *n*	- 7/16 *n*
2	-5.25	-5.25	-6.00	-6.00	-6.75	-6.75	-7.50	-7.50	-8.25	-8.25	-9.00	-9.00	-9.75	-9.75	- 12/15 *n*	- 12/15 *n*
3	-6.56	-6.56	-7.50	-7.50	-8.44	-8.44	-9.38	-9.38	-10.31	-10.31	-11.25	-11.25	-12.19	-12.19	- 15/16 *n*	- 15/16 *n*
4	-7.00	-7.00	-8.00	-8.00	-9.00	-9.00	-10.00	-10.00	-11.00	-11.00	-12.00	-12.00	-13.00	-13.00	- *n*	- *n*
5	-3.50	-2.50	-3.50	-2.50	-3.50	-2.50	-3.50	-2.50	-3.50	-2.50	-3.50	-2.50	-3.50	-2.50	- 7/2	- 5/2
6	-2.50	-1.50	-2.50	-1.50	-2.50	-1.50	-2.50	-1.50	-2.50	-1.50	-2.50	-1.50	-2.50	-1.50	- 5/2	- 3/2
7	-1.50	-0.50	-1.50	-0.50	-1.50	-0.50	-1.50	-0.50	-1.50	-0.50	-1.50	-0.50	-1.50	-0.50	- 3/2	-1/2
8	-1.00	0	-1.00	0	-1.00	0	-1.00	0	-1.00	0	-1.00	0	-1.00	0	- 1	0
9	4.65	4.65	4.95	4.95	5.27	5.27	5.60	5.60	5.95	5.95	6.31	6.31	6.68	6.68	7/16 *N*	7/16 *N*
10	3.32	3.32	3.54	3.54	3.76	3.76	4.00	4.00	4.25	4.25	4.51	4.51	4.77	4.77	5/16 *N*	5/16 *N*
11	1.99	1.99	2.12	2.12	2.26	2.26	2.40	2.40	2.55	2.55	2.70	2.70	2.86	2.86	3/16 *N*	3/16 *N*
12	0.66	0.66	0.71	0.71	0.75	0.75	0.80	0.80	0.85	0.85	0.90	0.90	0.95	0.95	1/16 *N*	1/16 *N*

13	3.06	3.06	3.50	3.50	3.94	3.94	4.38	4.38	4.81	4.81	5.25	5.25	5.69	5.69	7/16 n	7/16 n
14	5.25	5.25	6.00	6.00	6.75	6.75	7.50	7.50	8.25	8.25	9.00	9.00	9.75	9.75	12/16 n	12/16 n
15	6.56	6.56	7.00	7.00	8.44	8.44	9.38	9.38	10.31	10.31	11.25	11.25	12.19	12.19	15/16 n	15/16 n
16	0	0	0	0	0	0	0	0	0	0	0	0	0	0	0	0
17	-4.00	0	-4.00	0	-4.00	0	-4.00	0	-4.00	0	-4.00	0	-4.00	0	- 4	0
LENGTH COEFFICIENTS																
	K		K		K		K		K		K		K		K	
Horizontal	0.875000		1.000000		1.125000		1.250000		1.375000		1.500000		1.625000		1/8 n	
Diagonal	1.328768		1.414214		1.505199		1.600781		1.700184		1.802776		1.908042		1/8 N	
Vertical	1.000000		1.000000		1.000000		1.000000		1.000000		1.000000		1.000000		1	
ANGLES (IN DEGREES) BETWEEN MEMBERS																
Between horizontal and diagonal members	49		45		42		39		36		34		32		$8/n = \tan a$	
Between diagonal and vertical members	41		45		48		51		54		56		58		$90 - a$	
Between vertical and horizontal members	90		90		90		90		90		90		90		90	

Table 9 Flat Pratt 10 Panels at Top & Bottom

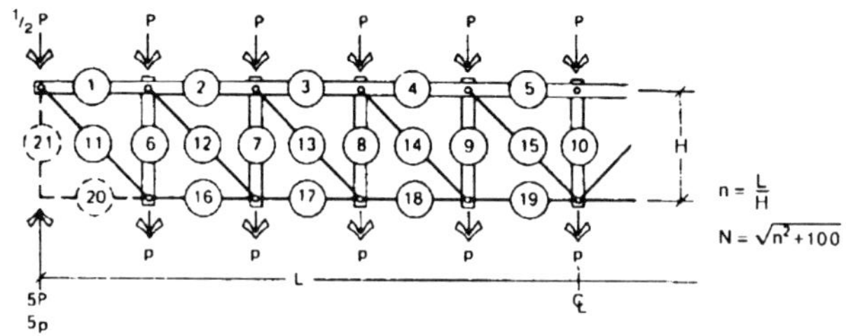

Member	VALUES OF n: 7		8		9		10		11		12		13		General Formulas	
	AXIAL FORCE COEFFICIENTS															
	K_1	K_2	K_1	K_2	K_1	K_2	K_1	K_2	K_1	K_2	K_1	K_2	K_1	K_2	K_1	K_2
1	-3.15	-3.15	-3.60	-3.60	-4.05	-4.05	-4.50	-4.50	-4.95	-4.95	-5.40	-5.40	-5.85	-5.85	$-9/20\,n$	$-9/20\,n$
2	-5.60	-5.60	-6.40	-6.40	-7.20	-7.20	-8.00	-8.00	-8.90	-8.90	-9.60	-9.60	-10.40	-10.40	$-16/20\,n$	$-16/20\,n$
3	-7.35	-7.35	-8.40	-8.40	-9.45	-9.45	-10.50	-10.50	-11.55	-11.55	-12.60	-12.60	-13.65	-13.65	$-21/20\,n$	$-21/20\,n$
4	-8.40	-8.40	-9.60	-9.60	-10.80	-10.80	-12.00	-12.00	-13.20	-13.20	-14.40	-14.40	-15.60	-15.60	$-24/20\,n$	$-24/20\,n$
5	-8.75	-8.75	-10.00	-10.00	-11.25	-11.25	-12.50	-12.50	-13.75	-13.75	-15.00	-15.00	-16.25	-16.25	$-25/20\,n$	$-25/20\,n$
6	-4.50	-3.50	-4.50	-3.50	-4.50	-3.50	-4.50	-3.50	-4.50	-3.50	-4.50	-3.50	-4.50	-3.50	$-9/2$	$-7/2$
7	-3.50	-2.50	-3.50	-2.50	-3.50	-2.50	-3.50	-2.50	-3.50	-2.50	-3.50	-2.50	-3.50	-2.50	$-7/2$	$-5/2$
8	-2.50	-1.50	-2.50	-1.50	-2.50	-1.50	-2.50	-1.50	-2.50	-1.50	-2.50	-1.50	-2.50	-1.50	$-5/2$	$-3/2$
9	-1.50	-0.50	-1.50	-0.50	-1.50	-0.50	-1.50	-0.50	-1.50	-0.50	-1.50	-0.50	-1.50	-0.50	$-3/2$	$-1/2$
10	-1.00	0	-1.00	0	-1.00	0	-1.00	0	-1.00	0	-1.00	0	-1.00	0	-1	0
11	5.49	5.49	5.76	5.76	6.05	6.05	6.36	6.36	6.69	6.69	7.03	7.03	7.38	7.38	$9/20\,N$	$9/20\,N$
12	4.27	4.27	4.48	4.48	4.71	4.71	4.95	4.95	5.20	5.20	5.47	5.47	5.74	5.74	$7/20\,N$	$7/20\,N$

13	3.05	3.05	3.20	3.20	3.36	3.36	3.54	3.54	3.72	3.72	3.91	3.91	4.10	4.10	5/20 *N*	5/20 *N*
14	1.83	1.83	1.92	1.92	2.02	2.02	2.12	2.12	2.23	2.23	2.34	2.34	2.46	2.46	3/20 *N*	3/20 *N*
15	0.61	0.61	0.64	0.64	0.67	0.67	0.71	0.71	0.74	0.74	0.78	0.78	0.82	0.82	1/20 *N*	1/20 *N*
16	3.15	3.15	3.60	3.60	4.05	4.05	4.50	4.50	4.95	4.95	5.40	5.40	5.85	5.85	9/20 *n*	9/20 *n*
17	5.60	5.60	6.40	6.40	7.20	7.20	8.00	8.00	8.80	8.80	9.60	9.60	10.40	10.40	16/20 *n*	16/20 *n*
18	7.35	7.35	8.40	8.40	9.45	9.45	10.50	10.50	11.55	11.55	12.60	12.60	13.65	13.65	21/20 *n*	21/20 *n*
19	8.40	8.40	9.60	9.60	10.80	10.80	12.00	12.00	13.20	13.20	14.40	14.40	15.60	15.60	24/20 *n*	24/20 *n*
20	0	0	0	0	0	0	0	0	0	0	0	0	0	0	0	0
21	-5.00	0	-5.00	0	-5.00	0	-5.00	0	-5.00	0	-5.00	0	-5.00	0	– 5	0
LENGTH COEFFICIENTS																
	K		*K*		*K*		*K*		*K*		*K*		*K*		*K*	
Horizontal	0.700000		0.800000		0.900000		1.000000		1.100000		1.200000		1.300000		1/10 *n*	
Diagonal	1.220656		1.280625		1.345362		1.414214		1.486607		1.562050		1.640122		1/10 *N*	
Vertical	1.000000		1.000000		1.000000		1.000000		1.000000		1.000000		1.000000		1	
ANGLES (IN DEGREES) BETWEEN MEMBERS																
Between horizontal and diagonal members	55		51		48		45		42		40		38		$10/n = \tan a$	
Between diagonal and vertical members	35		39		42		45		48		50		52		$90 - a$	
Between vertical and horizontal members	90		90		90		90		90		90		90		90	

Table 10 Flat Pratt 12 Panels at Top & Bottom

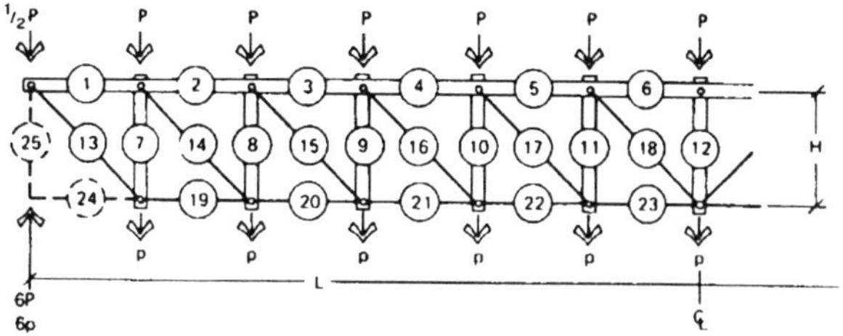

$n = \frac{L}{H}$

$N = \sqrt{n^2 + 144}$

Member	VALUES OF *n*														General Formulas	
	7		8		9		10		11		12		13			
AXIAL FORCE COEFFICIENTS																
	K_1	K_2	K_1	K_2	K_1	K_2	K_1	K_2	K_1	K_2	K_1	K_2	K_1	K_2	K_1	K_2
1	-3.21	-3.21	-3.67	-3.67	-4.13	-4.13	-4.58	-4.58	-5.04	-5.04	-5.50	-5.50	-5.96	-5.96	- 11/24 *n*	- 11/24 *n*
2	-5.83	-5.83	-6.67	-6.67	-7.50	-7.50	-8.33	-8.33	-9.17	-9.17	-10.00	-10.00	-10.83	-10.83	- 20/24 *n*	- 20/24 *n*
3	-7.88	-7.88	-9.00	-9.00	-10.13	-10.13	-11.25	-11.25	-12.38	-12.38	-13.50	-13.50	-14.63	-14.63	- 27/24 *n*	- 27/24 *n*
4	-9.33	-9.33	-10.67	-10.67	-12.00	-12.00	-13.33	-13.33	-14.67	-14.67	-16.00	-16.00	-17.33	-17.33	- 32/24 *n*	- 32/24 *n*
5	-10.21	-10.21	-11.67	-11.67	-13.13	-13.13	-14.58	-14.58	-16.04	-16.04	-17.50	-17.50	-18.96	-18.96	- 35/24 *n*	- 35/24 *n*
6	-10.50	-10.50	-12.00	-12.00	-13.50	-13.50	-15.00	-15.00	-16.50	-16.50	-18.00	-18.00	-19.50	-19.50	- 36/24 *n*	- 36/24 *n*
7	-5.50	-4.50	-5.50	-4.50	-5.50	-4.50	-5.50	-4.50	-5.50	-4.50	-5.50	-4.50	-5.50	-4.50	- 11/2	- 9/2
8	-4.50	-3.50	-4.50	-3.50	-4.50	-3.50	-4.50	-3.50	-4.50	-3.50	-4.50	-3.50	-4.50	-3.50	- 9/2	- 7/2
9	-3.50	-2.50	-3.50	-2.50	-3.50	-2.50	-3.50	-2.50	-3.50	-2.50	-3.50	-2.50	-3.50	-2.50	- 7/2	- 5/2
10	-2.50	-1.50	-2.50	-1.50	-2.50	-1.50	-2.50	-1.50	-2.50	-1.50	-2.50	-1.50	-2.50	-1.50	- 5/2	- 3/2
11	-1.50	-0.50	-1.50	-0.50	-1.50	-0.50	-1.50	-0.50	-1.50	-0.50	-1.50	-0.50	-1.50	-0.50	- 3/2	- 1/2
12	-1.00	0	-1.00	0	-1.00	0	-1.00	0	-1.00	0	-1.00	0	-1.00	0	- 1	0
13	6.37	6.37	6.61	6.61	6.88	6.88	7.16	7.16	7.46	7.46	7.78	7.78	8.11	8.11	11/24 *N*	11/24 *N*

14	5.21	5.21	5.41	5.41	5.63	5.63	5.86	5.86	6.10	6.10	6.36	6.36	6.63	6.63	9/24 N	9/24 N
15	4.05	4.05	4.21	4.21	4.38	4.38	4.56	4.56	4.75	4.75	4.95	4.95	5.16	5.16	7/24 N	7/24 N
16	2.89	2.89	3.00	3.00	3.13	3.13	3.25	3.25	3.39	3.39	3.54	3.54	3.69	3.69	5/24 N	5/24 N
17	1.74	1.74	1.80	1.80	1.88	1.88	1.95	1.95	2.03	2.03	2.12	2.12	2.21	2.21	3/24 N	3/24 N
18	0.58	0.58	0.60	0.60	0.63	0.63	0.65	0.65	0.68	0.68	0.71	0.71	0.74	0.74	1/24 N	1/24 N
19	3.21	3.21	3.67	3.67	4.13	4.13	4.58	4.58	5.04	5.04	5.50	5.50	5.96	5.96	11/24 n	11/24 n
20	5.83	5.83	6.67	6.67	7.50	7.50	8.33	8.33	9.17	9.17	10.00	10.00	10.83	10.83	20/24 n	20/24 n
21	7.88	7.88	9.00	9.00	10.13	10.13	11.25	11.25	12.38	12.38	13.50	13.50	14.63	14.63	27/24 n	27/24 n
22	9.33	9.33	10.67	10.67	12.00	12.00	13.33	13.33	14.67	14.67	16.00	16.00	17.33	17.33	32/24 n	32/24 n
23	10.21	10.21	11.67	11.67	13.13	13.13	14.58	14.58	16.04	16.04	17.50	17.50	18.96	18.96	35/24 n	35/24 n
24	0	0	0	0	0	0	0	0	0	0	0	0	0	0	0	0
25	-6.00	0	-6.00	0	-6.00	0	-6.00	0	-6.00	0	-6.00	0	-6.00	0	− 6	0
LENGTH COEFFICIENTS																
	K		K		K		K		K		K		K		K	
Horizontal	0.583333		0.666667		0.750000		0.833333		0.016667		1.000000		1.083333		1/12 n	
Diagonal	1.157704		1.201850		1.250000		1.301708		1.356562		1.414214		1.474317		1/12 N	
Vertical	1.000000		1.000000		1.000000		1.000000		1.000000		1.000000		1.000000		1	
ANGLES (IN DEGREES) BETWEEN MEMBERS																
Between horizontal and diagonal members	60		56		53		50		47		45		43		$12/n = \tan a$	
Between diagonal and vertical members	30		34		37		40		43		45		47		$90 - a$	
Between vertical and horizontal members	90		90		90		90		90		90		90		90	

Table 11 Flat Pratt 4 Panels at Top & Bottom

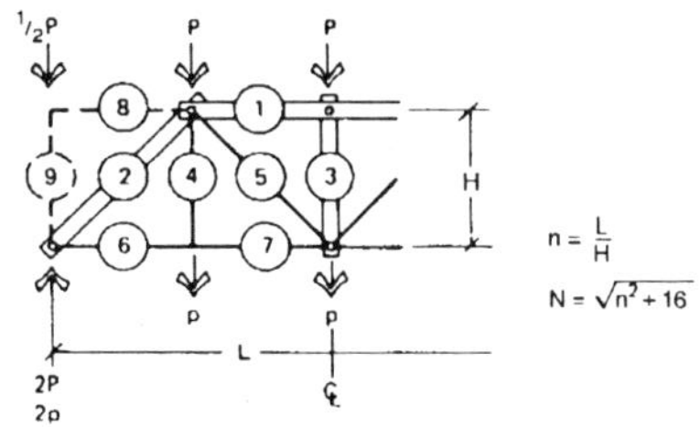

Member	VALUES OF *n*														General Formulas	
	7		8		9		10		11		12		13			
AXIAL FORCE COEFFICIENTS																
	K_1	K_2	K_1	K_2	K_1	K_2	K_1	K_2	K_1	K_2	K_1	K_2	K_1	K_2	K_1	K_2
1	-3.50	-3.50	-4.00	-4.00	-4.50	-4.50	-5.00	-5.00	-5.50	-5.50	-6.00	-6.00	-6.50	-6.50	$-1/2\,n$	$-1/2\,n$
2	-3.02	-3.02	-3.35	-3.35	-3.69	-3.69	-4.04	-4.04	-4.39	-4.39	-4.74	-4.74	-5.10	-5.10	$-3/8\,N$	$-3/8\,N$
3	-1.00	0	-1.00	0	-1.00	0	-1.00	0	-1.00	0	-1.00	0	-1.00	0	-1	0
4	0	1.00	0	1.00	0	1.00	0	1.00	0	1.00	0	1.00	0	1.00	0	1
5	1.01	1.01	1.12	1.12	1.23	1.23	1.35	1.35	1.46	1.46	1.58	1.58	1.70	1.70	$1/8\,N$	$1/8\,N$
6	2.63	2.63	3.00	3.00	3.38	3.38	3.75	3.75	4.13	4.13	4.50	4.50	4.88	4.88	$3/8\,n$	$3/8\,n$
7	2.63	2.63	3.00	3.00	3.38	3.38	3.75	3.75	4.13	4.13	4.50	4.50	4.88	4.88	$3/8\,n$	$3/8\,n$
8	0	0	0	0	0	0	0	0	0	0	0	0	0	0	0	0
9	-0.50	0	-0.50	0	-0.50	0	-0.50	0	-0.50	0	-0.50	0	-0.50	0	$-1/2$	0

LENGTH COEFFICIENTS								
	K	*K*	*K*	*K*	*K*	*K*	*K*	*K*
Horizontal	1.750000	2.000000	2.250000	2.500000	2.750000	3.000000	3.250000	$1/4\,n$
Diagonal	2.015564	2.236068	2.462214	2.692582	2.926175	3.162278	3.400368	$1/4\,N$
Vertical	1.000000	1.000000	1.000000	1.000000	1.000000	1.000000	1.000000	1
ANGLES (IN DEGREES) BETWEEN MEMBERS								
Between horizontal and diagonal members	30	27	24	22	20	18	17	$4/n = \tan a$
Between diagonal and vertical members	60	63	66	68	70	72	73	$90 - a$
Between vertical and horizontal members	90	90	90	90	90	90	90	90

Table 12 Modified Flat Pratt 6 Panels at Top & Bottom

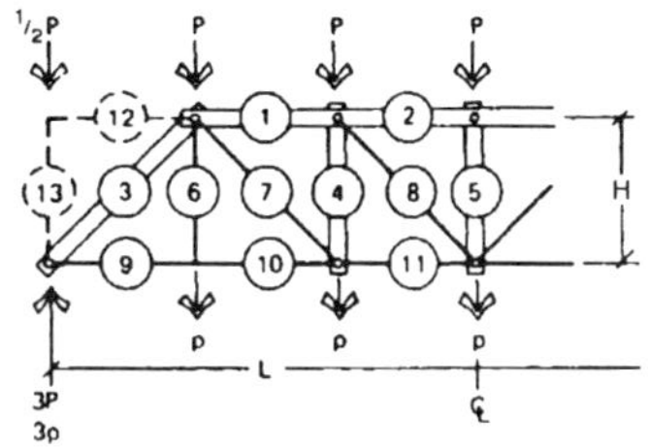

$n = \frac{L}{H}$

$N = \sqrt{n^2 + 36}$

Member	Values of *n*: 7		8		9		10		11		12		13		General Formulas	
AXIAL FORCE COEFFICIENTS																
	K_1	K_2	K_1	K_2	K_1	K_2	K_1	K_2	K_1	K_2	K_1	K_2	K_1	K_2	K_1	K_2
1	-4.67	-4.67	-5.33	-5.33	-6.00	-6.00	-6.67	-6.67	-7.33	-7.33	-8.00	-8.00	-8.67	-8.67	$-2/3\,n$	$-2/3\,n$
2	-5.25	-5.25	-6.00	-6.00	-6.75	-6.75	-7.50	-7.50	-8.25	-8.25	-9.00	-9.00	-9.75	-9.75	$-3/4\,n$	$-3/4\,n$
3	-3.84	-3.84	-4.17	-4.17	-4.51	-4.51	-4.86	-4.86	-5.22	-5.22	-5.59	-5.59	-5.97	-5.97	$-5/12\,N$	$-5/12\,N$
4	-1.50	-0.50	-1.50	-0.50	-1.50	-0.50	-1.50	-0.50	-1.50	-0.50	-1.50	-0.50	-1.50	-0.50	$-3/2$	$-1/2$
5	-1.00	0	-1.00	0	-1.00	0	-1.00	0	-1.00	0	-1.00	0	-1.00	0	-1	0
6	0	1.00	0	1.00	0	1.00	0	1.00	0	1.00	0	1.00	0	1.00	0	1
7	2.30	2.30	2.50	2.50	2.70	2.70	2.92	2.92	3.13	3.13	3.35	3.35	3.58	3.58	$1/4\,N$	$1/4\,N$
8	0.77	0.77	0.83	0.83	0.90	0.90	0.97	0.97	1.04	1.04	1.12	1.12	1.19	1.19	$1/12\,N$	$1/12\,N$
9	2.92	2.92	3.33	3.33	3.75	3.75	4.17	4.17	4.58	4.58	5.00	5.00	5.42	5.42	$5/12\,n$	$5/12\,n$
10	2.92	2.92	3.33	3.33	3.75	3.75	4.17	4.17	4.58	4.58	5.00	5.00	5.42	5.42	$5/12\,n$	$5/12\,n$
11	4.67	4.67	5.33	5.33	6.00	6.00	6.67	6.67	7.33	7.33	8.00	8.00	8.67	8.67	$2/3\,n$	$2/3\,n$
12	0	0	0	0	0	0	0	0	0	0	0	0	0	0	0	0
13	-0.50	0	-0.50	0	-0.50	0	-0.50	0	-0.50	0	-0.50	0	-0.50	0	$-1/2$	0

LENGTH COEFFICIENTS								
	K	K	K	K	K	K	K	K
Horizontal	1.166667	1.333333	1.500000	1.666667	1.833333	2.000000	2.166667	$1/6\,n$
Diagonal	1.536591	1.666667	1.802776	1.943651	2.088327	2.236068	2.386304	$1/6\,N$
Vertical	1.000000	1.000000	1.000000	1.000000	1.000000	1.000000	1.000000	1
ANGLES (IN DEGREES) BETWEEN MEMBERS								
Between horizontal and diagonal members	41	37	34	31	29	27	25	$6/n = \tan a$
Between diagonal and vertical members	49	53	56	59	61	63	65	$90 - a$
Between vertical and horizontal members	90	90	90	90	90	90	90	90

Table 13 Modified Flat Pratt 8 Panels at Top & Bottom

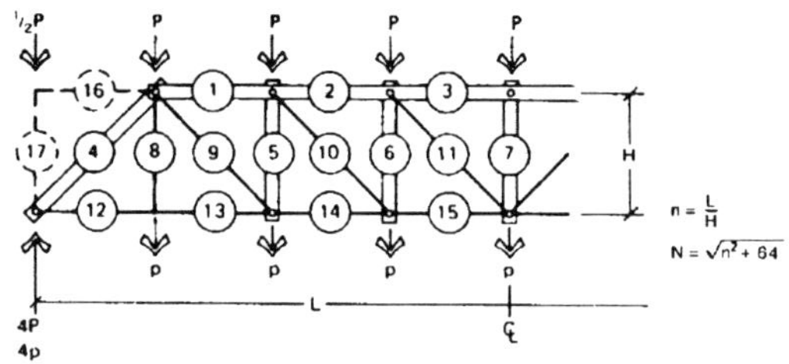

Member	VALUES OF *n*														General Formulas	
	7		8		9		10		11		12		13			
	AXIAL FORCE COEFFICIENTS															
	K_1	K_2	K_1	K_2	K_1	K_2	K_1	K_2	K_1	K_2	K_1	K_2	K_1	K_2	K_1	K_2
1	-5.25	-5.25	-6.00	-6.00	-6.75	-6.75	-7.50	-7.50	-8.25	-8.25	-9.00	-9.00	-9.75	-9.75	$-3/4\,n$	$-3/4\,n$
2	-6.56	-6.56	-7.50	-7.50	-8.44	-8.44	-9.38	-9.38	-10.31	-10.31	-11.25	-11.25	-12.19	-12.19	$-15/16\,n$	$-15/16\,n$
3	-7.00	-7.00	-8.00	-8.00	-9.00	-9.00	-10.00	-10.00	-11.00	-11.00	-12.00	-12.00	-13.00	-13.00	$-n$	$-n$
4	-4.65	-4.65	-4.95	-4.95	-5.27	-5.27	-5.60	-5.60	-5.95	-5.95	-6.31	-6.31	-6.68	-6.68	$-7/16\,N$	$-7/16\,N$
5	-2.50	-1.50	-2.50	-1.50	-2.50	-1.50	-2.50	-1.50	-2.50	-1.50	-2.50	-1.50	-2.50	-1.50	$-5/2$	$-3/2$
6	-1.50	-0.50	-1.50	-0.50	-1.50	-0.50	-1.50	-0.50	-1.50	-0.50	-1.50	-0.50	-1.50	-0.50	$-3/2$	$-1/2$
7	-1.00	0	-1.00	0	-1.00	0	-1.00	0	-1.00	0	-1.00	0	-1.00	0	-1	0
8	0	1.00	0	1.00	0	1.00	0	1.00	0	1.00	0	1.00	0	1.00	0	1
9	3.32	3.32	3.54	3.54	3.76	3.76	4.00	4.00	4.25	4.25	4.51	4.51	4.77	4.77	$5/16\,N$	$5/16\,N$
10	1.99	1.99	2.12	2.12	2.26	2.26	2.40	2.40	2.55	2.55	2.70	2.70	2.86	2.86	$3/16\,N$	$3/16\,N$
11	0.66	0.66	0.71	0.71	0.75	0.75	0.80	0.80	0.85	0.85	0.90	0.90	0.95	0.95	$1/16\,N$	$1/16\,N$
12	3.06	3.06	3.50	3.50	3.94	3.94	4.38	4.38	4.81	4.81	5.25	5.25	5.69	5.69	$7/16\,n$	$7/16\,n$

13	3.06	3.06	3.50	3.50	3.94	3.94	4.38	4.38	4.81	4.81	5.25	5.25	5.69	5.69	7/16 n	7/16 n
14	5.25	5.25	6.00	6.00	6.75	6.75	7.50	7.50	8.25	8.25	9.00	9.00	9.75	9.75	3/4 n	3/4 n
15	6.56	6.56	7.50	7.50	8.44	8.44	9.38	9.38	10.31	10.31	11.25	11.25	12.19	12.19	15/16 n	15/16 n
16	0	0	0	0	0	0	0	0	0	0	0	0	0	0	0	0
17	-0.50	0	-0.50	0	-0.50	0	-0.50	0	-0.50	0	-0.50	0	-0.50	0	- 1/2	0
LENGTH COEFFICIENTS																
	K		K		K		K		K		K		K		K	
Horizontal	0.875000		1.000000		1.125000		1.250000		1.375000		1.500000		1.625000		1/8 n	
Diagonal	1.328768		1.414214		1.505199		1.600781		1.700184		1.802776		1.908042		1/8 N	
Vertical	1.000000		1.000000		1.000000		1.000000		1.000000		1.000000		1.000000		1	
ANGLES (IN DEGREES) BETWEEN MEMBERS																
Between horizontal and diagonal members	49		45		42		39		36		34		32		$8/n = \tan a$	
Between diagonal and vertical members	41		45		48		51		54		56		58		$90 - a$	
Between vertical and horizontal members	90		90		90		90		90		90		90		90	

Table 14 Modified Flat Pratt 10 Panels at Top & Bottom

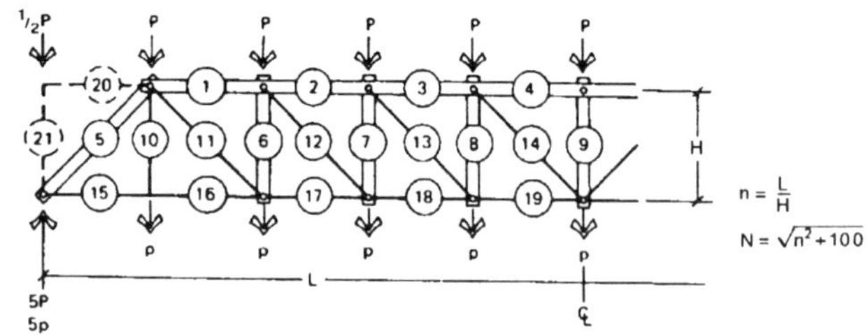

Member	VALUES OF n														General Formulas	
	7		8		9		10		11		12		13			
AXIAL FORCE COEFFICIENTS																
	K_1	K_2	K_1	K_2	K_1	K_2	K_1	K_2	K_1	K_2	K_1	K_2	K_1	K_2	K_1	K_2
1	-5.60	-5.60	-6.40	-6.40	-7.20	-7.20	-8.00	-8.00	-8.80	-8.80	-9.60	-9.60	-10.40	-10.40	$-4/5\ n$	$-4/5\ n$
2	-7.35	-7.35	-8.40	-8.40	-9.45	-9.45	-10.50	-10.50	-11.55	-11.55	-12.60	-12.60	-13.65	-13.65	$-21/20\ n$	$-21/20\ n$
3	-8.40	-8.40	-9.60	-9.60	-10.80	-10.80	-12.00	-12.00	-13.20	-13.20	-14.40	-14.40	-15.60	-15.60	$-6/5\ n$	$-6/5\ n$
4	-8.75	-8.75	-10.00	-10.00	-11.25	-11.25	-12.50	-12.50	-13.75	-13.75	-15.00	-15.00	-16.25	-16.25	$-5/4\ n$	$-5/4\ n$
5	-5.49	-5.49	-5.76	-5.76	-6.05	-6.05	-6.36	-6.36	-6.69	-6.69	-7.03	-7.03	-7.38	-7.38	$-9/20\ N$	$-9/20\ N$
6	-3.50	-2.50	-3.50	-2.50	-3.50	-2.50	-3.50	-2.50	-3.50	-2.50	-3.50	-2.50	-3.50	-2.50	$-7/2$	$-5/2$
7	-2.50	-1.50	-2.50	-1.50	-2.50	-1.50	-2.50	-1.50	-2.50	-1.50	-2.50	-1.50	-2.50	-1.50	$-5/2$	$-3/2$
8	-1.50	-0.50	-1.50	-0.50	-1.50	-0.50	-1.50	-0.50	-1.50	-0.50	-1.50	-0.50	-1.50	-0.50	$-3/2$	$-1/2$
9	-1.00	0	-1.00	0	-1.00	0	-1.00	0	-1.00	0	-1.00	0	-1.00	0	-1	0
10	0	1.00	0	1.00	0	1.00	0	1.00	0	1.00	0	1.00	0	1.00	0	1
11	4.27	4.27	4.48	4.48	4.71	4.71	4.95	4.95	5.20	5.20	5.47	5.47	5.74	5.74	$7/20\ N$	$7/20\ N$
12	3.05	3.05	3.20	3.20	3.36	3.36	3.54	3.54	3.72	3.72	3.91	3.91	4.10	4.10	$1/4\ N$	$1/4\ N$

13	1.83	1.83	1.92	1.92	2.02	2.02	2.12	2.12	2.23	2.23	2.34	2.34	2.46	2.46	3/20 N	3/20 N
14	0.61	0.61	0.64	0.64	0.67	0.67	0.71	0.71	0.74	0.74	0.78	0.78	0.82	0.82	1/20 N	1/20 N
15	3.15	3.15	3.60	3.60	4.05	4.05	4.50	4.50	4.95	4.95	5.40	5.40	5.85	5.85	9/20 n	9/20 n
16	3.15	3.15	3.60	3.60	4.05	4.05	4.50	4.50	4.95	4.95	5.40	5.40	5.85	5.85	9/20 n	9/20 n
17	5.60	5.60	6.40	6.40	7.20	7.20	8.00	8.00	8.80	8.80	9.60	9.60	10.40	10.40	4/5 n	4/5 n
18	7.35	7.35	8.40	8.40	9.45	9.45	10.50	10.50	11.55	11.55	12.60	12.60	13.65	13.65	21/20 n	21/20 n
19	8.40	8.40	9.60	9.60	10.80	10.80	12.00	12.00	13.20	13.20	14.40	14.40	15.60	15.60	6/5 n	6/5 n
20	0	0	0	0	0	0	0	0	0	0	0	0	0	0	0	0
21	-0.50	0	-0.50	0	-0.50	0	-0.50	0	-0.50	0	-0.50	0	-0.50	0	– 1/2	0
LENGTH COEFFICIENTS																
	K		K		K		K		K		K		K		K	
Horizontal	0.700000		0.800000		0.900000		1.000000		1.100000		1.200000		1.300000		1/10 n	
Diagonal	1.220656		1.280625		1.345362		1.414214		1.486607		1.562050		1.640122		1/10 N	
Vertical	1.000000		1.000000		1.000000		1.000000		1.000000		1.000000		1.000000		1	
ANGLES (IN DEGREES) BETWEEN MEMBERS																
Between horizontal and diagonal members	55		51		48		45		42		40		38		$10/n = \tan a$	
Between diagonal and vertical members	35		39		42		45		48		50		52		$90 - a$	
Between vertical and horizontal members	90		90		90		90		90		90		90		90	

Table 15 Modified Flat Pratt 12 Panels at Top & Bottom

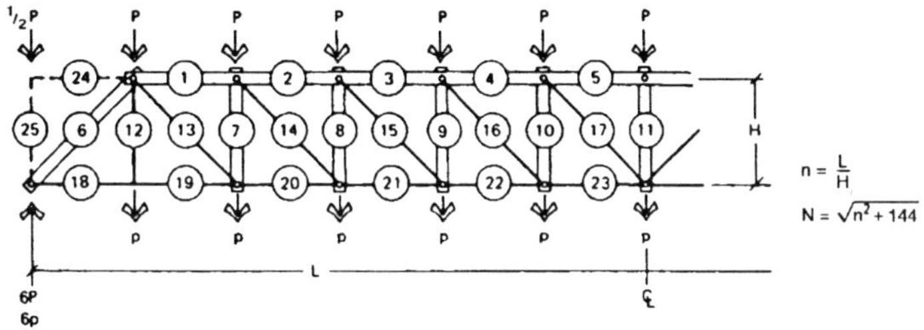

Member	VALUES OF *n*														General Formulas	
	7		8		9		10		11		12		13			
AXIAL FORCE COEFFICIENTS																
	K_1	K_2	K_1	K_2	K_1	K_2	K_1	K_2	K_1	K_2	K_1	K_2	K_1	K_2	K_1	K_2
1	-5.83	-5.83	-6.67	-6.67	-7.50	-7.50	-8.33	-8.33	-9.17	-9.17	-10.00	-10.00	-10.83	-10.83	$-20/24\,n$	$-20/24\,n$
2	-7.88	-7.88	-9.00	-9.00	-10.13	-10.13	-11.25	-11.25	-12.38	-12.38	-13.50	-13.50	-14.63	-14.63	$-27/24\,n$	$-27/24\,n$
3	-9.33	-9.33	-10.67	-10.67	-12.00	-12.00	-13.33	-13.33	-14.67	-14.67	-16.00	-16.00	-17.33	-17.33	$-32/24\,n$	$-32/24\,n$
4	-10.21	-10.21	-11.67	-11.67	-13.13	-13.13	-14.58	-14.58	-16.04	-16.04	-17.50	-17.50	-18.96	-18.96	$-35/24\,n$	$-35/24\,n$
5	-10.50	-10.50	-12.00	-12.00	-13.50	-13.50	-15.00	-15.00	-16.50	-16.50	-18.00	-18.00	-19.50	-19.50	$-36/24\,n$	$-36/24\,n$
6	-6.37	-6.37	-6.61	-6.61	-6.88	-6.88	-7.16	-7.16	-7.46	-7.46	-7.78	-7.78	-8.11	-8.11	$-11/24\,N$	$-11/24\,N$
7	-4.50	-3.50	-4.50	-3.50	-4.50	-3.50	-4.50	-3.50	-4.50	-3.50	-4.50	-3.50	-4.50	-3.50	$-9/2$	$-7/2$
8	-3.50	-2.50	-3.50	-2.50	-3.50	-2.50	-3.50	-2.50	-3.50	-2.50	-3.50	-2.50	-3.50	-2.50	$-7/2$	$-5/2$
9	-2.50	-1.50	-2.50	-1.50	-2.50	-1.50	-2.50	-1.50	-2.50	-1.50	-2.50	-1.50	-2.50	-1.50	$-5/2$	$-3/2$
10	-1.50	-0.50	-1.50	-0.50	-1.50	-0.50	-1.50	-0.50	-1.50	-0.50	-1.50	-0.50	-1.50	-0.50	$-3/2$	$-1/2$
11	-1.00	0	-1.00	0	-1.00	0	-1.00	0	-1.00	0	-1.00	0	-1.00	0	-1	0
12	0	1.00	0	1.00	0	1.00	0	1.00	0	1.00	0	1.00	0	1.00	0	1

13	5.21	5.21	5.41	5.41	5.63	5.63	5.86	5.86	6.10	6.10	6.36	6.36	6.63	6.63	9/24 *N*	9/24 *N*
14	4.05	4.05	4.21	4.21	4.38	4.38	4.56	4.56	4.75	4.75	4.95	4.95	5.16	5.16	7/24 *N*	7/24 *N*
15	2.89	2.89	3.00	3.00	3.13	3.13	3.25	3.25	3.39	3.39	3.54	3.54	3.69	3.69	5/24 *N*	5/24 *N*
16	1.74	1.74	1.80	1.80	1.88	1.88	1.95	1.95	2.03	2.03	2.12	2.12	2.21	2.21	3/24 *N*	3/24 *N*
17	0.58	0.58	0.60	0.60	0.63	0.63	0.65	0.65	0.68	0.68	0.71	0.71	0.74	0.74	1/24 *N*	1/24 *N*
18	3.21	3.21	3.67	3.67	4.13	4.13	4.58	4.58	5.04	5.04	5.50	5.50	5.96	5.96	11/24 *n*	11/24 *n*
19	3.21	3.21	3.67	3.67	4.13	4.13	4.58	4.58	5.04	5.04	5.50	5.50	5.96	5.96	11/24 *n*	11/24 *n*
20	5.83	5.83	6.67	6.67	7.50	7.50	8.33	8.33	9.17	9.17	10.00	10.00	10.83	10.83	20/24 *n*	20/24 *n*
21	7.88	7.88	9.00	9.00	10.13	10.13	11.25	11.25	12.38	12.38	13.50	13.50	14.63	14.63	27/24 *n*	27/24 *n*
22	9.33	9.33	10.67	10.67	12.00	12.00	13.33	13.33	14.67	14.67	16.00	16.00	17.33	17.33	32/24 *n*	32/24 *n*
23	10.21	10.21	11.67	11.67	13.13	13.13	14.58	14.58	16.04	16.04	17.50	17.50	18.96	18.96	35/24 *n*	35/24 *n*
24	0	0	0	0	0	0	0	0	0	0	0	0	0	0	0	0
25	-0.50	0	-0.50	0	-0.50	0	-0.50	0	-0.50	0	-0.50	0	-0.50	0	- 1/2	0
LENGTH COEFFICIENTS																
	K		*K*		*K*		*K*		*K*		*K*		*K*		*K*	
Horizontal	0.583333		0.666667		0.750000		0.833333		0.916667		1.000000		1.083333		1/12 n	
Diagonal	1.157704		1.201850		1.250000		1.301708		1.356568		1.414214		1.474317		1/12 *N*	
Vertical	1.000000		1.000000		1.000000		1.000000		1.000000		1.000000		1.000000		1	
ANGLES (IN DEGREES) BETWEEN MEMBERS																
Between horizontal and diagonal members	60		56		53		50		47		45		43		$12/n = \tan a$	
Between diagonal and vertical members	30		34		37		40		43		45		47		$90 - a$	
Between vertical and horizontal members	90		90		90		90		90		90		90		90	

Table 16 Flat Warren 2 Panels at Bottom

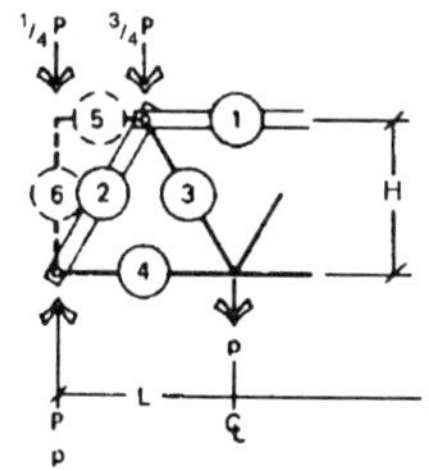

$n = \frac{L}{H}$

$N = \sqrt{n^2 + 16}$

Member	VALUES OF *n*														General Formulas	
	7		8		9		10		11		12		13			
AXIAL FORCE COEFFICIENTS																
	K_1	K_2	K_1	K_2	K_1	K_2	K_1	K_2	K_1	K_2	K_1	K_2	K_1	K_2	K_1	K_2
1	-1.31	-1.75	-1.50	-2.00	-1.69	-2.25	-1.88	-2.50	-2.05	-2.75	-2.25	-3.00	-2.44	-3.25	- 3/16 *n*	- 1/4 *n*
2	-1.51	-1.01	-1.68	-1.12	-1.85	-1.23	-2.02	-1.35	-2.19	-1.46	-2.37	-1.58	-2.55	-1.70	- 3/16 *N*	- 1/8 *N*
3	0	1.01	0	1.12	0	1.23	0	1.35	0	1.46	0	1.58	0	1.70	0	1/8 *N*
4	1.31	0.88	1.50	1.00	1.69	1.13	1.88	1.25	2.06	1.38	2.25	1.50	2.44	1.63	3/16 *n*	1/8 *n*
5	0	0	0	0	0	0	0	0	0	0	0	0	0	0	0	0
6	-0.25	0	-0.25	0	-0.25	0	-0.25	0	-0.25	0	-0.25	0	-0.25	0	- 1/4	0

LENGTH COEFFICIENTS								
	K	K	K	K	K	K	K	K
1, 4	3.500000	4.000000	4.500000	5.000000	5.500000	6.000000	6.500000	$1/2\,n$
2, 3	2.015564	2.236068	2.462214	2.692582	2.926175	3.162278	3.400368	$1/4\,N$
5	1.750000	2.000000	2.250000	2.500000	2.750000	3.000000	3.250000	$1/4\,n$
6	1.000000	1.000000	1.000000	1.000000	1.000000	1.000000	1.000000	1
ANGLES (IN DEGREES) BETWEEN MEMBERS								
Between horizontal and diagonal members	30	27	24	22	20	18	17	$4/n = \tan a$
Between diagonal members	120	126	132	136	140	144	146	$180 - 2a$
Between vertical and diagonal members	60	63	66	68	70	72	73	$90 - a$
Between vertical and horizontal members	90	90	90	90	90	90	90	90

Table 17 Flat Warren 3 Panels at Bottom

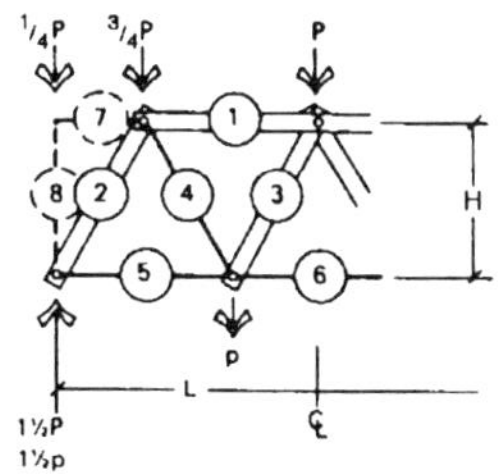

$n = \frac{L}{H}$

$N = \sqrt{n^2 + 36}$

Member	VALUES OF *n*														General Formulas	
	7		8		9		10		11		12		13			
	AXIAL FORCE COEFFICIENTS															
	K_1	K_2	K_1	K_2	K_1	K_2	K_1	K_2	K_1	K_2	K_1	K_2	K_1	K_2	K_1	K_2
1	-2.04	-2.33	-2.33	-2.67	-2.63	-3.00	-2.92	-3.33	-3.21	-3.67	-3.50	-4.00	-3.79	-4.33	$-7/24\,n$	$-1/3\,n$
2	-1.92	-1.54	-2.08	-1.67	-2.25	-1.80	-2.43	-1.94	-2.61	-2.09	-2.80	-2.24	-2.98	-2.39	$-5/24\,N$	$-1/6\,N$
3	-0.77	0	-0.83	0	-0.90	0	-0.97	0	-1.04	0	-1.12	0	-1.19	0	$-1/12\,N$	0
4	0.77	1.54	0.83	1.67	0.90	1.80	0.97	1.94	1.04	2.09	1.12	2.24	1.19	2.39	$1/12\,N$	$1/6\,N$
5	1.46	1.17	1.67	1.33	1.88	1.50	2.08	1.67	2.29	1.83	2.50	2.00	2.71	2.17	$5/24\,n$	$1/6\,n$
6	2.63	2.33	3.00	2.67	3.38	3.00	3.75	3.33	4.13	3.67	4.50	4.00	4.88	4.33	$3/8\,n$	$1/3\,n$
7	0	0	0	0	0	0	0	0	0	0	0	0	0	0	0	0
8	-0.25	0	-0.25	0	-0.25	0	-0.25	0	-0.25	0	-0.25	0	-0.25	0	$-1/4$	0

LENGTH COEFFICIENTS								
	K	K	K	K	K	K	K	K
1, 5, 6	2.333333	2.666667	3.000000	3.333333	3.666667	4.000000	4.333333	$1/3\,n$
2, 3, 4	1.536591	1.666667	1.802776	1.943651	2.088327	2.236068	2.386304	$1/6\,N$
7	1.166667	1.333333	1.500000	1.666667	1.833333	2.000000	2.166667	$1/6\,n$
8	1.000000	1.000000	1.000000	1.000000	1.000000	1.000000	1.000000	1
ANGLES (IN DEGREES) BETWEEN MEMBERS								
Between horizontal and diagonal members	41	37	34	31	29	27	25	$6/n = \tan a$
Between diagonal members	98	106	112	118	122	126	130	$180 - 2a$
Between vertical and diagonal members	49	53	56	59	61	63	65	$90 - a$
Between vertical and horizontal members	90	90	90	90	90	90	90	90

Table 18 Flat Warren 4 Panels at Bottom

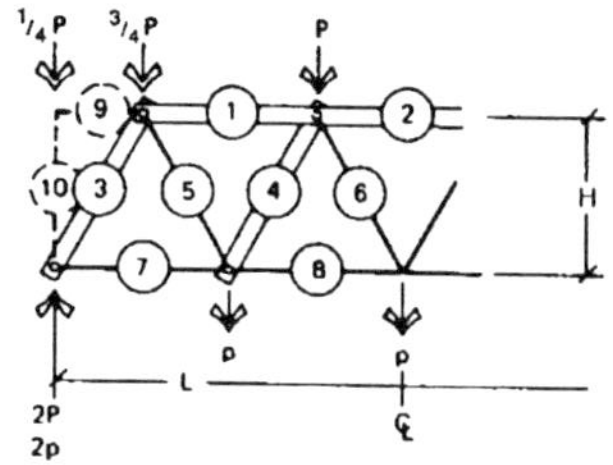

Member	VALUES OF *n*														General Formulas	
	7		8		9		10		11		12		13			
	AXIAL FORCE COEFFICIENTS															
	K_1	K_2	K_1	K_2	K_1	K_2	K_1	K_2	K_1	K_2	K_1	K_2	K_1	K_2	K_1	K_2
1	-2.41	-2.63	-2.75	-3.00	-3.09	-3.38	-3.44	-3.75	-3.78	-4.13	-4.13	-4.50	-4.47	-4.88	$-11/32\,n$	$-3/8\,n$
2	-3.28	-3.50	-3.75	-4.00	-4.22	-4.50	-4.69	-5.00	-5.16	-5.50	-5.68	-6.00	-6.09	-6.50	$-15/32\,n$	$-1/2\,n$
3	-2.33	-1.99	-2.47	-2.12	-2.63	-2.26	-2.80	-2.40	-2.98	-2.55	-3.15	-2.70	-3.34	-2.86	$-7/32\,N$	$-3/16\,N$
4	-1.33	-0.66	-1.41	-0.71	-1.51	-0.75	-1.60	-0.80	-1.70	-0.85	-1.80	-0.90	-1.91	-0.95	$-1/8\,N$	$-1/16\,N$
5	1.33	1.99	1.41	2.12	1.51	2.26	1.60	2.40	1.70	2.55	1.80	2.70	1.91	2.86	$1/8\,N$	$3/16\,N$
6	0	0.66	0	0.71	0	0.75	0	0.80	0	0.85	0	0.90	0	0.95	0	$1/16\,N$
7	1.53	1.31	1.75	1.50	1.97	1.69	2.19	1.88	2.41	2.06	2.63	2.25	2.84	2.44	$7/32\,n$	$3/16\,n$
8	3.28	3.06	3.75	3.50	4.22	3.94	4.69	4.38	5.16	4.81	5.63	5.25	6.09	5.69	$15/32\,n$	$7/16\,n$
9	0	0	0	0	0	0	0	0	0	0	0	0	0	0	0	0
10	-0.25	0	-0.25	0	-0.25	0	-0.25	0	-0.25	0	-0.25	0	-0.25	0	$-1/4$	0

LENGTH COEFFICIENTS								
	K	K	K	K	K	K	K	K
1, 2, 7, 8	1.750000	2.000000	2.250000	2.500000	2.275000	3.000000	3.250000	1/4 n
3, 4, 5, 6	1.328768	1.414214	1.505199	1.600781	1.700184	1.802776	1.908042	1/8 N
9	0.875000	1.000000	1.125000	1.250000	1.375000	1.500000	1.625000	1/8 n
10	1.000000	1.000000	1.000000	1.000000	1.000000	1.000000	1.000000	1
ANGLES (IN DEGREES) BETWEEN MEMBERS								
Between horizontal and diagonal members	49	45	42	39	36	34	32	$8/n = \tan a$
Between diagonal members	82	90	96	102	106	112	116	$180 - 2a$
Between vertical and diagonal members	41	45	48	51	54	56	58	$90 - a$
Between vertical and horizontal members	90	90	90	90	90	90	90	90

Table 19 Flat Warren 5 Panels at Bottom

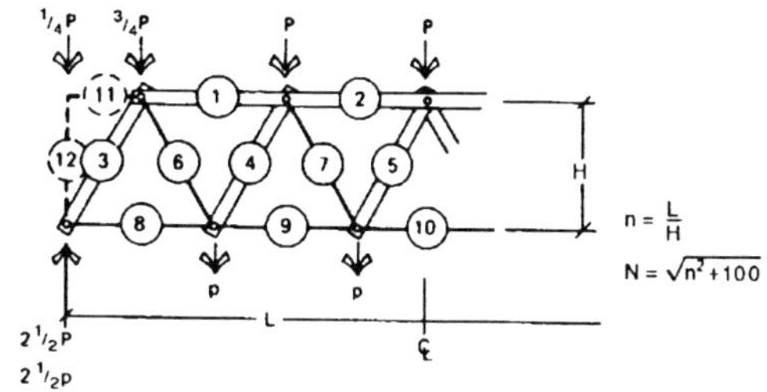

Member	VALUES OF n														General Formulas	
	7		8		9		10		11		12		13			
	AXIAL FORCE COEFFICIENTS															
	K_1	K_2	K_1	K_2	K_1	K_2	K_1	K_2	K_1	K_2	K_1	K_2	K_1	K_2	K_1	K_2
1	-2.63	-2.80	-3.00	-3.20	-3.38	-3.60	-3.75	-4.00	-4.13	-4.40	-4.50	-4.80	-4.88	-5.20	- 15/40 n	- 16/40 n
2	-4.03	-4.20	-4.60	-4.80	-5.18	-5.40	-5.75	-6.00	-6.33	-6.60	-6.90	-7.20	-7.48	-7.80	- 23/40 n	- 24/40 n
3	-2.75	-2.44	-2.88	-2.56	-3.03	-2.69	-3.18	-2.83	-3.35	-2.97	-3.52	-3.12	-3.69	-3.28	- 9/40 N	- 8/40 N
4	-1.83	-1.22	-1.92	-1.28	-2.02	-1.35	-2.12	-1.41	-2.23	-1.49	-2.34	-1.56	-2.46	-1.64	- 6/40 N	- 4/40 N
5	-0.61	0	-0.64	0	-0.67	0	-0.71	0	-0.74	0	-0.78	0	-0.82	0	- 2/40 N	0
6	1.83	2.44	1.92	2.56	2.02	2.69	2.12	2.83	2.23	2.97	2.34	3.12	2.46	3.28	6/40 N	8/40 N
7	0.61	1.22	0.64	1.28	0.67	1.35	0.71	1.41	0.74	1.49	0.78	1.56	0.82	1.64	2/40 N	4/40 N
8	1.58	1.40	1.80	1.60	2.03	1.80	2.25	2.00	2.48	2.20	2.70	2.40	2.93	2.60	9/40 n	8/40 n
9	3.68	3.50	4.20	4.00	4.73	4.50	5.25	5.00	5.78	5.50	6.30	6.00	6.83	6.50	21/40 n	20/40 n
10	4.38	4.20	5.00	4.80	5.63	5.40	6.25	6.00	6.88	6.60	7.50	7.20	8.13	7.80	25/40 n	24/40 n
11	0	0	0	0	0	0	0	0	0	0	0	0	0	0	0	0
12	-0.25	0	-0.25	0	-0.25	0	-0.25	0	-0.25	0	-0.25	0	-0.25	0	- 1/4	0

LENGTH COEFFICIENTS								
	K	K	K	K	K	K	K	K
1, 2, 8, 9, 10	1.400000	1.600000	1.800000	2.000000	2.200000	2.400000	2.600000	$1/5\,n$
3, 4, 5, 6, 7	1.220656	1.280625	1.345362	1.414214	1.486607	1.562050	1.640122	$1/10\,N$
11	0.700000	0.800000	0.900000	1.000000	1.100000	1.200000	1.300000	$1/10\,n$
12	1.000000	1.000000	1.000000	1.000000	1.000000	1.000000	1.000000	1
ANGLES (IN DEGREES) BETWEEN MEMBERS								
Between horizontal and diagonal members	55	51	48	45	42	40	38	$10/n = \tan a$
Between diagonal members	70	78	84	90	96	100	104	$180 - 2a$
Between vertical and diagonal members	35	39	42	45	48	50	52	$90 - a$
Between vertical and horizontal members	90	90	90	90	90	90	90	90

Table 20 Flat Warren 6 Panels at Bottom

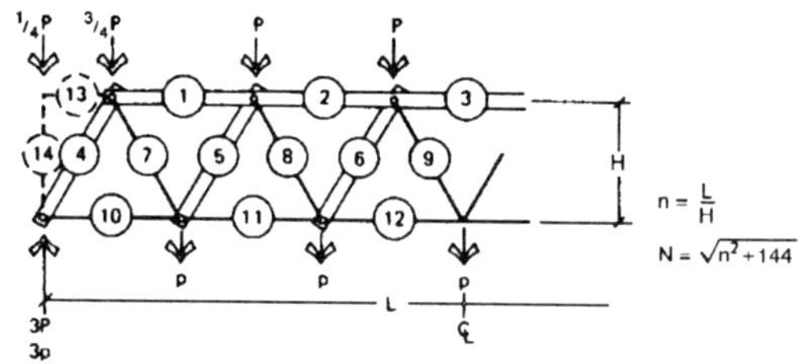

Member	VALUES OF *n*														General Formulas	
	7		8		9		10		11		12		13			
	AXIAL FORCE COEFFICIENTS															
	K_1	K_2	K_1	K_2	K_1	K_2	K_1	K_2	K_1	K_2	K_1	K_2	K_1	K_2	K_1	K_2
1	-2.77	-2.92	-3.17	-3.33	-3.56	-3.75	-3.96	-4.17	-4.35	-4.58	-4.75	-5.00	-5.15	-5.42	- 19/48 *n*	- 20/48 *n*
2	-4.52	-4.67	-5.17	-5.33	-5.81	-6.00	-6.46	-6.67	-7.10	-7.33	-7.75	-8.00	-8.40	-8.67	- 31/48 *n*	- 32/48 *n*
3	-5.10	-5.25	-5.83	-6.00	-6.56	-6.75	-7.29	-7.50	-8.02	-8.25	-8.75	-9.00	-9.48	-9.75	- 35/48 *n*	- 36/48 *n*
4	-3.18	-2.89	-3.31	-3.00	-3.44	-3.13	-3.58	-3.25	-3.73	-3.39	-3.89	-3.54	-4.05	-3.69	- 11/48 *N*	- 10/48 *N*
5	-2.32	-1.74	-2.40	-1.80	-2.50	-1.88	-2.60	-1.95	-2.71	-2.03	-2.83	-2.12	-2.95	-2.21	- 8/48 *N*	- 6/48 *N*
6	-1.16	-0.58	-1.20	-0.60	-1.25	-0.63	-1.30	-0.65	-1.36	-0.68	-1.41	-0.71	-1.47	-0.74	- 4/48 *N*	- 2/48 *N*
7	2.32	2.89	2.40	3.00	2.50	3.13	2.60	3.25	2.71	3.39	2.83	3.54	2.95	3.69	8/48 *N*	10/48 *N*
8	1.16	1.74	1.20	1.80	1.25	1.88	1.30	1.95	1.36	2.03	1.41	2.12	1.47	2.21	4/48 *N*	6/48 *N*
9	0	0.58	0	0.60	0	0.63	0	0.65	0	0.68	0	0.71	0	0.74	0	2/48 *N*
10	1.60	1.46	1.83	1.67	2.06	1.87	2.29	2.08	2.52	2.29	2.75	2.50	2.98	2.71	11/48 *n*	10/48 *n*
11	3.94	3.79	4.50	4.33	5.06	4.87	5.62	5.42	6.19	5.96	6.75	6.50	7.31	7.04	27/48 *n*	26/48 *n*

12	5.10	4.96	5.83	5.67	6.56	6.37	7.29	7.08	8.02	7.79	8.75	8.50	9.48	9.21	35/48 *n*	34/48 *n*
13	0	0	0	0	0	0	0	0	0	0	0	0	0	0	0	0
14	-0.25	0	-0.25	0	-0.25	0	-0.25	0	-0.25	0	-0.25	0	-0.25	0	− 1/4	0
LENGTH COEFFICIENTS																
	K		*K*		*K*		*K*		*K*		*K*		*K*		*K*	
1, 2, 3, 10, 11, 12	1.166667		1.333333		1.500000		1.666667		1.833333		2.000000		2.166667		1/6 *n*	
4, 5, 6, 7, 8, 9	1.157704		1.201850		1.250000		1.301708		1.356568		1.414214		1.474317		1/12 *N*	
13	0.583333		0.666667		0.750000		0.833333		0.916667		1.000000		1.083333		1/12 *n*	
14	1.000000		1.000000		1.000000		1.000000		1.000000		1.000000		1.000000		1	
ANGLES (IN DEGREES) BETWEEN MEMBERS																
Between horizontal and diagonal members	60		56		53		50		47		45		43		$12/n = \tan a$	
Between diagonal members	60		68		74		80		86		90		94		$180 - 2a$	
Between vertical and diagonal members	30		34		37		40		43		45		47		$90 - a$	
Between vertical and horizontal members	90		90		90		90		90		90		90		90	

Table 21 Flat Warren 7 Panels at Bottom

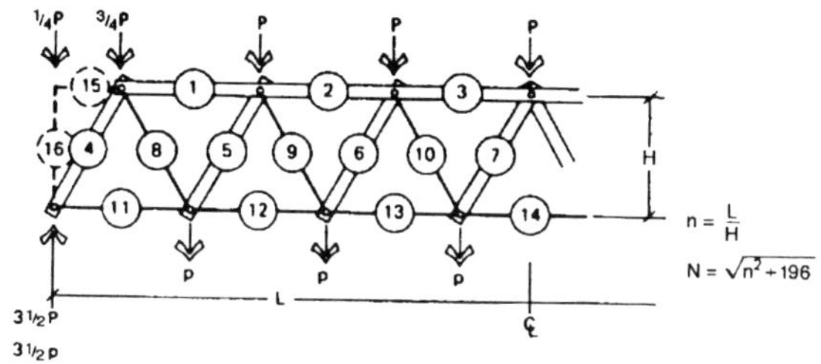

Member	VALUES OF n														General Formulas	
	7		8		9		10		11		12		13			
	AXIAL FORCE COEFFICIENTS															
	K_1	K_2	K_1	K_2	K_1	K_2	K_1	K_2	K_1	K_2	K_1	K_2	K_1	K_2	K_1	K_2
1	-2.88	-3.00	-3.29	-3.43	-3.70	-3.86	-4.11	-4.29	-4.52	-4.71	-4.93	-5.14	-5.34	-5.57	$-23/56\,n$	$-24/56\,n$
2	-4.88	-5.00	-5.57	-5.71	-6.27	-6.43	-6.96	-7.14	-7.66	-7.86	-8.36	-8.57	-9.05	-9.29	$-39/56\,n$	$-40/56\,n$
3	-5.88	-6.00	-6.71	-6.86	-7.55	-7.71	-8.39	-8.57	-9.23	-9.43	-10.07	-10.29	-10.91	-11.14	$-47/56\,n$	$-48/56\,n$
4	-3.63	-3.35	-3.74	-3.46	-3.86	-3.57	-3.99	-3.69	-4.13	-3.82	-4.28	-3.95	-4.44	-4.09	$-13/56\,N$	$-12/56\,N$
5	-2.80	-2.24	-2.88	-2.30	-2.97	-2.38	-3.07	-2.46	-3.18	-2.54	-3.29	-2.63	-3.41	-2.73	$-10/56\,N$	$-8/56\,N$
6	-1.68	-1.12	-1.73	-1.15	-1.78	-1.19	-1.84	-1.23	-1.91	-1.27	-1.98	-1.32	-2.05	-1.36	$-6/56\,N$	$-4/56\,N$
7	-0.56	0	-0.58	0	-0.59	0	-0.61	0	-0.64	0	-0.66	0	-0.68	0	$-2/56\,N$	0
8	2.80	3.35	2.88	3.46	2.97	3.57	3.07	3.69	3.18	3.82	3.29	3.95	3.41	4.09	$10/56\,N$	$12/56\,N$
9	1.68	2.24	1.73	2.30	1.78	2.38	1.84	2.46	1.91	2.54	1.98	2.63	2.05	2.73	$6/56\,N$	$8/56\,N$
10	0.56	1.12	0.58	1.15	0.59	1.19	0.61	1.23	0.64	1.27	0.66	1.32	0.68	1.36	$2/56\,N$	$4/56\,N$
11	1.63	1.50	1.86	1.71	2.09	1.93	2.32	2.14	2.55	2.36	2.79	2.57	3.02	2.79	$13/56\,n$	$12/56\,n$
12	4.13	4.00	4.71	4.57	5.30	5.14	5.89	5.71	6.48	6.29	7.07	6.86	7.66	7.43	$33/56\,n$	$32/56\,n$

13	5.63	5.50	6.43	6.29	7.23	7.07	8.04	7.86	8.84	8.64	9.64	9.43	10.45	10.21	45/56 n	44/56 n
14	6.13	6.00	7.00	6.86	7.88	7.71	8.75	8.57	9.63	9.43	10.50	10.29	11.38	11.14	49/56 n	48/56 n
15	0	0	0	0	0	0	0	0	0	0	0	0	0	0	0	0
16	−0.25	0	−0.25	0	−0.25	0	−0.25	0	−0.25	0	−0.25	0	−0.25	0	− 1/4	0
LENGTH COEFFICIENTS																
	K		K		K		K		K		K		K		K	
1, 2, 3, 11, 12, 13, 14	1.000000		1.142857		1.285714		1.428571		1.571429		1.714286		1.857143		1/7 n	
4, 5, 6, 7, 8, 9, 10	1.118034		1.151751		1.188808		1.228904		1.271750		1.317078		1.364641		1/14 N	
15	0.500000		0.571429		0.642857		0.714286		0.785714		0.857143		0.928571		1/14 n	
16	1.000000		1.000000		1.000000		1.000000		1.000000		1.000000		1.000000		1	
ANGLES (IN DEGREES) BETWEEN MEMBERS																
Between horizontal and diagonal members	63		60		57		54		52		49		47		$14/n = \tan a$	
Between diagonal members	54		60		66		72		76		82		86		$180 - 2a$	
Between diagonal and vertical members	27		30		33		36		38		41		43		$90 - a$	
Between vertical and horizontal members	90		90		90		90		90		90		90		90	

Table 22 Flat Warren 8 Panels at Bottom

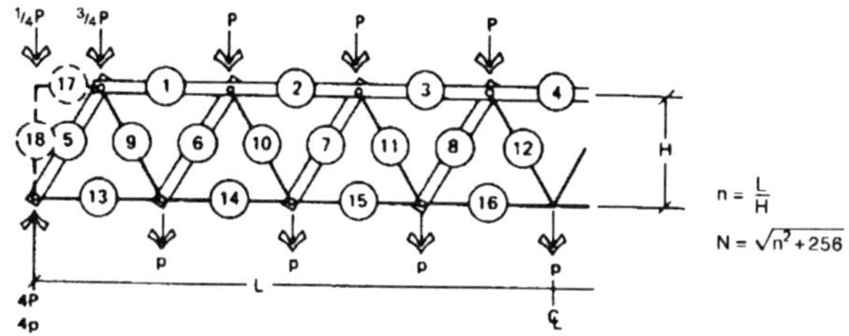

Member	VALUES OF n: 7		8		9		10		11		12		13		General Formulas	
	AXIAL FORCE COEFFICIENTS															
	K_1	K_2	K_1	K_2	K_1	K_2	K_1	K_2	K_1	K_2	K_1	K_2	K_1	K_2	K_1	K_2
1	-2.95	-3.06	-3.38	-3.50	-3.80	-3.94	-4.22	-4.38	-4.64	-4.81	-5.06	-5.25	-5.48	-5.69	- 27/64 n	- 28/64 n
2	-5.14	-5.25	-5.88	-6.00	-6.61	-6.75	-7.34	-7.50	-8.08	-8.25	-8.81	-9.00	-9.55	-9.75	- 47/64 n	- 48/64 n
3	-6.45	-6.56	-7.38	-7.50	-8.30	-8.44	-9.22	-9.38	-10.14	-10.31	-11.06	-11.25	-11.98	-12.19	- 59/64 n	- 60/64 n
4	-6.89	-7.00	-7.88	-8.00	-8.86	-9.00	-9.84	-10.00	-10.83	-11.00	-11.81	-12.00	-12.80	-13.00	- 63/64 n	- 64/64 n
5	-4.09	-3.82	-4.19	-3.91	-4.30	-4.02	-4.42	-4.13	-4.55	-4.25	-4.69	-4.38	-4.83	-4.51	- 15/64 N	- 14/64 N
6	-3.27	-2.73	-3.35	-2.80	-3.44	-2.87	-3.54	-2.95	-3.64	-3.03	-3.75	-3.13	-3.87	-3.22	- 12/64 N	- 10/64 N
7	-2.18	-1.64	-2.24	-1.68	-2.29	-1.72	-2.36	-1.77	-2.43	-1.82	-2.50	-1.88	-2.58	-1.93	- 8/64 N	- 6/64 N
8	-1.09	-0.55	-1.12	-0.56	-1.15	-0.57	-1.18	-0.59	-1.21	-0.61	-1.25	-0.63	-1.29	-0.64	- 4/64 N	- 2/64 N
9	3.27	3.82	3.35	3.91	3.44	4.02	3.54	4.13	3.64	4.25	3.75	4.38	3.87	4.51	12/64 N	14/64 N
10	2.18	2.73	2.24	2.80	2.29	2.87	2.36	2.95	2.43	3.03	2.50	3.13	2.58	3.22	8/64 N	10/64 N
11	1.09	1.64	1.12	1.68	1.15	1.72	1.18	1.77	1.21	1.82	1.25	1.88	1.29	1.93	4/64 N	6/64 N
12	0	0.55	0	0.56	0	0.57	0	0.59	0	0.61	0	0.63	0	0.64	0	2/64 N
13	1.64	1.53	1.88	1.75	2.11	1.97	2.34	2.19	2.58	2.41	2.81	2.63	3.05	2.84	15/64 n	14/64 n

14	4.27	4.16	4.88	4.75	5.48	5.34	6.09	5.94	6.70	6.53	7.31	7.13	7.92	7.72	39/64 n	38/64 n
15	6.02	5.91	6.88	6.75	7.73	7.59	8.59	8.44	9.45	9.28	10.31	10.13	11.17	10.97	55/64 n	54/64 n
16	6.89	6.78	7.88	7.75	8.86	8.72	9.84	9.69	10.83	10.66	11.81	11.63	12.80	12.59	63/64 n	62/64 n
17	0	0	0	0	0	0	0	0	0	0	0	0	0	0	0	0
18	−0.25	0	−0.25	0	−0.25	0	−0.25	0	−0.25	0	−0.25	0	−0.25	0	− 1/4	0

LENGTH COEFFICIENTS

	K	K	K	K	K	K	K	K
1, 2, 3, 4, 13, 14, 15, 16	0.875000	1.000000	1.125000	1.250000	1.375000	1.500000	1.625000	1/8 n
5, 6, 7, 8, 9, 10, 11, 12	1.091516	1.118034	1.147347	1.179248	1.213530	1.250000	1.288471	1/16 N
17	0.437500	0.500000	0.562500	0.625000	0.687500	0.750000	0.812500	1/16 n
18	1.000000	1.000000	1.000000	1.000000	1.000000	1.000000	1.000000	1

ANGLES (IN DEGREES) BETWEEN MEMBERS

Between horizontal and diagonal members	66	63	61	58	55	53	51	$16/n = \tan a$
Between diagonal members	48	54	58	64	70	74	78	$180 - 2a$
Between diagonal and vertical members	24	27	29	32	35	37	39	$90 - a$
Between vertical and horizontal members	90	90	90	90	90	90	90	90

Table 23 Flat Warren 9 Panels at Bottom

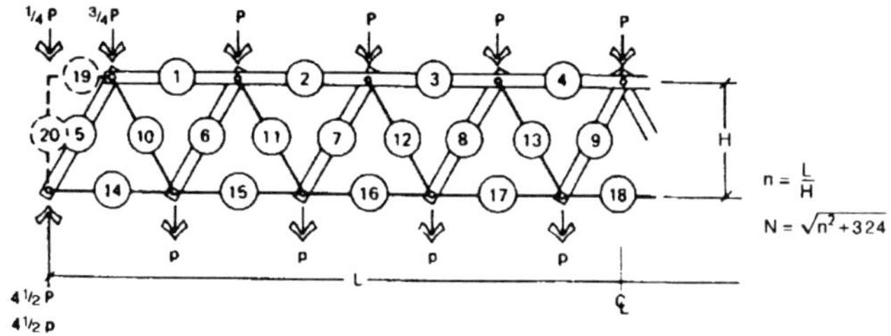

Member	VALUES OF n														General Formulas	
	7		8		9		10		11		12		13			
	AXIAL FORCE COEFFICIENTS															
	K_1	K_2	K_1	K_2	K_1	K_2	K_1	K_2	K_1	K_2	K_1	K_2	K_1	K_2	K_1	K_2
1	-3.01	-3.11	-3.44	-3.56	-3.88	-4.00	-4.31	-4.44	-4.74	-4.89	-5.17	-5.33	-5.60	-5.78	$-31/72\,n$	$-32/72\,n$
2	-5.35	-5.44	-6.11	-6.22	-6.88	-7.00	-7.64	-7.78	-8.40	-8.56	-9.17	-9.33	-9.93	-10.11	$-55/72\,n$	$-56/72\,n$
3	-6.90	-7.00	-7.89	-8.00	-8.88	-9.00	-9.86	-10.00	-10.85	-11.00	-11.83	-12.00	-12.82	-13.00	$-71/72\,n$	$-72/72\,n$
4	-7.68	-7.78	-8.78	-8.89	-9.88	-10.00	-10.97	-11.11	-12.07	-12.22	-13.17	-13.33	-14.26	-14.44	$-79/72\,n$	$-80/72\,n$
5	-4.56	-4.29	-4.65	-4.38	-4.75	-4.47	-4.86	-4.58	-4.98	-4.69	-5.11	-4.81	-5.24	-4.93	$-17/72\,N$	$-16/72\,N$
6	-3.76	-3.22	-3.83	-3.28	-3.91	-3.35	-4.00	-3.43	-4.10	-3.52	-4.21	-3.61	-4.32	-3.70	$-14/72\,N$	$-12/72\,N$
7	-2.68	-2.15	-2.74	-2.19	-2.80	-2.24	-2.86	-2.29	-2.93	-2.34	-3.00	-2.40	-3.08	-2.47	$-10/72\,N$	$-8/72\,N$
8	-1.61	-1.07	-1.64	-1.09	-1.68	-1.12	-1.72	-1.14	-1.76	-1.17	-1.80	-1.20	-1.85	-1.23	$-6/72\,N$	$-4/72\,N$
9	-0.54	0	-0.55	0	-0.56	0	-0.57	0	-0.59	0	-0.60	0	-0.62	0	$-2/72\,N$	0
10	3.76	4.29	3.83	4.38	3.91	4.47	4.00	4.58	4.10	4.69	4.21	4.81	4.32	4.93	$14/72\,N$	$16/72\,N$
11	2.68	3.22	2.74	3.28	2.80	3.35	2.86	3.43	2.93	3.52	3.00	3.61	3.08	3.70	$10/72\,N$	$12/72\,N$
12	1.61	2.15	1.64	2.19	1.68	2.24	1.72	2.29	1.76	2.34	1.80	2.40	1.85	2.47	$6/72\,N$	$8/72\,N$

13	0.54	1.07	0.55	1.09	0.56	1.12	0.57	1.14	0.59	1.17	0.60	1.20	0.62	1.23	2/72 *N*	4/72 *N*
14	1.65	1.56	1.89	1.78	2.13	2.00	2.36	2.22	2.60	2.44	2.83	2.67	3.07	2.89	17/72 *n*	16/72 *n*
15	4.38	4.28	5.00	4.89	5.63	5.50	6.25	6.11	6.88	6.72	7.50	7.33	8.13	7.94	45/72 *n*	44/72 *n*
16	6.32	6.22	7.22	7.11	8.13	8.00	9.03	8.89	9.93	9.78	10.83	10.67	11.74	11.56	65/72 *n*	64/72 *n*
17	7.49	7.39	8.56	8.44	9.63	9.50	10.69	10.56	11.76	11.61	12.83	12.67	13.90	13.72	77/72 *n*	76/72 *n*
18	7.88	7.78	9.00	8.89	10.13	10.00	11.25	11.11	12.38	12.22	13.50	13.33	14.63	14.44	81/72 *n*	80/72 *n*
19	0	0	0	0	0	0	0	0	0	0	0	0	0	0	0	0
20	−0.25	0	−0.25	0	−0.25	0	−0.25	0	−0.25	0	−0.25	0	−0.25	0	− 1/4	0

LENGTH COEFFICIENTS

	K	*K*	*K*	*K*	*K*	*K*	*K*	*K*
1, 2, 3, 4, 14, 15, 16, 17, 18	0.777778	0.888889	1.000000	1.111111	1.222222	1.333333	1.444444	1/9 *n*
5, 6, 7, 8, 9, 10, 11, 12, 13	1.072956	1.094318	1.118034	1.143959	1.171946	1.201850	1.233534	1/18 *N*
19	0.388889	0.444444	0.500000	0.555556	0.611111	0.666667	0.722222	1/18 *n*
20	1.000000	1.000000	1.000000	1.000000	1.000000	1.000000	1.000000	1

ANGLES (IN DEGREES) BETWEEN MEMBERS

Between horizontal and diagonal members	69	66	63	61	59	56	54	$18/n = \tan a$
Between diagonal members	42	48	54	58	62	68	72	$180 - 2a$
Between diagonal and vertical members	21	24	27	29	31	34	36	$90 - a$
Between vertical and horizontal members	90	90	90	90	90	90	90	90

Table 24 Flat Warren 10 Panels at Bottom

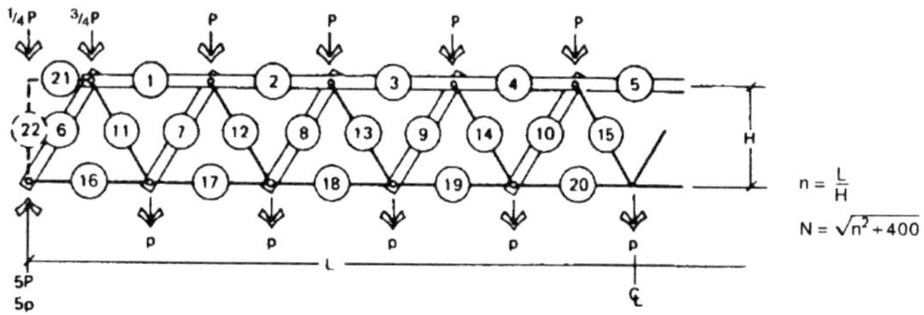

Member	VALUES OF *n*														General Formulas	
	7		8		9		10		11		12		13			
	AXIAL FORCE COEFFICIENTS															
	K_1	K_2	K_1	K_2	K_1	K_2	K_1	K_2	K_1	K_2	K_1	K_2	K_1	K_2	K_1	K_2
1	-3.06	-3.15	-3.50	-3.60	-3.94	-4.05	-4.38	-4.50	-4.81	-4.95	-5.25	-5.40	-5.69	-5.85	- 35/80 *n*	- 36/80 *n*
2	-5.51	-5.60	-6.30	-6.40	-7.09	-7.20	-7.88	-8.00	-8.66	-8.80	-9.45	-9.60	-10.24	-10.40	- 63/80 *n*	- 64/80 *n*
3	-7.26	-7.35	-8.30	-8.40	-9.34	-9.45	-10.38	-10.50	-11.41	-11.55	-12.45	-12.60	-13.49	-13.65	- 83/80 *n*	- 84/80 *n*
4	-8.31	-8.40	-9.50	-9.60	-10.69	-10.80	-11.88	-12.00	-13.06	-13.20	-14.25	-14.40	-15.44	-15.60	- 95/80 *n*	- 96/80 *n*
5	-8.66	-8.75	-9.90	-10.00	-11.14	-11.25	-12.38	-12.50	-13.61	-13.75	-14.85	-15.00	-16.09	-16.25	- 99/80 *n*	-100/80 *n*
6	-5.03	-4.77	-5.12	-4.85	-5.21	-4.93	-5.31	-5.03	-5.42	-5.14	-5.54	-5.25	-5.67	-5.37	- 19/80 *N*	- 18/80 *N*
7	-4.24	-3.71	-4.31	-3.77	-4.39	-3.84	-4.47	-3.91	-4.57	-3.99	-4.66	-4.08	-4.77	-4.17	- 16/80 *N*	- 14/80 *N*
8	-3.18	-2.65	-3.23	-2.69	-3.29	-2.74	-3.35	-2.80	-3.42	-2.85	-3.50	-2.92	-3.58	-2.98	- 12/80 *N*	- 10/80 *N*
9	-2.12	-1.59	-2.15	-1.62	-2.19	-1.64	-2.24	-1.68	-2.28	-1.71	-2.33	-1.75	-2.39	-1.79	- 8/80 *N*	- 6/80 *N*
10	-1.06	-0.53	-1.08	-0.54	-1.10	-0.55	-1.12	-0.56	-1.14	-0.57	-1.17	-0.58	-1.19	-0.60	- 4/80 *N*	- 2/80 *N*
11	4.24	4.77	4.31	4.85	4.39	4.93	4.47	5.03	4.57	5.14	4.66	5.25	4.77	5.37	16/80 *N*	18/80 *N*
12	3.18	3.71	3.23	3.77	3.29	3.84	3.35	3.91	3.42	3.99	3.50	4.08	3.58	4.17	12/80 *N*	14/80 *N*
13	2.12	2.65	2.15	2.69	2.19	2.74	2.24	2.80	2.28	2.85	2.33	2.92	2.39	2.98	8/80 *N*	10/80 *N*
14	1.06	1.59	1.08	1.62	1.10	1.64	1.12	1.68	1.14	1.71	1.17	1.75	1.19	1.79	4/80 *N*	6/80 *N*

15	0	0.53	0	0.54	0	0.55	0	0.56	0	0.57	0	0.58	0	0.60	0	2/80 *N*
16	1.66	1.58	1.90	1.80	2.14	2.03	2.38	2.25	2.61	2.48	2.85	2.70	3.09	2.93	19/80 *n*	18/80 *n*
17	4.46	4.38	5.10	5.00	5.74	5.63	6.38	6.25	7.01	6.88	7.65	7.50	8.29	8.13	51/80 *n*	50/80 *n*
18	6.56	6.48	7.50	7.40	8.44	8.33	9.38	9.25	10.31	10.18	11.25	11.10	12.19	12.03	75/80 *n*	74/80 *n*
19	7.96	7.88	9.10	9.00	10.24	10.13	11.38	11.25	12.51	12.38	13.65	13.50	14.79	14.63	91/80 *n*	90/80 *n*
20	8.66	8.58	9.90	9.80	11.14	11.03	12.38	12.25	13.61	13.48	14.85	14.70	16.09	15.93	99/80 *n*	98/80 *n*
21	0	0	0	0	0	0	0	0	0	0	0	0	0	0	0	0
22	−0.25	0	−0.25	0	−0.25	0	−0.25	0	−0.25	0	−0.25	0	−0.25	0	− 1/4	0
LENGTH COEFFICIENTS																
	K		*K*		*K*		*K*		*K*		*K*		*K*		*K*	
1, 2, 3, 4, 5, 16, 17, 18, 19, 20	0.700000		0.800000		0.900000		1.000000		1.100000		1.200000		1.300000		1/10 *n*	
6, 7, 8, 9, 10, 11, 12, 13, 14, 15	1.059481		1.077033		1.096586		1.118034		1.141271		1.166190		1.192686		1/20 *N*	
21	0.350000		0.400000		0.450000		0.500000		0.550000		0.600000		0.650000		1/20 *n*	
22	1.000000		1.000000		1.000000		1.000000		1.000000		1.000000		1.000000		1	
ANGLES (IN DEGREES) BETWEEN MEMBERS																
Between horizontal and diagonal members	71		68		66		63		61		59		57		$20/n = \tan a$	
Between diagonal members	38		44		48		54		58		62		66		$180 - 2a$	
Between diagonal and vertical members	19		22		24		27		29		31		33		$90 - a$	
Between vertical and horizontal members	90		90		90		90		90		90		90		90	

APPENDIX 13: BRIDGE INSPECTION DOCUMENTS

In the step-by-step procedure discussed in the text, the gathering of all data is an essential part of the rehabilitation process. Bridges through the course of their usable life are periodically inspected by governing authorities, most of which are the departments of transportation of various states. The general procedures and documents for inspections employed by the North Carolina Department of Transportation are typical of other states. Examples are reproduced here by permission:

- North Carolina Department of Transportation, Division of Highways Bridge Maintenance Unit—Structure Data File
- Bridge Inspection Record And Summary (Supplement Record for Trusses)
- Bridge Inspection Record And Summary (R.C. Box Culverts)
- Bridge Inspection Record And Summary (Pipe and Arch Culverts)
- Field Inspection Report Bridge Inspection and Analysis
- Field Sketch for Vertical and Horizontal Clearances
- Stream Bed Soundings
- Regulatory and Warning Sign Notice—Bridge Maintenance Unit
- Prompt Action Notice—Bridge Inspection and Analysis
- Bridge Inspector's Recommendations for Maintenance Repairs

Design teams engaged in the preliminary design phase of rehabilitation projects should be familiar with these documents. They describe in practical terms the condition of the structure, furnishing therefore the basic data necessary for evaluating the existing damage and eventually formulating a preliminary assessment of the procedure to be proposed. By knowing the existence of such documents, the design team is better able to formulate questions for department of transportation of the state involved and thus able to expedite the design process.

NORTH CAROLINA
DEPARTMENT OF TRANSPORTATION
DIVISION OF HIGHWAYS
BUILDING MAINTENANCE UNIT

STRUCTURE DATA FILE

COUNTY __________ BRIDGE NO ______

CONTENTS: DATA CARD
STRUCTURE DATA WORK SHEETS
STREAM BEAD SOUNDINGS & PROFILE SKETCH
FORM 501 OR 502
FORM BMD—9
RETURNED PROMPT ACTION NOTICE SHOWING REPAIRS
Other sketches and notes showing structure details
PHOTOGRAPHS

STRUCTURE ANALYSIS (CHECK) _____ YES or NO ______
WHEN PLANS ARE AVAILABLE ________________________

FIELD INSPECTOR
____
____
____
____
____
____
____
____
____
____
____

date by

Figure A13-1 North Carolina Department of Transportation, Division of Highways Bridge Maintenance Unit—Structure Data File form (reproduced courtesy of North Carolina Department of Transportation).

BRIDGE INSPECTION RECORD AND SUMMARY

(SUPPLEMENT RECORD FOR TRUSSES)

INSPECTION TYPE________________ INSPECTION DATE________________

BRIDGE NO.________________ COUNTY________________ TRUSS TYPE________________

LENGTH________________ FOR SPAN________________

EVALUATION CODE: 0-2 is critical, 3-4 is poor, 5 and 6 are fair, 7,8 & 9 are good

EVALUATION FOR SI&A ITEM 59

INSPECTION ITEM	GRADE	BRIEF DESCRIPTION
18. LOWER CHORDS & CONNECTIONS		
19. END FLOOR BEAMS & CONNECTIONS		
20. INTERMEDIATE FLOOR BEAMS & CONNECTIONS		
21. BOTTOM LATERAL STRUTS-SWAY BRACING, & CONNECTIONS		
22. LONGITUDINAL STRINGERS		
23. STRINGER CONNECTIONS OR BEARINGS		
24. END POSTS		
25. VERTICALS		
26. DIAGONALS & CONNECTIONS		
27. PORTAL STRUTS-BRACING & CONNECTIONS		
28. TOP CHORDS & CONNECTIONS		
29. TOP LATERAL STRUTS-SWAY BRACING & CONNECTIONS		
30. TOP LATERALS & CONNECTIONS		
31. TRUSS ALIGNMENT		
32. BEARING ASSEMBLIES (INCLUDE MISALIGNMENT)		

TRUSS NOMENCLATURE AND TYPES

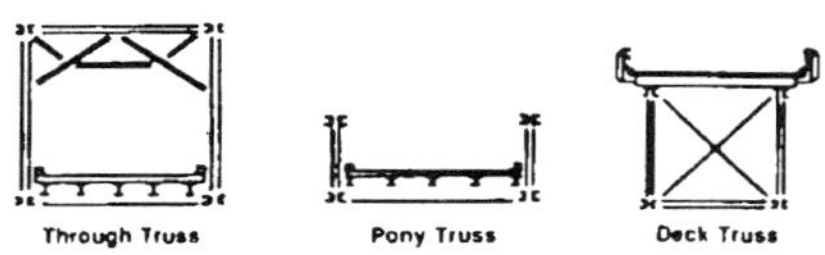

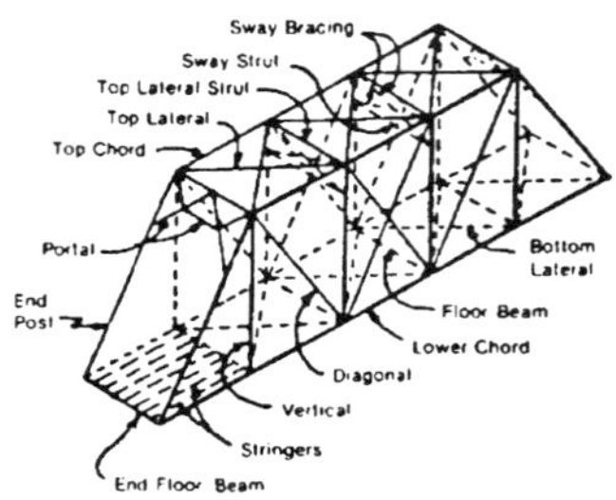

NOTES: ________________

64. SPECIAL INSPECTION REQUESTED FOR ________________

SHEET____ of________

Figure A13-2 Bridge Inspection Record And Summary (Supplement Record for Trusses) form (reproduced courtesy of North Carolina Department of Transportation).

BRIDGE I&A
FORM 1(90)C
Rev. 10-91

BRIDGE INSPECTION RECORD AND SUMMARY
(R.C. BOX CULVERTS)

STRUCTURE No. ________ COUNTY ________ ROUTE ________
OVER ________ ROUTE ORIENTATION ________ YEAR BUILT ________
INSPECTION TYPE ________ INSPECTION DATE ________

RECORD THE FOLLOWING ITEMS:

1. NUMBER & SIZE OF BARRELS (W X H) ____________
2. LENGTH ALONG C\L ROADWAY ____________
3. LENGTH ALONG C\L CULVERT ____________
4. DISTANCE FROM CROWN ROADWAY TO BED ____________
5. SKEW (+-) ________
6. APPROACH RD'WY. WIDTH (SHOW-PAVED OR UNPAVED) ________
7. APPROACH SHOULDER WIDTH (SHOW LEFT OR RIGHT) ____________
8. INCLUDE SKETCHES TO SHOW CULVERT SITUATIONS AND ROADWAY CLEARANCES THAT CANNOT BE DETAILED ON THIS SHEET.

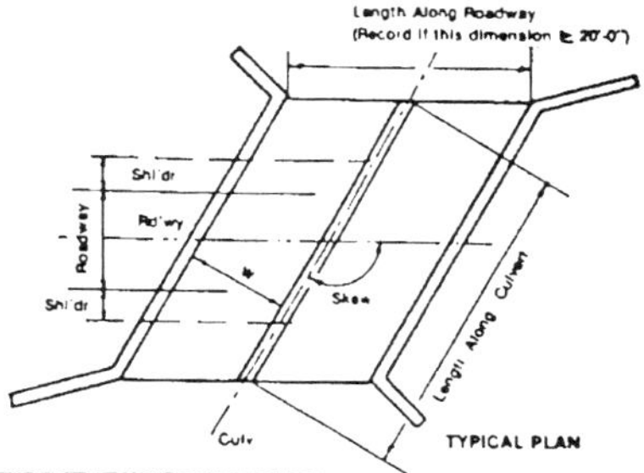

TYPICAL PLAN

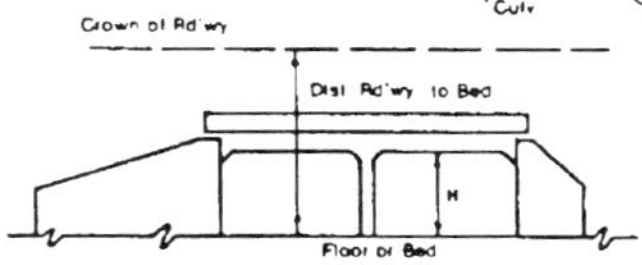

TYPICAL END ELEVATION

EVALUATION CODE: 0 - 2 IS CRITICAL, 3 & 4 IS POOR, 5 AND 6 ARE FAIR, 7 - 9 ARE GOOD

	INSPECTION ITEM		GRADE	BRIEF DESCRIPTION
EVALUATE FOR SI&A ITEM 62	1. TOP SLAB			
	2. BOTTOM SLAB			
	3. EXT. & INT. WALLS			
	4 WINGWALLS - RETAINING WALLS			
	5. HEADWALLS - TOEWALLS - FLUMES			
	6. STRUCTURE ALIGNMENT - SETTLEMENT (INCLUDE SCOUR)			
	7. DRAINAGE SYSTEMS (ON STRUCTURE)			
EVAL. FOR SI&A ITEM 61	8. CHANNEL & CHANNEL PROT.	a. WATERWAY		
		b. ALIGNMENT		
		c. SCOUR		
		d. SLOPE PROT. (RIP-RAP, DIKES, ETC.)		
	9. APPROACH ROADWAY CONDITION			
	10. ESTIMATED REMAINING LIFE			

11. SI&A RATINGS			
	11. CHANNEL & CHANNEL PROT.	ITEM 61	
	12. CULVERT & RETAINING WALLS	ITEM 62	
	13. WATERWAY ADEQUACY	ITEM 71	
	14. APPROACH ROADWAY ALIGNMENT	ITEM 72	
	15. FIELD SCOUR EVALUATION		

16. PRESENTLY POSTED ________
17 REGULATORY SIGN NOTICE ISSUED ________
18 PROMPT - ACTION NOTICE ISSUED ________
19 TOTAL FIELD INSPECTION TIME ________
20. INSPECTED BY: ________ (TEAM LEADER)

21. REVIEWED BY: ________

22. SPECIAL INSPECTION REQUESTED FOR: ________

SHEET ____ OF ____

Figure A13-3 Bridge Inspection Record And Summary (R.C. Box Culverts) form (reproduced courtesy of North Carolina Department of Transportation).

BRIDGE I & A FORM I <90/U
REV. 10-91

BRIDGE INSPECTION RECORD AND SUMMARY
<PIPE AND ARCH CULVERTS>

STRUCTURE NO. ______ COUNTY ______ ROUTE ______

OVER ______ ROUTE ORIENTATION ______ YEAR BUILT ______

INSPECTION TYPE ______ INSPECTION DATE ______

RECORD THE FOLLOWING ITEMS:

1. NUMBER, SIZE, & TYPE OF CULVERT ______
2. LENGTH ALONG C/L ROADWAY ______
3. LENGTH ALONG C/L CULVERT ______
4. DISTANCE FROM CROWN OF ROADWAY TO BED ______
5. SKEW ______
6. APPROACH RD'WY. WIDTH ______
7. APPROACH SHL'DR. WIDTH ______
8. INCLUDE SKETCHES TO SHOW CULVERT SITUATIONS AND ROADWAY CLEARANCES THAT CANNOT BE DETAILED ON THIS SHEET.

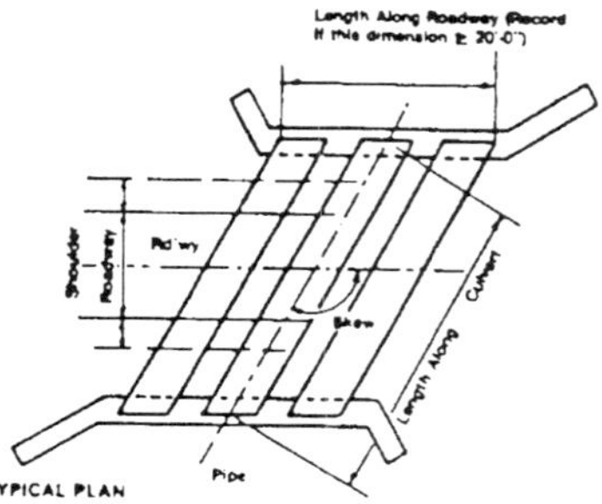

TYPES:

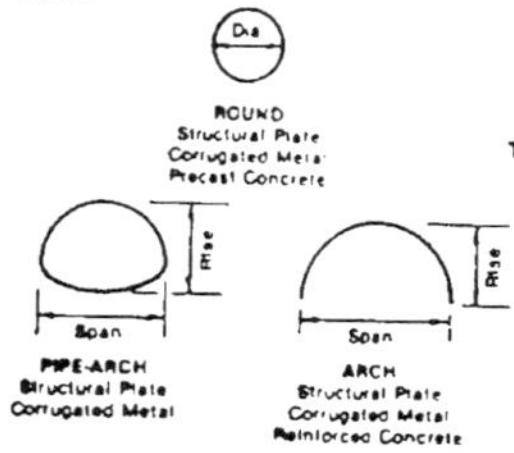

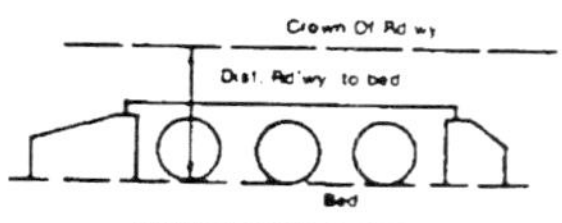

EVALUATION CODE: 0 - 2 IS CRITICAL, 3 & 4 IS POOR, 5 AND 6 ARE FAIR, 7 - 9 ARE GOOD

	INSPECTION ITEM		GRADE	
EVAL. FOR SI&A ITEM 62	1. STRUCTURE CONDITION			
	2. WINGWALLS - RETAINING WALLS			
	3. HEADWALLS - TOEWALLS - FLUMES			
	4. STRUCTURE ALIGNMENT - SETTLEMENT <INCLUDE SCOUR>			
EVAL. FOR SI&A ITEM 61	5. CHANNEL & CHANNEL PROT.	a. WATERWAY		
		b. ALIGNMENT		
		c. SCOUR		
		d. SLOPE PROT. <RIP RAP - DIKES, ETC.>		
	6. APPROACH ROADWAY CONDITION			
	7. ESTIMATED REMAINING LIFE			

SI&A RATINGS	8. CHANNEL & CHANNEL PROTECTIONS	ITEM 61	
	9. CULVERTS & RETAINING WALLS	ITEM 62	
	10. WATERWAY ADEQUACY	ITEM 71	
	11. APPROACH ROADWAY ALIGNMENT	ITEM 72	
	12. FIELD SCOUR EVALUATION		

13. PRESENTLY POSTED ______
14. REGULATORY SIGN NOTICE ISSUED ______
15. PROMPT-ACTION NOTICE ISSUED ______
16. TOTAL FIELD INSPECTION TIME ______
<INCLUDE TRAVEL & WRITE-UP>
17. INSPECTED BY: ______
<TEAM LEADER>
18. REVIEWED BY: ______

19 SPECIAL INSPECTION REQUESTED FOR ______ SHEET ____ OF ____

Figure A13-4 Bridge Inspection Record And Summary (Pipe and Arch Culverts) form (reproduced courtesy of North Carolina Department of Transportation).

Figure A13-6 Field Inspection Report Bridge Inspection and Analysis form (reproduced courtesy of North Carolina Department of Transportation).

BIRDGE MAINT. WORKSHEET 1-1-90

FIELD SKETCH FOR VERTICAL AND HORIZONTAL CLEARANCES

BRIDGE NO. ______________ COUNTY __________________ DATE ____________ BY ________________________

APPROACH ROADWAY WIDTH (SI&A ITEM 32)

UNDERCLEARANCE (SI&A ITEMS 10, 28, 38, 39, 40, 47, 54, 55 & 56)

SHEET _____ OF ________

Figure A13-7 Field Sketch for Vertical and Horizontal Clearances form (reproduced courtesy of North Carolina Department of Transportation).

BR. INSP. WORKSHEET(1-1-90)

STREAM BED SOUNDINGS

(SEE NEXT SHEET FOR PROFILE SKETCH)

BRIDGE NO. ______ COUNTY ______ DATE ______ BY ______

RECORD SOUNDINGS FROM TOP OF RAIL. OTHER LOCATION IF NEEDED: ______

DISTANCE H. W. MARK TO TOP OF RAIL ______ LOCATION H. W. MARK ______

NOTE: DRAFT A STREAM BED PROFILE SKETCH AND RECORD SOUNDINGS AT THE STREAM FACE OF ABUTMENTS, C/L OF PIERS, WE, & WS , AND MAJOR CONTOUR CHANGES ON THE DOWNSTREAM SIDE. RECORD SOUNDINGS AT THE STREAM FACE OF ABUTMENTS AND C/L OF PIERS ON THE UPSTREAM SIDE.

DOWNSTREAM			UPSTREAM		
STATION	SOUNDING	DESCRIPTION	STATION	SOUNDING	DESCRIPTION

SHEET___ OF______

Figure A13-8 Stream Bed Soundings form (reproduced courtesy of North Carolina Department of Transportation).

Bridge I & A Form 10 (86)F STATE of NORTH CAROLINA DEPT. of TRANSPORTATION & HIGHWAY SAFETY BRIDGE MAINTENANCE UNIT BRIDGE INSPECTION	PROMPT ACTION NOTICE BRIDGE INSPECTION & ANALYSIS	BRIDGE NO. ____________ COUNTY ____________ ROUTE ____________

Team Leader: ______________________________ Date ____________

Assisted By: ______________________________

MEMORANDUM TO: ______________________________
Bridge Maintenance Supervisor

A condition inspection of the subject bridge was made ____________. It is recommended that the conditions listed below be given prompt attention. The original notice and two (2) copies of this memorandum are furnished for your use. Please note on the original notice and both copies the action taken as a result of these recommendations. Send the original notice to the Head of Bridge Maintenance and one (1) copy to your Superintendent. Retain one (1) copy for your files.

CC: Head of Bridge Maintenance
Bridge Maintenance Superintendent
File Copy

Figure A13-9 Prompt Action Notice—Bridge Inspection and Analysis form (reproduced courtesy of North Carolina Department of Transportation).

BRIDGE INSPECTOR'S RECOMMENDATIONS FOR MAINTENANCE REPAIRS

BRIDGE ______________ COUNTY ______________ DIVISION ______________

INSPECTED BY ______________ DATE ______________ CHECKED BY ______________ DATE ______________

THESE REPAIRS SHOULD BE MADE WITHIN TWELVE MONTHS FROM THE DATE OF THIS INSPECTION.

FUNCTION CODE	DESCRIPTION OF FUNCTION	UNIT	QUANITY	REMARKS	EST. COST
COPY – NCDOT					

Sheet _____ of _____

Figure A13-10 Bridge Inspector's Recommendations for Maintenance Repairs (reproduced courtesy of North Carolina Department of Transportation).

APPENDIX 14: CABLES AND WIRE ROPES

Data on cables and wire ropes demand some efforts to be found in the large scope of technical publications. Therefore, to shorten the time to have access to the manufacturing companies of cables and wire ropes, the following information will be useful to the reader who tries to get to the source of these products to attain the available literature covering not only the cables and the wire ropes *per se* but also the related hardware, including fittings, and other details.

At present, major companies manufacturing cables and wire ropes have formed a board referred to as *The Wire Rope Technical Board*. Members of this board are several companies: Bridon American Corporation, The Rochester Corporation, Wire Rope Corporation of America, Paulsen Wire Rope Corporation, Broderick & Bascom Rope Company, and Williamsport Wirerope Works. Of these, the following four companies are those that deal specifically with applications to bridge structures:

Bridon American Corporation
Stevens Lane
Exeter, PA 18643
Phone: 717-822-3349 x110
Fax: 717-654-5873

Wire Rope Corporation of America
609 N. Second Street
St. Joseph, MO 64501
Phone: 816-236-5055
Fax: 816-236-5040

Williamsport Wirerope Works
P.O. Box 3188
100 Maynard Street
Williamsport, PA 17701
Phone: 717-327-4232
Fax: 717-327-4274

Paulsen Wire Rope Corporation
880 South Second Street
Sunbury, PA 17801
Phone: 717-286-7141
Fax: 717-286-0791

For further information concerning cables and wire ropes the following are recommended:

- *Wire Rope Users Manual,* Third Edition
- *Wire Rope Sling Users Manual,* First Edition
- *Wire Rope Sling Users Manual,* Second Edition

These publications are available from:

WRTB Fulfillment
3246 Fall Creek Highway, Suite 160
Granbury, TX 76049-7979

Another publication of major interest is: *Structural Applications of Steel Cables for Buildings*, (ASCE 19-96) Standards Committee of the ASCE, 1997, a manual of standards available from the American Society of Civil Engineers (ASCE).

APPENDIX 15: U. S. COAST GUARD PERMIT APPLICATION FORMS FOR BRIDGE CONSTRUCTION

The following are forms and a sample of required drawings used to attain the U. S. Coast Guard permit for bridge construction as required by law. While this topic has been addressed in the text, it is useful for designers to see the documents typically required for bridge construction.

Sample Letter of Application for Permit

Commander (obr)
Ninth Coast Guard District
1240 East Ninth Street

Cleveland, Ohio 44199

Dear Sir:

Application is hereby made by *(name of consultant or engineering firm) (name of applicant)* at *(address)* for approval by the Commandant, U. S. Coast Guard, of the location and plans of a *(type of bridge structure)* to be constructed *(or modified)* across the *(name of waterway)* at *(city), (state),* _____________ miles above the mouth of the waterway as shown on the attached plans.

Federal funds will/will not be utilized and have been/are being applied for. Federal agencies which must grant approvals, easements, or other actions for this project include

____

The bridge will have a/no significant impact on the human environment. The impacts on the human environment are as follows: *(briefly describe key issues)*

____

An environmental analysis describing these effects has been prepared in the format of Section 102(2)(C) of the National Environmental Policy Act of 1969 and is enclosed.

There are/are no wildlife and waterfowl refuges, recreational areas, public parks or historic sites in the vicinity or in the way of the *(type of structure)* or its approaches *(as appropriate)*.

Legal authority for the *(type of structure)* is found in the General Bridge Act of 1946. The laws of the State of ____________ do not require us to obtain a state permit for this work *(or enclose state permits)*.

Enclosed herewith is a letter of authorization from *(structure owner)* and an extract of *(the motions from Corporate meetings, etc.)* authorizing me to make this application.

This *(type of structure)* will replace the existing *(type of structure)* at *(city, state)*, will be removed when the new *(type of structure)* is completed. *(If applicable)*

Sincerely,

Encls:
(1) Original and three copies of map of the vicinity and plans of *(type of structure)*
(2) Environmental Analysis *(or two copies of FHWA Final EIS or FONSI)* as appropriate
(3) Authorization for applicant make application
(4) Evidence of ownership of old *(type of structure)*
(5) Water quality certification under 33 U.S.C. 1251 *(or copy of letter requesting same)*
(6) CZM consistency statement
(7) State agency concurrence in CZM consistency certification

Figure A15-1 Sample Letter of Bridge Permit Application (reproduced courtesy U. S. Coast Guard).

Checklist

ENCLOSURES TO APPLICATION FORM (as applicable)

( ) letter authorizing agent to act in applicant's behalf
( ) letter authorizing modification or removal of another's bridge
( ) State license to construct the proposed bridge
( ) extract from corporation's charter
( ) proof of ownership of the land that the proposed bridge will be located on
( ) extracts of motions from meetings authorizing construction of the proposed bridge
( ) water quality certification
( ) CZM consistency certification
( ) state concurrence with consistency certification
( ) environmental document
( ) other federal, state and local permits
( ) identification of property owners adjacent to the proposed bridge site

Drawings

LOCATION MAP

( ) show the location of the proposed bridge in red
( ) show the location of existing bridge
( ) show wildlife and waterfow refuges, historical and archaeological sites, public parks and recreation areas
( ) show graphic scale
( ) show north arrow
( ) show direction of streamflow by use of an arrow
( ) show towns in project vicinity

PLAN VIEW

( ) show properties adjacent to the proposed bridge and identify respective owners
( ) show length and width of bridge (proposed and, as appropriate, existing)
( ) show fendering system and indicate type of construction materials
( ) show banks of the waterway
( ) show navigation channel limits
( ) show and identify structures immediately adjacent to the proposed bridge and their pier alignment in relation to the proposed bridge
( ) show graphic scale
( ) show north arrow
( ) show horizontal clearance normal to the channel axis
( ) show channel axis

ELEVATION VIEW (looking upstream)

( ) show the navigational opening in red
( ) show the horizontal clearance normal to the channel
( ) show the vertical clearance above the appropriate datum
( ) show the elevation of the waterway bottom
( ) show the amount of fill
( ) show the graphic scale

TITLE BLOCKS

( ) Indicate applicant
( ) Indicate waterway name and mile point
( ) Indicate the location of project (city, county, state)
( ) Indicate date of plans
( ) Indicate sheet number

Figure A15-2 Bridge Permit Application Check List (reproduced courtesy U. S. Coast Guard).

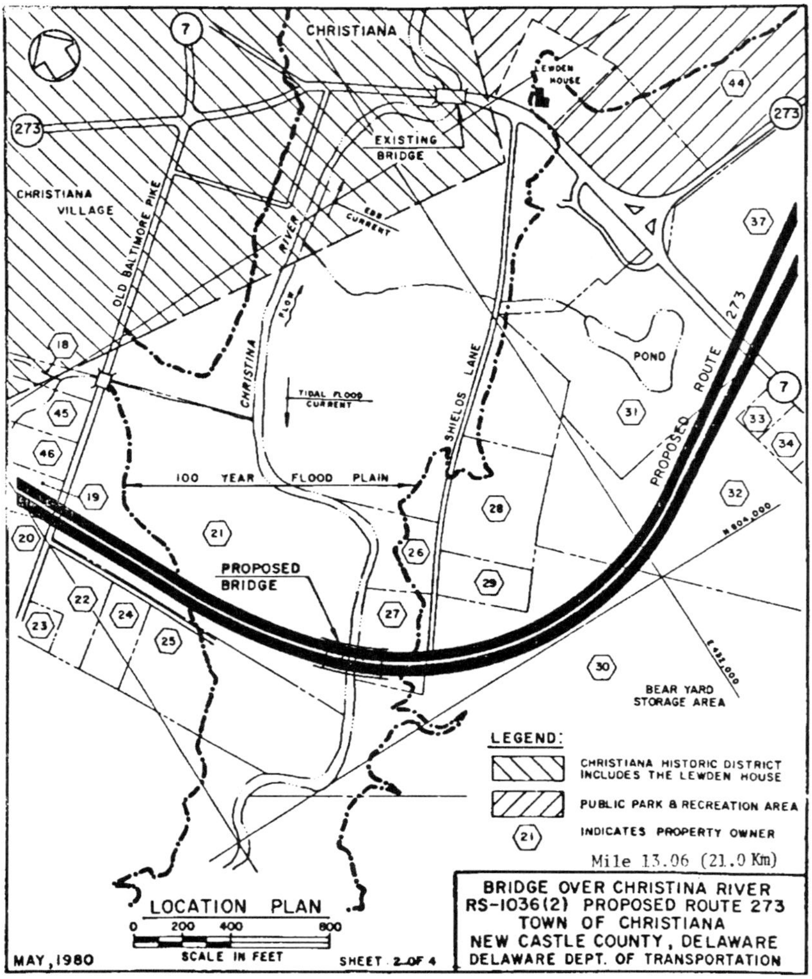

Figure A15-3 Sample of Drawings Required for Bridge Permit Application (reproduced courtesy U. S. Coast Guard).

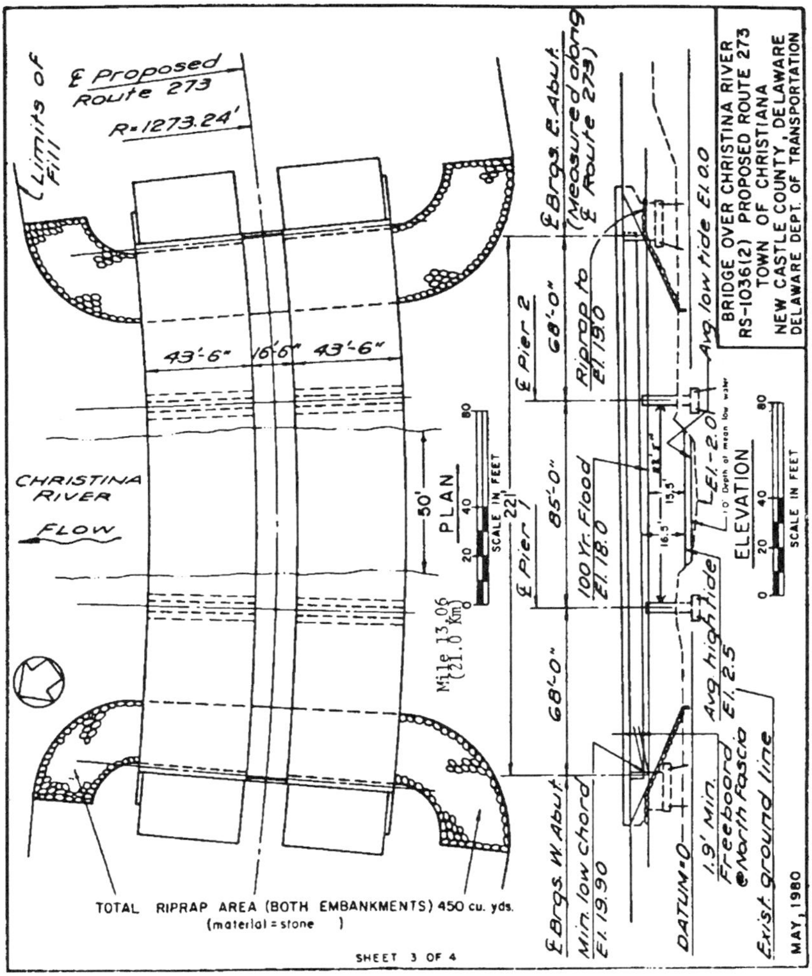

Figure A15-3 (continued) Sample of Drawings Required for Bridge Permit Application (reproduced courtesy U. S. Coast Guard).

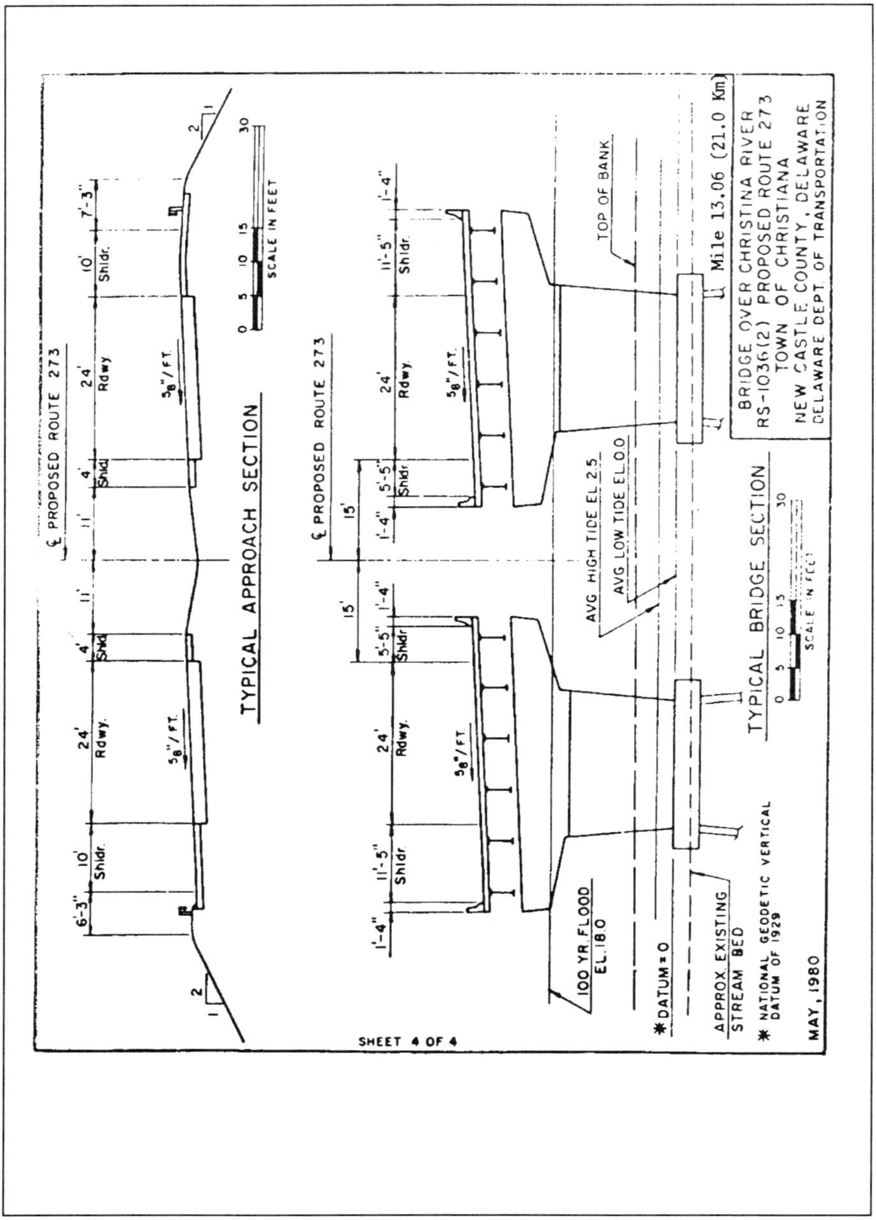

Figure A15-3 (continued) Sample of Drawings Required for Bridge Permit Application (reproduced courtesy U. S. Coast Guard).

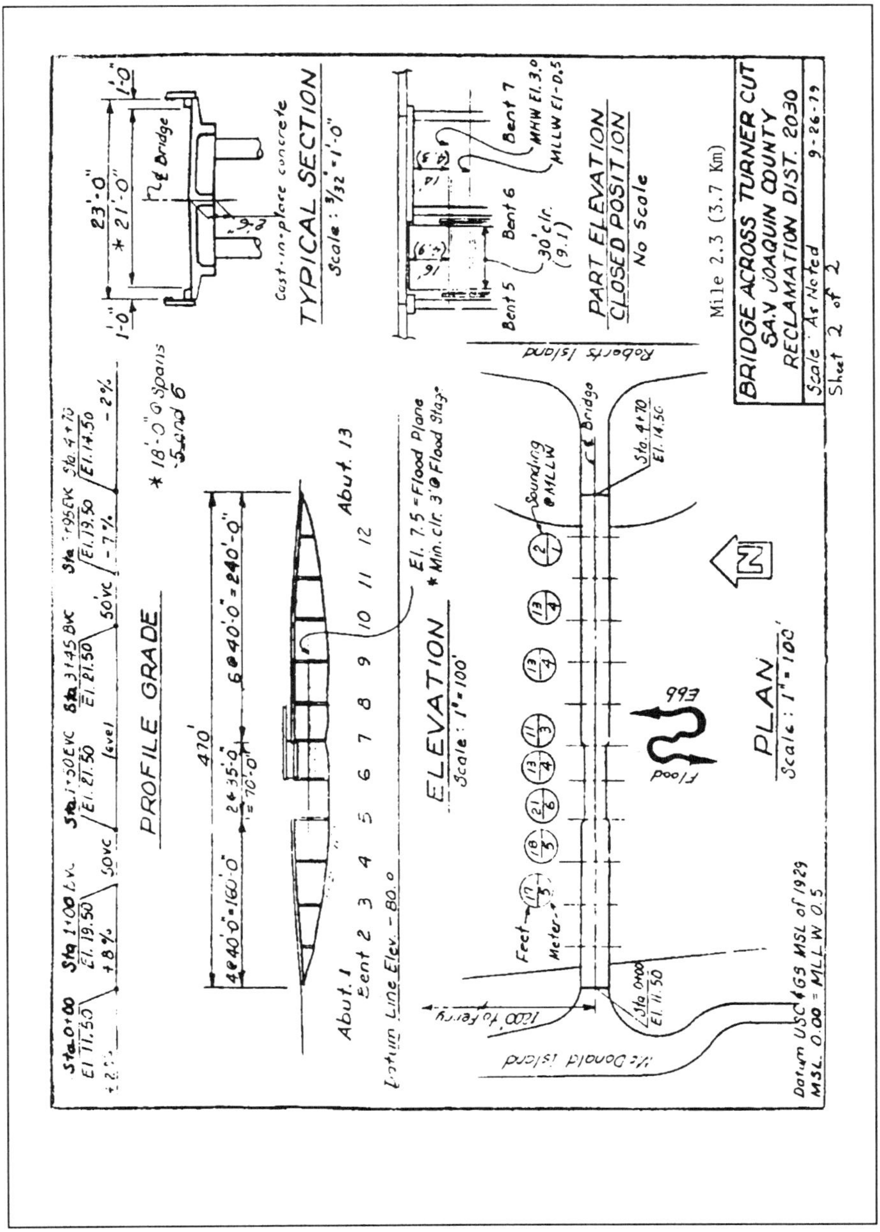

Figure A15-3 (continued) Sample of Drawings Required for Bridge Permit Application (reproduced courtesy U. S. Coast Guard).

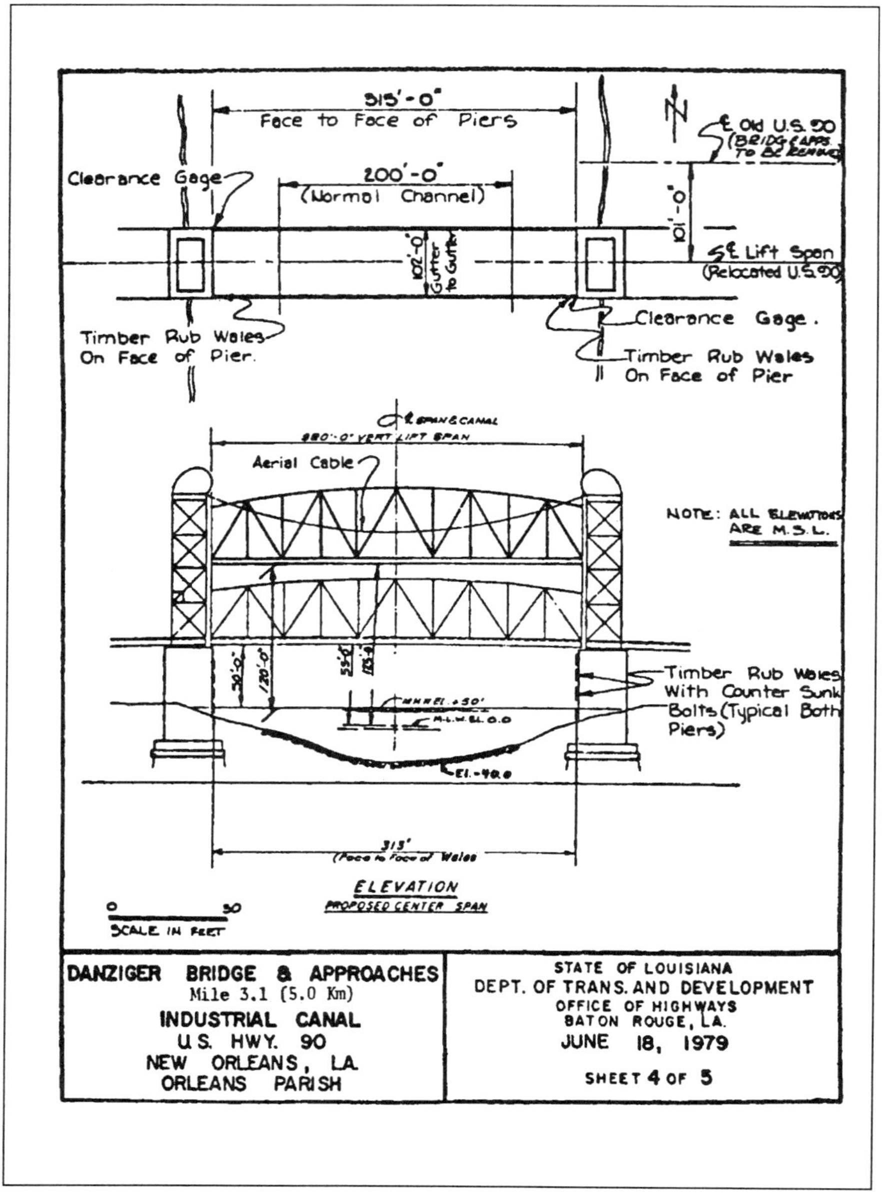

Figure A15-3 (continued) Sample of Drawings Required for Bridge Permit Application (reproduced courtesy U. S. Coast Guard).

GENERAL BIBLIOGRAPHY*

AASHTO, *Standard Specifications for Highway Bridges*. American Association of State Highway and Transportation Officials, 1996.

ACI, "Analysis and design of reinforced concrete bridge structures," ACI 343R-88, 1988, pp. 343-162.

ACI, "Esthetics in concrete bridge design," MP-1, 1990, p. 334.

ACI, "Standard practice for concrete highway bridge construction," ACI 345-82, 1982, pp. 345-82.

API, "Recommended practice," API RP2A, January 1980.

Allan, B. J., "Aesthetics in bridge engineering," *Jour. Inst. of Highway Engineers*, May 1971, pp. 7-16.

______, "Aesthetics in highway engineering," *Jour. Inst. of Highway Engineers*, November 1968, pp. 17-24.

______, "Bridgework in the new town of Cumbernauld," *Proceedings of the Institution of Civil Engineers*, November 1964, pp. 497-534.

______, "Some notes on significance of form in bridge engineering," *Proceedings of the Institution of Civil Engineers*, Vol. 60, part 1, February 1976, pp. 79-94.

---

*See also Special Bibliography on Bridge Aesthetics at end of Chapter 4.

Ahlskog, J. J., "Bridge rehabilitation, strengthening and seismic retrofit with a new lightweight aluminum deck system," *Proceedings of the Second Symposium on Practical Solutions for Bridge Strengthening and Rehabilitation Bsar ll,* March 24-25, 1997, Kansas City, MO, pp. 33-45.

American Institute of Steel Construction, Design Manual for Orthotropic Steel Plate Deck Bridges. American Institute of Steel Construction, 1964.

Asakawa, K., Kawakami, H., Yoshioka, T., and Matsumoto, N., Hidaka, I., "Dynamic response of steel and composite girders under high-speed train," *Q. Rep. Railway Technical Research Institution,* Vol. 24, no. 3, September 1983, pp. 121-122.

Bell, D., "Over Stony Creek," *Proceedings Symposium on Practical Solutions for Bridge Strengthening and Rehabilitation,* April 5-6, 1993, Des Moines, IA, pp. 203-207.

Billington, D. P., "Bridge design and regional esthetics," *Journal of the Structural Division,* ASCE, Vol. 107, no. 3, March 1981, pp. 473-486.

______, "History and esthetics in concrete arch bridges," *Journal of the Structural Division,* ASCE, Vol. 103, no. 11, November 1977, pp. 2129-2143.

______, *The Tower and the Bridge.* 1983.

______, *Robert Maillart and the Art of Reinforced Concrete.* New York, Architectural History Foundation and The MIT Press, 1990, p. 151.

Biliszczuk, J., et al., "Problems of rehabilitation of precast concrete bridges in Poland," *Proceedings of theSecond Symposium on Practical Solutions for Bridge Strengthening and Rehabilitation Bsar ll,* March 24-25, 1997, Kansas City, MO, pp. 45-55.

Brown, David, *Bridges.* New York, Macmillan, 1993.

Burke, Martin P., Jr., "Bridge aesthetics bibliography," *Bridge Aesthetics Around the World,* Transportation Research Board, Washington, D.C., 1988.

______, "Bridge aesthetics-rules, formulas, and principles-the Negative View," ACI, MP-1, 1990, pp. 71-78.

California Department of Transportation, *Bridge Design Practice.* June 1993, pp. 7-i-7-26.

Clifford, D., *Art and Understanding.* London, Evelyn, Adams and Mackay, 1968, pp. 91-166.

Coates, A. C., "Repair of the Chelsea Street Bridge after collision with a tanker," *Proceedings of the Second Symposium on Practical Solutions for Bridge Strengthening and Rehabilitation Bsar ll,* March 24-25, 1997, Kansas City, MO, pp. 315-322.

Cope, R.J., ed., *Concrete Bridge Engineering: Performance and Advances*. Elsevier Applied Science, 1987, pp. 346.

Corven, J. A., "Strengthening and rehabilitation of three highway bridges," *Proceedings of the Symposium on Practical Solutions for Bridge Strengthening and Rehabilitation* , April 5-6, 1993, Des Moines, IA, pp. 55-64.

Crouch, A. G. D. "Bridge aesthetics: a sociological approach," *Transactions*. Institution of Engineers Australia. Vol. CE16, no. 2, 1969, pp. 138-142.

______, *The Aesthetics of Bridges*. Thesis (B. E.), University of Queensland, 1969.

Delgado, R., and dos Santos, S., "Modelling of railway bridge-vehicle interaction on high-speed tracks," *Computers and Structures*, Vol. 63, n o. 3, May 1997. pp. 511-523.

Det Norske Veritas, *Rules for the Design, Construction and Inspection of Fixed Offshore Structures*. 1977.

Dober, R. P., *Environmental Design*. New York, Van Nostrand, 1969.

Editor, "Assemblages soudes de profile creux circulaires avec decoupes d'intersection conception et verification des assemblages," *NF,* June 1978, pp. 22-250.

______, "Dutch regulations for the design of tubular structures," SG-TC-18, Draft, 1977.

______ , *Proceedings*. Second Symposium on Practical Solutions for Bridge Strengthening and Rehabilitation Bsar ll , March 24-25, 1997, Kansas City, MO, pp.□167-176.

Elliott, A. L., "Aesthetics of highway bridges," *Civil Engineering,* June, 1968, pp. 64-69.

______, "Esthetic development of California's bridges," American Society of Civil Engineers Structural Division. Vol. 109. no. 9, 1983.

Fendrich, L., "Feste Fahrbahn fuer Gleise und Weichen," *Betonwerk und Fertigteil-Technik/Concrete Precasting Plant,* Vol. 61, no. 5, May 1995, p. 8.

Frampton, K., Webster A., and Tishchhause A., *Calatrava Bridges*. Zurich, Artemis, 1993, pp. 219.

Fu, C. C., and Jausurawong, N., "Boulevard bridge rehabilitation - upgrading the load capacity of an aged truss bridge," *Proceedings of the Second Symposium on Practical Solutions for Bridge Strengthening and Rehabilitation Bsar II,* March 24-25, 1997, Kansas City, MO, pp. 409-414.

Fyfe, E. R., and Kuruvilla, S. J., "Column seismic retrofit using high strength fiber/epoxy jackets," *Proceedings of the Symposium on Practical Solutions for Bridge Strengthening and Rehabilitation,* April 5-6, 1993, Des Moines, IA, pp.□17-122.

Gagnaire, M., and Uso, T., "Performance evaluation of frame relay based on the packet trains model," *Proceedings of the IFIP TC6 Second International Conference on Broadband Communications*, 1994.

Gibstein, M. B., "The static strength of T-joints subjected to in-plane bending moments," *Det Norske Veritas,* Report no. 76-137, Oslo, July, 1976.

Gimsing, N., *Cable Supported Bridges; Concept and Design*. John Wiley and Sons, 1983, pp. 400.

Ghosn, M., Moses, F., and Gobieski, J., "Evaluation of steel bridges using in-service testing," *Structural Design of Bridges,* Transportation Research Board, Washington, DC, 1986, pp. 71-78.

Green, R., and Strevel, D., "Composite box girders during construction," paper presented at 56th Annual Meeting, Transportation Research Board, Washington, DC, January 1997.

Harichandran, R. S., Hatfield F. J., and Kifle, B. "Shear reinforcement of wood beams with polymer composite materials," *Proceedings of the Second Symposium on Practical Solutions for Bridge Strengthening and Rehabilitation Bsar II,* March 24-25, 1997, Kansas City, MO, pp. 353-362.

Hedgren, A. W. Jr., and Schmitt W., "Remodeling of the Julien Dubuque Bridge," *Proceedings*. Symposium on Practical Solutions for Bridge Strengthening and Rehabilitation, April 5-6, 1993, Des Moines, IA, pp. 3-12.

Heins, C. P., "Box-girder bridge design - state of the art," paper presented at 1978 National Engineering Conference, American Institute of Steel Construction, May 1978, Los Angeles, CA.

Highway Research Board, "Investigation of wind forces on highway bridges," *Proceedings*. Thirty-Second Annual Meeting, National Academy of Sciences, National Research Council, January 13-16, 1953, Special Report 10, publication 272, p. 30.

Hogg, J., ed., *Psychology and the Visual Arts*. Harmondsworth, England, Penguin, 1969.

Homaee, H. V., et al., "Rehabilitation of the historic Strawberry Mansion Bridge," *Proceedings of the Second Symposium on Practical Solutions for Bridge Strengthening and Rehabilitation Bsar II,* March 24-25, 1997, Kansas City, MO, pp. 111-119.

______, "Replacement of eyebar linkages and pins in the anchor span of the Summit Bridge," *Proceedings of the Second Symposium on Practical Solutions for Bridge Strengthening and Rehabilitation Bsar II,* March 24-25, 1997, Kansas City, MO, pp. 323-331.

Ichimaru, Y., Ishii, N., and Shinohara, O., "Structural design of S-curve bridges," *Proceedings*. Conceptual Design of Structures International Symposium, University of Stuttgart, October 7-11, 1996, Stuttgart, pp. 616-623.

Institution of Civil Engineers, "Steel box girders," *Proceedings,* from conference held on April 25, 1972, p. 98.

Ito, M., "Selection of bridge types: from Japanese experiences," *Proceedings*. Conceptual Design of Structures, International Symposium, University of Stuttgart, October 7-11, 1996, Stuttgart, pp. 65-72.

Jackson, C., "NSB'S first 200 km/h route nears opening," *Railway Gazette International*, Vol. 150, no. 2, February 1994. pp. 98-99.

Jones, C. J., *Design Methods Seeds of Human Futures*. John Wiley and Sons, 1970.

Joyce, J. R., and Pryor, A. J., "Development of the roadside and aesthetics in road design," *Jour. I. E. Aust.*, Vol. 40, no. 7-8, July-August, 1968, pp. 139-153.

Karshenas, M., and Kaspar, I. I., "Application of isolation bearings to bridges in low to moderate seismic areas," *Proceedings*. Second Symposium on Practical Solutions for Bridge Strengthening and Rehabilitation Bsar II, March 24-25, 1997, Kansas City, Missouri, pp. 343-352.

Kim, J. B., Grenci, T. A., Klingler, J. C., and Kwacz, K., "Rehabilitation and upgrading of an historic truss bridge in Stuyvesant Falls, New York," *Proceedings*. Symposium on Practical Solutions for Bridge Strengthening and Rehabilitation, April 5-6, 1993, Des Moines, IA, pp. 195-201.

Kline, T. R., O'Connor, J. P., and Yates, G. R., "Repair or replacement evaluation Manayunk Bridge, Pennsylvania," *Proceedings Symposium on Practical Solutions for Bridge Strengthening and Rehabilitation,* April 5-6, 1993, Des Moines, IA, pp. 285-290.

Kurobane, K., Makino, Y., and Mitsui, Y., "Ultimate strength design formulae for simple tubular joints," *IABSE,* Tokyo, September 1976.

______ , "Ultimate strength formulae for simple tubular joints," IIW Doc. XV-385-76, Department of Architecture, Kumemoto University, May 1976.

Kurobane, Y., "New developments and practices in the tubular joint design," IIW Doc. XV-488-81, Oporto, Portugal, September 1981.

Lally, A., "Steel box girder erection," *Proceedings.* I.C.E. International Conference on Steel Box Girder Bridges, February 13-14, 1973.

Lanyi, R. S., "Field prestressing repairs to precast concrete bridge griders," *Proceedings.* Symposium on Practical Solutions for Bridge Strengthening and Rehabilitation, April 5-6, 1993, Des Moines, IA, pp. 291-299.

Lawrie, R., and Seim, C., "The strengthening of the steel girder approach to the Berkely Bridge in Norfolk, Virginia," *Proceedings.* Symposium on Practical Solutions for Bridge Strengthening and Rehabilitation, April 5-6, 1993, Des Moines, IA, pp. 41-44.

______ , "Aesthetics of bridge design," *P. C. I. Journal,* February 1968, pp. 14-31.

______ , *Brucken/ Bridges,* Deutsche Verlags-Anstalt GmbH. Stuttgart, 1982.

LeRoy, D. H., and Kauhl N. E., "Rehabilitation of the Macombs Dam Bridge," *Proceedings.* Second Symposium on Practical Solutions for Bridge Strengthening and Rehabilitation Bsar ll, March 24-25, 1997, Kansas City, MO pp. 155-164.

Malek, M., "Begins work on upgrading first corridor," *Railway Gazette International,* June 6, 1995, p. 151.

Marshall, P. W., "A review of american criteria for tubular structures and proposed revisions," IIW Doc. XV-405-77, London, 1977, rev. March 1978.

Matsura, A., Asakawa, K., Kawakami, H., and Ito, H., "Dynamic response of concrete-steel composite girder under running train," *Q Rep Railway Technical Research Institution,* Vol. 25, no. 3, September 1984. , pp. 85-88.

Mattock, A.H., et al., *Commentary on Criteria for Design of Steel-Concrete Composite Box Girder Highway Bridges,* August 1967.

Medeot, R., and Romagnolo, M., "Seismic retrofitting and strengthening of Aglio Viaduct," *Proceedings.* Symposium on Practical Solutions for Bridge Strengthening and Rehabilitation, April 5-6, 1993, Des Moines, IA, pp. 163-172.

Mee, B., Whitten, B., and Neijikovsky, B., "Railway testing using a portable ride quality and vibration measurement system with GPS," *Proceedings*. SPIE-The Ineternational Society for Optical Engineering, Vol. 2458, 1995, pp.138-147.

Miller, D. K., "Ensuring weld quality under field repair and retrofit conditions," *Proceeding*. Symposium on Practical Solutions for Bridge Strengthening and Rehabilitation, April 5-6, 1993, Des Moines, IA, pp. 259-262.

Ministry of Transport, *The Appearance of Bridges*. Her Majesty's Stationery Office, London, 1964.

Moore, H., as cited by Read, H., *Philosophy of Modern Art*, London, Faber and Faber, 1969, p. 209.

Moreau, W. J., Warshaw, R., and Gilstad, D. E., "Waking Rip Van Winkle," *Proceedings*. Symposium on Practical Solutions for Bridge Strengthening and Rehabilitation, April 5-6, 1993, Des Moines, IA, pp. 175-184.

Moritoh, Y., Zenda, Y., and Nagakura, K., "Noise control of high speed shinkansen," *Journal of Sound and Vibration*, Vol. 193, no. 1, May 30, 1996, pp. 319-334.

Moser, C. A., *Survey Methods in Social Investigation*. London, Heinemann, 1969.

Naka, T., Kato, B., and Kanatani, H., "Experimental study on welded tubular connections," report, Research Institute of Welding, University of Tokyo, 1964.

National Research Council, *Proceedings 7*. Fourth International Bridge Engineering Conference. San Francisco, CA, August 28-30, 1995, p. 408.

Nickerson, R. L., and Veshosky, D., "Criteria for bridge selection.," *Modern Steel Construction,* Vol. 32, no. 9, 1992, pp. 39-44.

Niemierko, A., "Some design and performance problems related to behaviour of bridge structures on high-speed railway lines," *Archives of Civil Engineering*, Vol. 42, no. 4, 1997. pp. 401-423.

O'Conner, C., Burgess, M. D., and Egan, M. J., "Bridge appearance," *American Society of Civil Engineers, Structural Division,* Vol. 106, no. 1, 1980, pp.□1678-1687.

______, "Bridge appearance," *Journal of the Structural Division*, ASCE, Vol. 106, no. 1, January 1980, pp. 331-349.

Ohta, T., Takahashi, N., and Yamane, T., "Aesthetic design method for bridges," *Journal of Structural Engineering*, ASCE, Vol. 113, no. 8, 1987, pp. 1678-87.

______, "Studies on aesthetic design method of double-deck bridges with different types constructed contiguously and its application," *Proceedings of the Japan Society of Civil Engineers*, Vol. 344, no. I-1, April 1984, pp. 205-213 (in Japanese).

Ohtsuka, H., et al., "Drift capacity and failure mechanism of toll reinforced concrete piers," *Proceedings of the 25th Japan Congress on Materials Research*, Kyoto, Japan, 1982, pp. 261-265.

Pan, B. P., Plummer, F. B., and Kuang, J. G., "Ultimate strength design of tubular joints," OTC 2644, 1976.

Parr, J. G., "The social impact of engineers," *Engineering Journal*, December 1968.

Plato. *Philebus*. Quoted by Herbert Read, *Art Now*. Faber and Faber, London, 1960, p. 72.

Podolny, W., and Scalzi, J., *Construction and Design of Cable-Stayed Bridges*. John Wiley and Sons, 1976.

Potzl, M., "Robustness in bridge design," *Proceedings*. Conceptual Design of Structures, International Symposium, University of Stuttgart, October 7-11, 1996, Stuttgart, pp. 127-135.

Leto, I. V., "Preliminary design of the Messina Strait Bridge," *Proceedings of the Institution of Civil Engineers*, Civil Engineering, American Society of Civil Engineers Italy International Group, Vol. 102, no. 3, August 1994. pp. 122-129.

Pritchard, B., "Shock transmission units for bridge substructure strengthening with minimum traffic disruption," *Proceedings*. Symposium on Practical Solutions for Bridge Strengthening and Rehabilitation, April 5-6, 1993, Des Moines, IA, pp. 221-230.

Read, H., *Art and Industry*. Faber and Faber, London, 1956, pp. 40-41.

______, *Art Now*. Faber and Faber, London, 1960, p. 35.

______, *The Meaning of Art*. Harmondsworth, England, Penguin, 1967.

Reber, J. B., *Ultimate Strength Design of Tubular Joints*. October 1972, pp. 1664.

Reina, P., Normile, D., and Green, P., "High-speed rail ready to boom," *Engineering News-Record*, Vol. 233, no. 8, August 22, 1994, pp. 24-28.

Reekie, F. R., *Design in the Built Environment*. London, Edward Arnold, 1972.

Richardson, G., and Associates, "Curved girder workshop textbook," Federal Highway Administration, September 1976.

Robinson, H., *The Principles of Architectural Composition*. 9th Impression, London, The Architectural Press, 1963.

Rogers, H. D., and Biggs D. T., "Rehabilitation of the Red Kill Bridge," *Proceedings*. Symposium on Practical Solutions for Bridge Strengthening and Rehabilitation, April 5-6, 1993, Des Moines, IA, pp. 33-40.

Ruskin, J., *Modern Painters*. London, George Allen, 1904.

Shimokawa, H., et al., "Hitsuishijima and Iwakurojima Bridges," paper, 12th Congress of IABSE, Vancouver, Canada, September 1984.

Sparrow, K. D., *Ultimate Strengths of Welded Joints in Tubular Steel Structures*. PhD Thesis, Kingston Polytechnic, September 1979.

Speigel, E., Bisdikian, C., and Tantawi, A., "Characterization of the traffic on high-speed token-ring networks," *Performance Evaluation*, Vol. 19, no. 1, January 1994, pp. 47-72.

Stamenkovic, A., Sparrow, K. D., and Irving, D. J., "Existing methods for calculating the static strength of welded T-, Y-, N-, K-, and X-joints in CHS sections," Technical Report, Parts 1, 2 and 3, Kingston Polytechnic, June 1980.

Stamenkovic, A., and Sparrow, K. D., "Experimental determination of the ultimate static strength of T-joints in circular hollow steel sections subject to axial load and Moment," *Proceedings*. First International Conference on Joints and Structural Steelwork, Teeside Polytechnic, UK, April 7, 1981, Pentech Press, London.

______, "A review of existing methods for the determination of the static axial strength of welded T-, Y-, N-, K-, and X-joints in circular hollow steel sections," *Proceedings*. First International Conference on Joints and Structural Steelwork, Teeside Polytechnic, UK, April 7, 1981, Pentech Press, London.

State of California, Department of Public Works, *Manual of Bridge Design Practice*. 3rd ed., Sacramento, CA, 1971.

Strobel, G. C., "Steel box girders highlight design of I-480 in Omaha," *Civil Engineering*, Vol. 39, no. 12, December 1969.

Tanabe, M., Wakui, H., Matsumoto, N., "Efficient numercial method for dynamic interaction analysis of high-speed Shimkansen vehicles, rail and bridge," *Proceedings*. 1994 ASME International Computers Conference and Exhibition, 1994, pp. 567-572.

Taylor, A. W., and Stone, W. C., "Jacket thickness requirements for seismic retrofitting of circular bridge columns," *Proceedings*. Symposium on Practical Solutions for Bridge Strengthening and Rehabilitation, April 5-6, 1993, Des Moines, IA, pp. 249-258.

Tanner, D. L., "Steel box girders and steel bents support double deck roadways," *Acier-Stahl-Steel*, January 1970.

Tessiatore, R., and Nusekabel, K., "Tied-arch redecking - Rock Island Centennial Bridge," *Proceedings*. Second Symposium on Practical Solutions for Bridge Strengthening and Rehabilitation Bsar II, March 24-25, 1997, Kansas City, MO, pp. 387-396.

Transportation Research Board. *Structural Design of Bridges*. Transportation Research Record 1072, Washington, DC, 1986, pp. 78.

Troitsky, M. S., *Cable-Stayed Bridges: an Approach to Modern Bridge Design*. (second Edition), Van Nostrand Reinhold, New York, 1977, pp. 469.

Tung, D. H. H., "Analysis of curved twin box-girder bridges," reprint from *ASCE Structural Engineering Conference*, Seattle, WA, May 8-12, 1967.

______, and Fountain, R. S., "Approximate torsional analysis of curved box girders by the M/R method," *Engineering Journal*, AISC, Volume 7, no. 3, July 1970.

U. S. Steel, *Steel/Concrete Composite Box-Girder Bridges, A Construction Manual*. 1978.

Vitek, J., Verma R., and Banan M. R., "Seismic design criteria and retrofit schemes for the San Mateo-Hayward toll bridge," *Proceedings*. Second Symposium on Practical Solutions for Bridge Strengthening and Rehabilitation Bsar II, March 24-25, 1997, Kansas City, MO, pp. 197-206.

Wardenier, J. "Design rules for predominantly statically loaded welded joints in CHS," IIW Doc. XV-79, Bratislawa, 1979.

Wardenier, J., and Stark, J. W. B., "The static strength of welded lattice girder joints in structural hollow sections," *Part 3: Joints made of Circular Hollow Sections*, Stevin Report 6-78-4, Delft University of Technology and TNO-IBBC.

Watson, S., and Hurd, M. K., "Esthetics in concrete bridge design," American Concrete Institute MP-1, 1990

Washio, K., Togo, T., and Mitsui, N., "Experimental study on local failure of chords in tubular truss joints," *Technology Reports of Osaka University*, Vol. 18, Parts I and II, no. 850, 1968, and Vol. 19, no. 874, 1969.

Wengenroth, R. H., "Bridge engineer looks at aesthetics of structures," *Journal of the Structural Division*, ASCE, Vol. 97, no. ST4, April, 1971.

______ , "Bridge engineer looks at esthetics of structures," *Journal of the Structural Division*, ASCE, Vol. 97, no. 4, Apr., 1971, pp. 1227-1237.

Wright, R. N., "Design of box girders of deformable cross section," *Highway Research Record,* no. 295, Bridges - 10 Reports, 1969.

______ , Abdel-Samad, S. R., and Robinson, A. R., "BEF analogy for analysis of box girders," *Journal of the Structural Division*, ASCE, Vol. 94, ST7, Proc. Paper 6025, July 1968, pp. 1719-1743.

Xiao, W., and Olson B., "Seismic retrofit of bridge columns using prefabricated composite jackets," *Proceedings*. Second Symposium on Practical Solutions for Bridge Strengthening and Rehabilitation Bsar II, March 24-25, 1997, Kansas City, MO, pp. 45-54.

Yang, Y., and Yau, Y., "Vehicle-bridge interaction element for dynamic analysis," *Journal of Structural Engineering*, Vol. 123, no. 11, November 1997. pp. 1512-1518.

Zuk, W., "Methodology for evaluating the aesthetic appeal of bridge designs," *Highway Research Report* no. 428.

INDEX

B

M

N

O

Q

R

S

T

U

V